W9-DIN-015

REPLACING DARWIN

The New *Origin of Species*

Nathaniel T. Jeanson

First printing: October 2017
Second printing: November 2017

Master Books®, P.O. Box 726, Green Forest, AR 72638

Master Books® is a division of the New Leaf Publishing Group, Inc.

ISBN: 978-1-68344-075-8
Library of Congress Number: 2017913476

Cover by Intelligent Designs Creative

Please consider requesting that a copy of this volume be purchased by your local library system.

Printed in the United States of America

Please visit our website for other great titles:
www.masterbooks.com

For information regarding author interviews,
please contact the publicity department at (870) 438-5288.

A portion of the proceeds from this book funds
continuing research on the origin of species.

To
Patrick, Jason, and José
And many more like them

Contents

Introduction

Why Now?

Chapter 1

Inevitable

This book makes a bold claim — that the events of the last 130 years have rewritten the history of life on this planet. On the surface, this statement might seem outrageous. A development this provocative would create at least as much upheaval as Darwin's publication did.

Yet, upon deeper reflection, both revolutions become less surprising. In fact, in the wider context of the history of biology, you could argue that both paradigm shifts were inevitable.

To see how, an analogy is helpful. If we think of the origin of species* as a scientific jigsaw puzzle, each species is one piece of the bigger puzzle. Clues to their origins represent additional pieces. Paradigm shifts are major revisions in how the puzzle is put together.

This latter concept requires some explanation. For jigsaw puzzles that come in a box, paradigm shifts are rare. The box cover guides the progress, and the total number of pieces constrains the possible arrangements. Even if a puzzle is large, these two factors streamline and smooth the assembly process.

Unlike typical jigsaw puzzles, the puzzle of the origin of species does not come in a box. No cover exists. The final number of pieces are unknown. In fact, nearly all pieces must be actively sought. Consequently, with each new discovery, the potential for massive overhaul lurks in the background.

* Unless otherwise noted in this book, when I use the term *species*, I am using it in the biological sense — in other words, as a formal unit of classification.

For example, consider the state of the puzzle prior to Darwin's day. Just a century before Darwin published *On the Origin of Species*, the first pieces were discovered. In the late 1700s, Carolus Linnaeus entered the modern concept of *species* into our scientific vocabulary.[1] Since Linnaeus began the modern discipline of *taxonomy* (the identification and classification of life), the pieces of Linnaeus' puzzle were a fraction of the total known today. As an illustration, of the more than 5,400 mammal species that exist today, Linnaeus documented fewer than 200.[2] Linnaeus and his contemporaries had very little data with which to work.

Nevertheless, despite this dearth, the available pieces suggested themselves into a plausible puzzle image. As a modern illustration of this old image, consider the polar bear species (Color Plate 1). It occupies a snowy environment along with its fellow Arctic residents, the Arctic fox (Color Plate 2) and Arctic hare (Color Plate 3). In all three species, white coats[3] camouflage them against the stark Arctic environment.

If you move south to warmer climates, bears, foxes, and rabbits lose their white coats in favor of something more suitable.[4] In the North American forests and mountain ranges, black bears lumber much more secretively than if their coats where white (Color Plate 4). In the grasslands and forests of Eurasia, red foxes dart more clandestinely than the Arctic fox would (Color Plate 5). In the blazing heat of the desert, jackrabbits don't need white coats; instead, they need to stay cool. Unlike arctic hares, jackrabbits possess large ears — efficient thermoregulators (Color Plate 6). The challenges of each environment are matched by the unique features of each species.

This pattern is true globally. From the frigid ice floes of the Arctic to the tropical lagoons of the Great Barrier Reef; from the expansive Serengeti savanna to the dense Black Forest of Germany; from the airless slopes of the Himalayas to the soundless depths of the Pacific; from the deathly dry Sahara to the humid and lush jungles of the Amazon, species seem to have been made for the environments in which they reside.

The natural implications of these observations are clear. Going back to William Paley[5] in 1802, scholars recognized that purpose implies design. Design implies a designer. With an analogy to the familiar realm of human design, Paley illustrated his reasoning. For example, if a watch were discovered lying on the ground, no one would explain its origin by the action of wind and rain over millions of years. Rather, they would observe the intricate fit of the parts to one another and conclude the obvious — that a designer made the watch for a specific purpose. Similarly, the match between species and their environments suggested deliberate purpose — as if, by the purposeful action of a Designer, species were designed for the environments in which they reside.

In technical terms, this view led to very specific conclusions in two arenas. Since species fit their locales so well, it would seem that they were made for their individual habitats — that is, that they were created separate and distinct from other species. This implied that species do not become other species — a view known as *species' fixity*. Conversely, since species appear to have been made for their individual habitats, it would seem that they have always been in their current locations. Thus, in the arena of geography, the design argument implied the fixity of species' locations.

In jigsaw puzzle terms, it was as if each species formed its own isolated puzzle. Environmental clues might decorate the outer edges of each species' puzzle. But, rather than connect different species' puzzles to one another, these clues seemed to separate the puzzles.

With few pieces in hand, this view was easy to maintain. The absence of obvious connecting pieces between puzzles produced a convincing set of isolated images. Nevertheless, due to the fact that many pieces were still waiting to be discovered, the potential for massive overhaul lingered.

By 1859, Darwin and his contemporaries had discovered many new pieces to these puzzles. Drawing on the growing knowledge in the fields of biogeography (i.e., the geographic distribution of species around the globe), anatomy, physiology, embryology, geology, and paleontology, Darwin began to see connections where prior investigators saw only empty space.[6] Eventually, Darwin proposed that all species evolved from one or a few common ancestors — a massive paradigm shift.

Because of the unusual nature of the puzzle of the origin of species, paradigm shifts are inevitable.

Like the 18th century, the scope of species diversity in Darwin's day was a fraction of today's variety. In 1859, the scientific community had no knowledge of the majority of species we have now documented. With over 1.6 million[7] plant, animal, fungal, and bacterial species currently known, hundreds of thousands of pieces were missing from Darwin's puzzle.[8]

With Darwin barely 100 years removed from Linnaeus' foundational work, this fact shouldn't be surprising.

Darwin didn't compensate for this ignorance of species diversity with any special abilities. His lifespan wasn't any longer than the average lifespan today. He observed the world for 73 years. And then he died. Furthermore, in those 73 years, he was subject to the technology of the 1800s. He couldn't travel the globe as easily as we do. Without the information exchange facilitated by the Internet, he couldn't benefit as easily from the travels and discoveries of others. Yet Darwin tried to tackle one of the biggest questions in biology.

Since 1859, we've had time to reevaluate his picture — much more time than he had to propose and appraise it. We've also had more space. Today, travel is virtually unrestricted. Few corners of the world have remained recalcitrant to scientific exploration. Furthermore, the Internet makes information sharing faster than ever before. A global community of millions[9] of scientists can pool their resources and build on one another's work. Though lifespans have changed little, the cumulative observations of these scientists have built an unprecedented body of knowledge on the diversity and operation of life.

Consequently, the puzzle image has changed.

Three developments have led the way. First, after Darwin wrote *On the Origin of Species*, an entire field of science was born — and then matured. Unlike any other field of science, this field directly constrains and guides the answer to the origin of species. Consequently, it's the most relevant field to our question. You could say that the edge pieces of the puzzle have finally been found.

Second, premature conclusions were corrected. Anyone who has gotten stuck trying to put together a puzzle without a box cover and without edge pieces would do what Darwin did — they would test piece after piece until they found a plausible connection. However, without the constraints of a box cover image and of edge pieces, it's easy to link pieces where no link exists. Connections that initially appear plausible eventually give way to the correct links, once additional pieces are found. Darwin made many such premature links. Since 1859, we've been able to unlock some of the connections that he erroneously made, while cementing correct ones.

Third, in the last few years, the critical corner pieces were found. Several remarkable scientific discoveries were made — ones entirely unanticipated by the trajectory of discoveries prior. With these pieces in hand, the framework of the puzzle and the existing connections among pieces have been reoriented.

Individually, each of these developments carried minor significance. By analogy, if you were trying to put a puzzle together, the discovery of a few edge pieces would be helpful. But it wouldn't be earth-shattering. Conversely, if you found a corner piece, this discovery would be fantastic. But if the remaining pieces have been forced together in clumsy and incorrect ways, the corner would do little good. Finally, if all you did was unlock a few misconnected pieces, you would rejoice in the removal of barriers to progress. But the happiness of this success would soon be outweighed by the intimidating scope of the remaining task. In isolation, these discoveries would do little to reveal the final image.

Together, they produced a compelling picture of how species came to be.

To be sure, large chunks of the puzzle still need to be filled in. Having the corner pieces, edge pieces, and a couple of correctly connected center pieces is a huge step forward. But significant holes in the puzzle remain.[10] Explaining in detail the origin of every species that ever lived is a monstrous undertaking. Much work remains to be done.

Nevertheless, the puzzle picture that we possess today is far different from the one that Darwin created. And it is far superior. It puts the far reaches of the globe — and the species that they contain — into an image that is as captivating as it is surprising.

This book tells the story of how this picture came to be.

Part One

A Field Is Born

Chapter 2
The Secret of Life

Sweating over a large, unassembled jigsaw puzzle, I take deliberate steps to simplify the vexing challenge. First, I search for the edge pieces. Once I've found them all, it's a fairly simple task of trial and error to connect them. Far fewer possible connections exist among these pieces than among the center pieces. Furthermore, once connected, they provide an enormously helpful framework in which to connect the rest of the puzzle.

If my puzzle came without a box cover, the edges would be doubly useful. Edges limit the amount of possible horizontal and vertical connections among the center pieces. This saves me the enormous frustration that follows endless trial and error of unassembled center pieces. Edges also naturally suggest how the final image will look. Even though each piece contains a tiny part of the whole image, I can sense the final subject matter just by looking at the edge pieces. In the assembly of a jigsaw puzzle, the identification of edge pieces is a major step forward.

Since species are not literal jigsaw puzzles, the biological analogy for edge pieces might not be obvious. The parallel becomes clearer upon brief reflection. Consider species with which we are familiar. We recognize zebras by their stripes, elephants by their trunks, giraffes by their long necks, bald eagles by the color of the feathers on their heads, and monarch butterflies by the patterns on their gossamer wings. Species are defined by their traits.

This is true across all life. Mammals, reptiles, birds, amphibians, fish, starfish, sea urchins, crustaceans, arachnids, insects, worms of all sorts, shellfish, octopi, snails, corals, jellyfish, sponges, mosses, ferns, conifers, grasses, orchids, fruit trees, fungi, algae, bacteria, and all the other life forms on earth possess unique combinations of traits. Though some features require a microscope to visualize, traits define species.

Therefore, the question of the origin of species is a question of the origin of traits. If you want to know the origin of zebras, you need to discover the origin of stripes. The origin of elephants is bound up in the origin of trunks. Giraffe origins are inextricable from the origin of long necks. The origin of eagles goes hand in hand with the origin of white feathers. Butterfly history is read off the history of butterfly wing patterns. The origin of the rest of the species is found in the origin of the traits that define them.

The solution to the origin of traits represents the hard constraints on the origin of species — the edge pieces of the puzzle.

In 1859, zero edge pieces were known. With careful reflection, the reason for this is easy to grasp. For example, if you wanted to discover the origin of these traits, you could begin by watching how they behave each generation. However, unlike humans, species don't keep written records of their family trees. Thus, as a first step, you might start by investigating human family trees.

The simplest place to start is your own family tree. Perhaps you recognize your father's chin in your jaw. You might investigate how far back on your family tree you can trace this chin shape. However, if you're like me, your family tree is probably small. Going back further than a few generations, I don't know who my relatives are. If your tree is like mine, your ability to examine trait behavior is severely limited by your ignorance of your ancestors.

Your attempts to track traits might encounter a second hurdle. If your family tree is small, you might compensate by tracing additional family trees. In doing so, you'd probably have to follow more traits than chin shape. For example, you might document the behavior of red hair and freckles. If you did, you might observe that, on occasion, the trait disappears from a family tree. A red-headed parent might have no red-headed offspring. Or a great-grandparent might have red hair, but several generations of descendants might not. As the scope of your investigation expands, you might find several traits that behave in odd and inexplicable ways.

These idiosyncrasies apply to both living and extinct species. In fact, when fossils are part of the equation, the problems multiply. Unlike recorded family trees, fossils have no explicit genealogical connection to anything alive today. Ancestral relationships have to first be inferred from indirect data before trait behavior can be tracked. Even more troubling, the placement of fossils on family trees requires implicit assumptions about how traits behave. Assuming a mode of inheritance to prove a mode of inheritance is circular reasoning. In other words, fossils cannot inform how traits behave.

If you had access to a microscope, you'd discover the most inexplicable behavior of all. *All* traits are erased each generation — and then rebuilt. When sperm meets egg, the visible features that define multicellular species are not present. Instead, these characteristics arise via the process of development.

In summary, if you rigorously tracked the behavior of visible traits, you'd discover an intimidating number of paradoxes. These paradoxes would raise a host of perplexing questions. Do traits form spontaneously? Can they be destroyed? Can they be changed? If so, how much can they be changed? Are traits blended? Particulate? Inherited as a whole? Separated into units? Independent? Interdependent?[1]

Consider the ramifications of this uncertainty. If traits can appear and disappear, how could you trace species' ancestry? What markers would you use to fill in the family tree? Furthermore, if all traits are rebuilt every generation, can any species become any other species? Do any constraints on change exist? Might a fish spontaneously spawn a spider? Without an answer to the mystery of heredity, the origin of species would be an enigma.

When Darwin wrote *On the Origin of Species*, paradoxes — not edge pieces of the puzzle — were all that the scientific community possessed.

The first steps toward resolving these paradoxes were taken in 1865 — six years after Darwin's seminal publication. An Austrian monk, Gregor Mendel (Figure 2.1), solved the paradoxes of family trees. Like nearly every other species, his subject of study — pea plants — did not keep written records of inheritance. So Mendel did it for them.

Figure 2.1 — Gregor Mendel

Mendel watched and documented inheritance in pea plants over several generations. His tedious labors involved counting pea plant offspring and recording the traits that appeared in each generation. Over the span of nearly a decade, Mendel made hundreds of crosses and counted thousands of peas. The mathematical precision with which Mendel documented his results laid the groundwork for several revolutionary inferences.[2]

One of the first discoveries that Mendel made was the discrete nature of genetic information. For example, Mendel crossed pea plants with pure-breeding* yellow seeds to pea plants with pure-breeding green seeds. All of the offspring (i.e., the F_1 generation) of this union bore yellow seeds (Color Plate 7). He didn't observe yellowish-green seeds or some other blend between the two colors. Instead, the traits remained distinct. Mendel's experiments demonstrated the fact of *particulate* inheritance rather than *blended* inheritance.

Mendel called these particulate units of inheritance *unit factors.*

Applied more broadly to species, Mendel's discoveries were a tremendous step forward. When traits appear and disappear on family trees, it must have something to do with the unit

* In other words, they always bred true for a particular trait. For example, pea plants that are pure-breeding for yellow seeds produce offspring that always have yellow seeds.

factors behind these traits. For example, red hair and freckles might be encoded by unit factors. Your chin shape might be as well. Consequently, if we can identify and track unit factors, we might be able to reconstruct the family trees for each species.

When the offspring of Mendel's cross (i.e., the plants bearing strictly yellow seeds) were self-fertilized, about 75 percent of this second generation (i.e., the F_2 generation) bore yellow seeds. Twenty-five percent bore green seeds — a 3:1 ratio (Color Plate 7). In other words, in Mendel's experiments, green seed color disappeared in the offspring of the cross between pure-breeding plants. Yet it reappeared in subsequent generations. Thus, the instructions for green didn't get eliminated. They were just hidden by yellow for a single generation. Mendel referred to the behavior of the yellow seeds as *dominant* since the instructions for yellow dominated over the instructions for green. He referred to the behavior of the green seeds as *recessive.*

The mathematical proportions of these offspring yielded another discovery about the nature of inheritance. Mendel self-fertilized the second-generation plants (Color Plate 7). In the offspring (the third generation, or F_3), he again observed consistent mathematical ratios in the seed color trait. If second-generation (F_2) green seed plants were self-fertilized, they always bore green seeds. When second generation (F_2) yellow seed plants were self-fertilized, one-third of the offspring (F_3) bore strictly yellow seeds, while the other two-thirds bore yellow seeds and green seeds in a 3:1 ratio (Color Plate 7). Without going into the details of the math, Mendel was able to infer the behavior of a single trait. In this case, the differing versions of the seed color trait *segregated* from one another each generation. The paternal instructions and the maternal instructions were not combined; they stayed distinct.

These observations begin to explain why red hair appears and disappears irregularly on family trees. Red hair behaves in a recessive manner. When two red-haired individuals bear children, the offspring all possess red hair. But if one of the parents lacks red hair, the offspring are mixed — some might have red hair, some might not. In fact, if the parent without red hair has no red-headed ancestors, none of the children might possess red hair. If both parents lack red hair — but if both have a red-headed ancestor — a red-headed offspring might still arise. But the red-haired offspring will likely represent a minority of the children.

Take zebras as another example. Though zebras, horses, and donkeys are all separate species, they can hybridize to produce sterile offspring. When striped zebras are crossed to unstriped donkeys, the offspring bear stripes (Color Plate 8). They aren't spotted or strictly solid color throughout. Instead, the striping pattern is still distinct. The same phenomenon occurs when zebras and horses hybridize (Color Plate 8). In short, it seems that the unit factor for stripes is dominant.

However, in the offspring of these crosses, the striping pattern varies. Sometimes stripes occur primarily on the legs of the hybrids (Color Plate 9), rather than both on the legs and on the sides of the torso (Color Plate 8). Perhaps several unit factors control coat color patterning. The stripes trait might dominate only one aspect of the process of creating the coat color patterning in the adult. Regardless, stripes in zebras behave largely according to the principles of inheritance that Mendel uncovered in plants.

Together, the discovery of unit factors, of dominant and recessive traits, and of the segregation of genetic information began to define clear rules for the ways in which traits behave each generation.

Mendel performed even more complex crosses, which revealed yet another fundamental principle of inheritance. In addition to seed color, Mendel crossed plants bearing variety in other traits, such as seed form. For example, some of his plants bore seeds that were wrinkled while others bore seeds that were round (Color Plate 10). Just like with seed color, one form was dominant (e.g., round), and the other recessive (e.g., wrinkled).

In one series of experiments, Mendel crossed plants that differed in seed color and seed form. The first cross involved pure-breeding yellow-colored, round form seed plants. These were crossed with pure-breeding green-colored, wrinkled form seed plants. As expected, all offspring (F_1 generation) had yellow seeds that were round. When these second-generation plants (F_1) were self-fertilized, every possible combination appeared in their offspring (F_2 generation). Some F_2 plants had yellow and round seeds (about 9/16th of the offspring); some had yellow and wrinkled seeds (about 3/16th of the offspring); some had green and round seeds (about 3/16th of the offspring); and some had green and wrinkled seeds (about 1/16th of the offspring) (Color Plate 10). In other words, the behavior of one trait was not tied to the behavior of another. If the seed color trait behaved one way, it had no effect on the behavior of the seed shape trait.

Mathematically, the proportions of the offspring corresponded to what might be predicted from a model of two traits inherited independent of one another. Diagrammatically, all of these probabilities can be derived with a *Punnett square* (Color Plate 10). Since we're dealing with two traits that each have dominant and recessive forms, the predicted frequencies are slightly skewed. Due to their recessive nature, recessive forms are predicted to appear less frequently than the dominant ones. But by assuming that traits are independent of one another, and that the differing versions of each trait segregate from one another (i.e., this is what the Punnett square diagram does in Color Plate 10), we can make mathematical sense of the results.

In contrast, if the behavior of one trait was dependent upon the behavior of another, then the offspring would show ratios that did *not* agree with the probabilities derived from the Punnett square.

Since the ratios agreed, Mendel inferred that the instructions for different traits *sorted independently of one another.*

This discovery opened up a whole new world of possibilities. Think of all the traits that define your external features. Think of all the traits that define each species. Unit factors must exist for each of these traits. Some might be dominant, while others are recessive. Since the dominant and recessive versions of unit factors segregate from one another, traits can appear and disappear each generation. In a sense, these genetic discoveries define some of the limits to trait behavior. Conversely, since unit factors sort independently of one another, an overwhelming number of potential combinations exist. Together, Mendel's principles uncovered both the boundaries and the enormous opportunities for variety that exist within each species.

In the years following Mendel's work, other principles of inheritance have been discovered. In other words, exceptions to Mendel's rules were found. For example, some unit factors are linked and do not sort independently of one another. As another example, other unit factors do indeed act in concert to produce a blended trait outcome. Nevertheless, Mendel's findings laid a major foundation for our modern understanding of trait behavior.

For reasons unknown, Darwin appears to have been unaware of Mendel's work. Conversely, for equally unknown reasons, when Mendel died in 1884, his discoveries died with him, not to be resurrected until the turn of the century.

Despite the strength and rigor of Mendel's conclusions, his results never answered the question of *why* unit factors behave in the way that they do. In other words, Mendel's rules successfully predicted how traits interact and combine. Yet these predictions didn't reveal whether the rules could change — or whether they have changed in the past.

If the rules could change, then traits might behave in entirely unexpected ways. Perhaps a fish could become a spider. Maybe it did in the distant past. Or maybe this change is impossible. Without the *why* of the rules of inheritance, these speculations would exist unfettered by reality. Without these fetters, the origin of traits would remain a mystery. Without an explanation for the origin of traits, the origin of species would remain an unsolved puzzle.

Several decades after Mendel's death, meticulous observation of cells during cell division put Mendel's principles in more concrete, subcellular, and potentially mechanistic terms. Under the microscope, *somatic cells* (i.e., non-reproductive cells) were observed to divide through a process of cell division termed *mitosis*. In both animals and plants, before the nucleus breaks down, structures that look like flexible noodles — *chromosomes* — appear in a period of time termed *prophase* (Figure 2.2). By *prometaphase*, the membrane surrounding the nucleus breaks down (Figure 2.2).

During *metaphase* (Figure 2.2), the chromosomes line up in the center of the cell. These chromosomes appear as x-like structures because they represented two identical (replicated) chromosomes that are still partially joined. In *anaphase* (Figure 2.2), the replicated chromosomes are separated from one another and are pulled toward opposite ends of the cell. By *telophase* (Figure 2.2), the cell begins to split into two cells, each cell containing a chromosome content identical to the other cell, and the nuclear membrane begins to reappear. *Cytokinesis* completes the splitting process.

Leaving Mendel's discoveries aside for the moment, consider the ramifications of what we just observed. The fact that the process of nuclear division was so complex suggested a functional role for chromosomes. If chromosomes were inert and irrelevant to heredity, why would cells take such care to pass them on via such a convoluted cycle?

The behavior of somatic cells was just the beginning of the answer to the question of inheritance. More relevant was the behavior of *germ cells* or *gametes* — sperm and egg. Chromosomes are also distributed among these cells during normal development in a cell division process termed *meiosis*. This process bears strong resemblance to mitosis, but also differs in key ways. In both mitosis and meiosis, the chromosome number and behavior is very predictable. In the early 1900s, the American scientist Walter Sutton provided the documentation. During the process of meiosis, Sutton observed that the chromosomes occurred in recognizable *pairs* (Figure 2.3). For example, in Figure 2.3, for each long x-like structure, there exists another long x-like structure. For each short x-like structure, there exists another short x-like structure. Technically, each x-like structure consists of two chromosomes — one a replicated copy of another. Furthermore, these pairs were separated during meiosis, and the individual chromosomes in each x-like structure were distributed among the various products of *spermatogenesis* and *oogenesis* — the processes which give rise to sperm and egg (Figure 2.3). Sutton suggested that one member of each chromosome pair was ultimately maternal in origin and the other, paternal (Figure 2.3). In other words, an individual inherits one

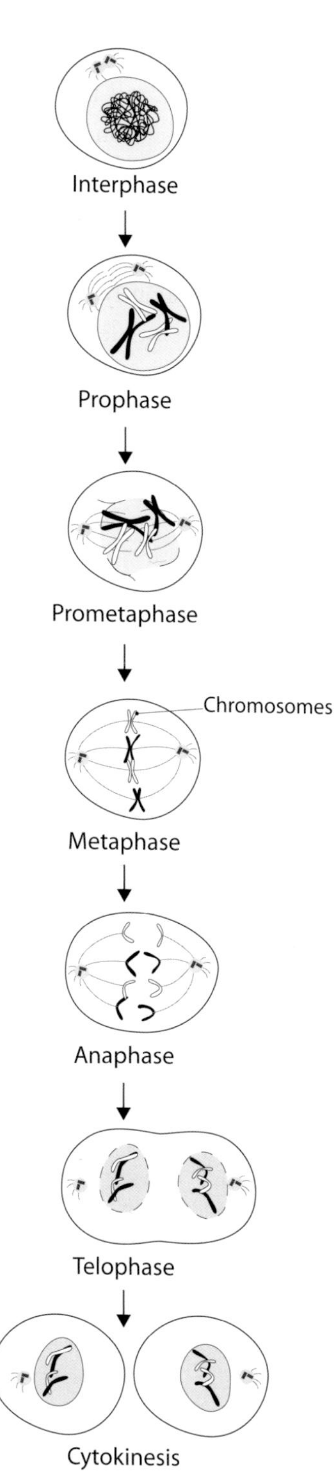

Figure 2.2. Cell division in non-reproductive cells. Somatic cells (non-reproductive cells) undergo a very strict series of cell division events in a process termed mitosis. In early phases (e.g., prophase), replicated chromosomes become visible. Until anaphase, the replicated chromosomes remain attached, leading to the familiar x-like shape seen in each of the four chromosomes in the metaphase display shown here. After metaphase, the remaining phases of the cell cycle are dedicated to separating these replicated chromosomes and divvying them up into two daughter cells.

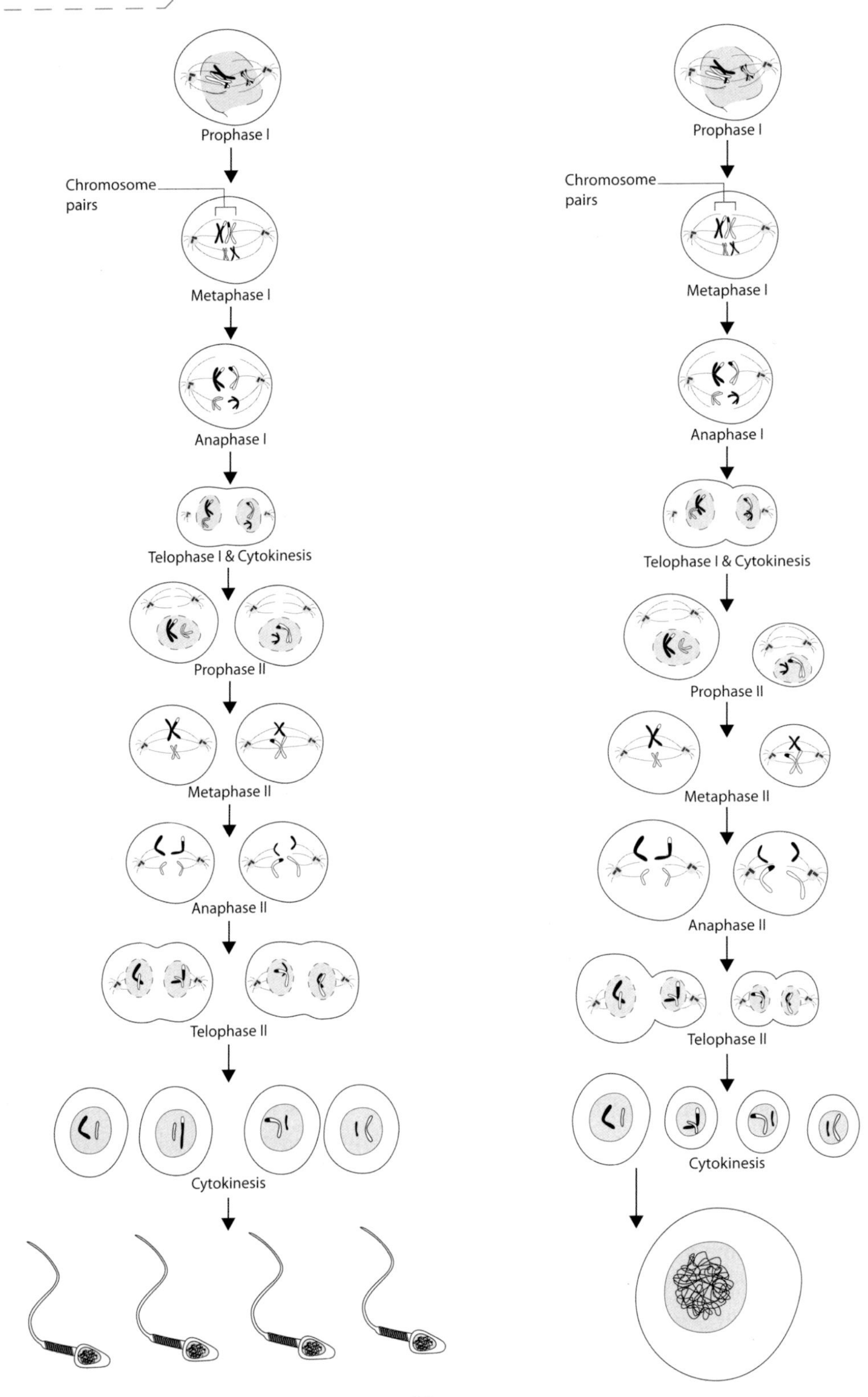
Prophase I
Chromosome pairs
Metaphase I
Anaphase I
Telophase I & Cytokinesis
Prophase II
Metaphase II
Anaphase II
Telophase II
Cytokinesis
Prophase I
Chromosome pairs
Metaphase I
Anaphase I
Telophase I & Cytokinesis
Prophase II
Metaphase II
Anaphase II
Telophase II
Cytokinesis

Figure 2.3 (previous page). Cell division in sperm and egg. To produce sperm and egg, cells undergo a process of two cell divisions termed meiosis. One of the major differences between mitosis and meiosis is the arrangement of the chromosome pairs in the first cell division. In mitosis, chromosomes line up vertically (i.e., with respect to the display in figure 2.2) during metaphase, and the replicated chromosomes are separated to produce identical chromosome content in the resultant daughter cells. In meiosis, during the first metaphase, chromosome pairs line up next to one another (i.e., with respect to the display in this figure), and pairs — not individual chromosomes — are separated from one another. Then, during the second round of cell division, in which chromosomes line up vertically during metaphase II, the replicated chromosomes are separated from one another. Because the second round of cell division during meiosis proceeds without additional DNA replication, the chromosome content in sperm and egg is half of that in somatic cells (a condition known as haploid; somatic cells have the full — diploid — chromosome content).

Of the four products of meiosis during the formation of egg cells, only one of the four products eventually becomes a mature egg. In contrast, all four products of meiosis during the formation of sperm cells result in mature sperm.

During prophase I of meiosis, a chromosomal diversification process happens ("crossing over"), which is discussed in more detail in chapter 9.

member of each pair from each parent — as implied by the chromosome numbers and arrangement in Figure 2.3.

Immediately, Sutton's discovery suggested a link to Mendel's findings. Since chromosomes physically segregate from one another during the formation of sperm and egg, perhaps chromosomes contain the unit factors which segregate over successive generations.

For example, pea plants contain chromosomes. If two pure-breeding plants were crossed to one another, each pure-breeding parent would possess a pair of chromosomes. However, in the process of meiosis, even though the parents would each contain two chromosomes, they wouldn't produce gametes with *pairs* of chromosomes. They would produce reproductive cells with *individual* chromosomes (Color Plate 11).

If we treat chromosomes as the repositories for unit factors, we can make sense of Mendel's results. For example, a pure-breeding yellow seed color parent would pass on a chromosome containing only yellow seed color instructions. A pure-breeding green seed color parent would pass on a chromosome containing only the green seed color instructions. In the offspring (F_1), the gametes would fuse and restore the chromosomal pairing arrangement, resulting in a chromosome for yellow seed color paired with a chromosome for green seed color. Since yellow dominates green, the seeds of the offspring (F_1) would all be yellow, despite having a chromosome for yellow seed color and a chromosome for green seed color (Color Plate 11).

Self-fertilization of these chromosomally mixed individuals to produce the next generation (F_2) illustrates the chromosome-Mendel link as well (Color Plate 11). Since the process of meiosis occurs again in the F_1 individuals, individual chromosomes — not pairs of chromosomes — would be produced in the gametes of the F_1 individuals. This entails that the chromosome for yellow seed color and the chromosome for green seed color would segregate from one another. To produce the F_2 individuals, these gametes with their individual chromosomes could be combined in a variety of ways — resulting in both yellow and green seeded individuals in the F_2 generation (Color Plate 11).

Treating chromosomes as the repositories for unit factors also makes sense of Mendel's more complex crosses (Color Plate 12). For example, in the parents, let's postulate that one set of chromosomes had the unit factors for seed color and another set of chromosomes the unit factors for seed shape.[3] Since chromosomes come in pairs, the pure-breeding pea plants would have chromosome pairs containing the same information. Pea plants with smooth yellow seeds would possess a chromosome pair with both copies specifying yellow color. In another chromosome pair, both copies would specify smooth shape (Color Plate 12). The pure-breeding pea plants with rough green seeds would possess a similar state. One chromosome pair would consist of both copies specifying green color. Another chromosome pair would consist of both copies specifying rough shape (Color Plate 12).

Chromosome number		Combinations in germ cells	Combinations in somatic cells
In somatic cells	In germ cells		
2	1	2	4
4	2	4	16
6	3	8	64
8	4	16	256
10	5	32	1,024
12	6	64	4,096
14	7	128	16,384
16	8	256	65,536
18	9	512	262,144
20	10	1,024	1,048,576
22	11	2,048	4,194,304
24	12	4,096	16,777,216
26	13	8,192	67,108,864
28	14	16,384	268,435,456
30	15	32,768	1,073,741,824
32	16	65,536	4,294,967,296
34	17	131,072	17,179,869,184
36	18	262,144	68,719,476,736

Table 2.1. Tremendous potential for combinatorial diversity in species with high chromosome numbers. Adapted and redrawn from W.S. Sutton, "The Chromosomes in Heredity," *Biological Bulletin*, 1903, 4:231–251.

Since each parent would pass on only one member of each chromosome pair during the formation of gametes, the fusion of these gametes to produce the offspring (F_1) would lead to a very predictable outcome (Color Plate 12). In the F_1 individuals, one member of the chromosome pair specifying seed color would have the unit factor for yellow. The other member of the pair would have the unit factor for green. For the chromosome pair specifying seed shape, one member of the pair would have the unit factor for smooth. The other member of the pair would have the unit factor for rough. Since yellow and smooth dominate green and rough, only pea plants with smooth yellow seeds would be seen.

Chromosome number		Combinations in germ cells	Combinations in somatic cells
In somatic cells	In germ cells		
46	23	8,388,608	70,368,744,177,664

Table 2.2. Tremendous potential for combinatorial diversity in humans.

When these individuals (F_1) were self-fertilized, the offspring (F_2) would each receive a single member of each chromosome pair from their parents. The individual members of each chromosome pair would segregate from each other. The two pairs of chromosomes would sort independently of each other. Thus, all four combinations of traits — due to differing combinations of chromosome pairs — would be visible in the offspring. The mathematical proportions of these traits would follow the predicted probabilities outlined by the way in which chromosomes could be inherited (Color Plate 12).

In sum, predicting and tallying results either by visible appearance or by chromosome distribution yields the same conclusion — the ratios that Mendel documented in 1865.

Along with a number of critical functional tests in the early 1900s, Sutton's observations began to sketch a more comprehensive picture of inheritance.

The synthesis between these chromosomal observations and Mendel's conclusions raised a number of intriguing possibilities. If chromosomes do indeed encode Mendel's unit factors, and if unit factors specify traits, then the number of traits encoded by chromosomes is mind-boggling. In 1903, Sutton published a table[4] (adapted in Table 2.1) showing the theoretical number of chromosome combinations[5] (and, by extension, trait combinations) that were possible from varying chromosome numbers. In humans, 46 chromosomes exist, implying a bewildering possibility of trait combinations[6] (Table 2.2). In zebras, 32–46 chromosomes exist, a fact which

Species	Common name	Chromosome number		Combinations in germ cells	Combinations in somatic cells
		In somatic cells	In germ cells		
Caenorhabditis elegans	Roundworm	12	6	64	4,096
Apis mellifera	Honeybee	32	16	65,536	4,294,967,296
Xenopus laevis	African clawed frog	36	18	262,144	68,719,476,736
Anolis carolinensis	Green anole	36	18	262,144	68,719,476,736
Danio rerio	Zebrafish	50	25	33,554,432	1,125,899,906,842,620
Equus caballus	Horse	64	32	4,294,967,296	18,446,744,073,709,600,000
Gallus gallus	Chicken	78	39	549,755,813,888	302,231,454,903,657,000,000,000

Table 2.3. Tremendous potential for combinatorial diversity across the animal kingdom.

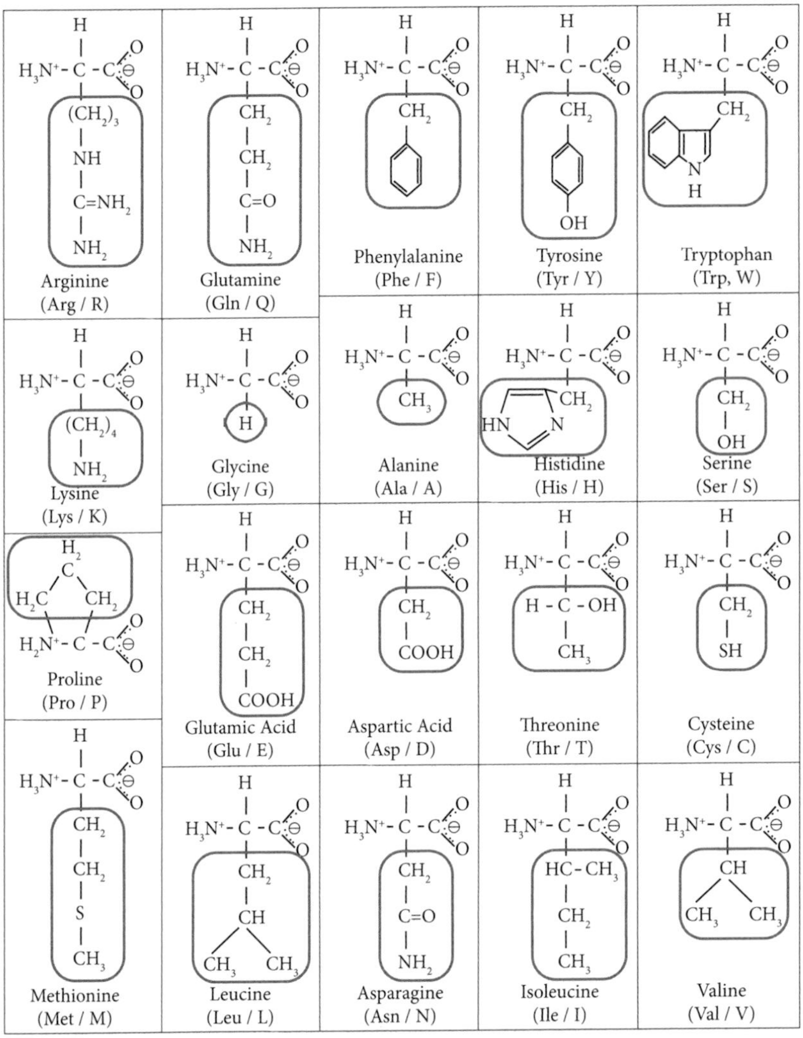

Arginine (Arg / R)
Glutamine (Gln / Q)
Phenylalanine (Phe / F)
Tyrosine (Tyr / Y)
Tryptophan (Trp, W)
Lysine (Lys / K)
Glycine (Gly / G)
Alanine (Ala / A)
Histidine (His / H)
Serine (Ser / S)
Proline (Pro / P)
Glutamic Acid (Glu / E)
Aspartic Acid (Asp / D)
Threonine (Thr / T)
Cysteine (Cys / C)
Methionine (Met / M)
Leucine (Leu / L)
Asparagine (Asn / N)
Isoleucine (Ile / I)
Valine (Val / V)

Figure 2.4. Chemical diagrams of 20 standard amino acids. In the chemical diagrams themselves, chemical elements are represented with single letter abbreviations: carbon (C), nitrogen (N), oxygen (O), hydrogen (H), sulfur (S). Five-sided and six-sided chemical structures consist entirely of carbon and hydrogen, unless otherwise indicated. Chemical bonds indicated by lines (single bond by one line; double bond by two lines). Subscripts denote number of atoms of the adjacent element. Below each chemical diagram (i.e., at the base of each box), the name of each amino acid is given three ways — as a full name, as a three-letter abbreviation, and as a single-letter abbreviation (the latter two ways are in parentheses). Rounded gray boxes indicate the various identities of the "R" placeholder in Figure 2.6.

also implies that a tremendous diversity of traits could be encoded on chromosomes. Across the animal kingdom[7] (Table 2.3), chromosome diversity suggested that Sutton's discoveries[8] were applicable to virtually all multicellular life.[9]

Sutton's findings began to explain *why* traits behave in the manner that they do, which moved the scientific community one step closer to understanding the origin of traits — and to discovering the critical edge pieces to the puzzle of the origin of species.

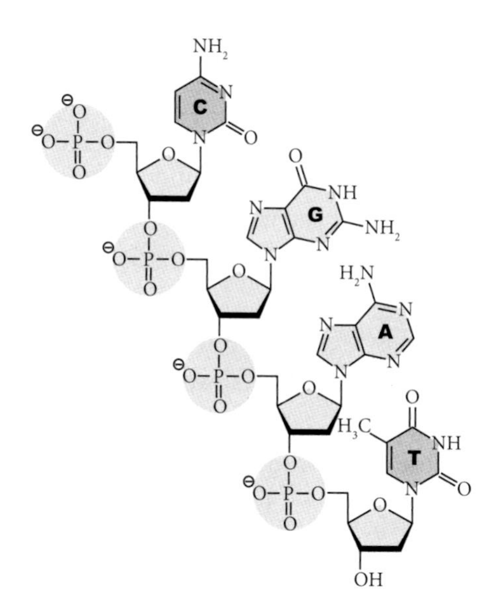

Figure 2.5. Chemical diagram of 4 linked DNA nucleotides. Elements are represented with single letter abbreviations: carbon (C), nitrogen (N), oxygen (O), hydrogen (H), phosphorus (P). Five-sided and six-sided chemical structures consist entirely of carbon and hydrogen, unless otherwise indicated. Chemical bonds indicated by lines (single bond by one line; double bond by two lines). Subscripts denote number of atoms of the adjacent element. The four different nucleotide subunits are designated by the bold, single-letter abbreviations for the defining bases (i.e., the six-sided chemical structures, or the joined five-sided and six-sided chemical structures): Adenine (A), cytosine (C), thymine (T), guanine (G).

Once inheritance was firmly connected to chromosomes, a new question greeted geneticists (scientists who study genetics). At a surface level of explanation, chromosome behavior described trait behavior. But the behavior of chromosomes didn't explain enough detail to be useful. How *specifically* did chromosomes contain the information for traits? How could a microscopic cellular structure possess the instruction manual for building the traits that define each species? How do chromosomes act like a molecular code?[10]

Before these questions could be broached, a much more mundane — but inescapable — problem had to be solved. In multicellular species, chromosomes are chemically composed of at least two major biological molecules, protein and deoxyribonucleic acid (DNA). Proteins are composed of chemical substances termed *amino acids*. Twenty different amino acids occur normally in most species (Figure 2.4). DNA is composed of chemical substances termed *nucleotides*, of which four different kinds occur (Figure 2.5). Which substance — DNA or protein — was the primary carrier of the genetic information? And how did it do so?

In a sense, the discovery of the chromosome-trait link was like connecting the content of this book to the pages between the covers — but without identifying whether the paper or the ink contained the information. It would

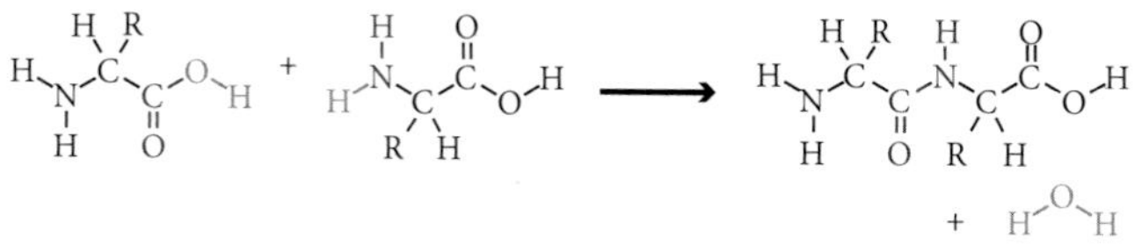

Figure 2.6. Linking individual amino acid subunits into a chain. Elements are represented with single letter abbreviations: carbon (C), nitrogen (N), oxygen (O), hydrogen (H). The single letter R is a placeholder for various amino-acid-specific chemical links (see Figure 2.4 for examples). Chemical bonds indicated by lines (single bond by one line; double bond by two lines).

be impossible to determine the book's contents without knowing whether the fabric of the paper or the arrangement of the ink into letters carried the information.

In the cellular realm, the chemical details naturally suggested a candidate. Both proteins and DNA are composed of chains of their respective subunits, amino acids and nucleotides (Figures 2.5–2.6). Theoretically, proteins made better candidates than DNA for carrying genetic information. With 20 amino acids to choose from (Figure 2.4), many protein chain combinations were possible — many more than for DNA chains of nucleotides.

For example, let's say that the information in chromosomes is contained in the chemical equivalent of three-letter words. If DNA carried the information, only 64 total words would be possible (4 * 4 * 4 = 64). In contrast, if proteins carried the information, 8,000 total words would be possible (20 * 20 * 20 = 8,000). The contrast grows stronger as the word size grows bigger.

At this stage in our discussion, these speculations might seem completely disconnected from reality. How could a chain of chemicals carry information for two eyes, two ears, a nose, and a mouth? Of all the things to hypothesize, these scientific pursuits might appear irrational. Yet this is where prior investigations led. The hypothesis that information was encoded in chains of chemicals was the best guess possible in light of the known facts of the time.

Furthermore, we regularly encode information in chains other than printed Roman letters. Braille is language that uses raised dots. A chain of physical bumps encodes complex information. Chemicals represent a kind of braille — but at a much tinier level than the dots we touch.

The theoretical contrast between the information-carrying capacity of protein and DNA chains prompted investigators in the early 1900s to propose that DNA acted like a scaffold — like the paper on which braille dots are printed. With just four nucleotides to choose from, it seemed straightforward to erect a repeating DNA structure upon which protein information could be hung. Furthermore, researchers measured the amount of the four different nucleotides in cells and found them to be equal. Together, these facts seemed to support a model termed the *tetranucleotide hypothesis* — the view that DNA is a passive, structural edifice in the process of inheritance, not an active repository of heritable information.[11]

Other observations seemed consistent with this idea. In terms of dry weight in cells, proteins represent 50 percent — they are major components. By contrast, early chemical analyses of DNA suggested that it was a small molecule — too small to carry much information. In addition, when nucleotide material was compared across various species, the nucleotide ratios appeared to be the same. If nucleotides contained the heritable information, you might expect their ratios to vary across creatures whose heritable features are diverse. Yet no simple to complex hierarchy was apparent.

In the early 1900s, traits such as red hair, chin shape, zebra stripes, and the elephant's trunk were thought to be encoded by chains of amino acids on chromosomes.

In 1928, a British biologist, Fred Griffith, described an intriguing finding in tiny bacterial cells — with large ramifications for the relationship between chromosomes and traits. In his experi-

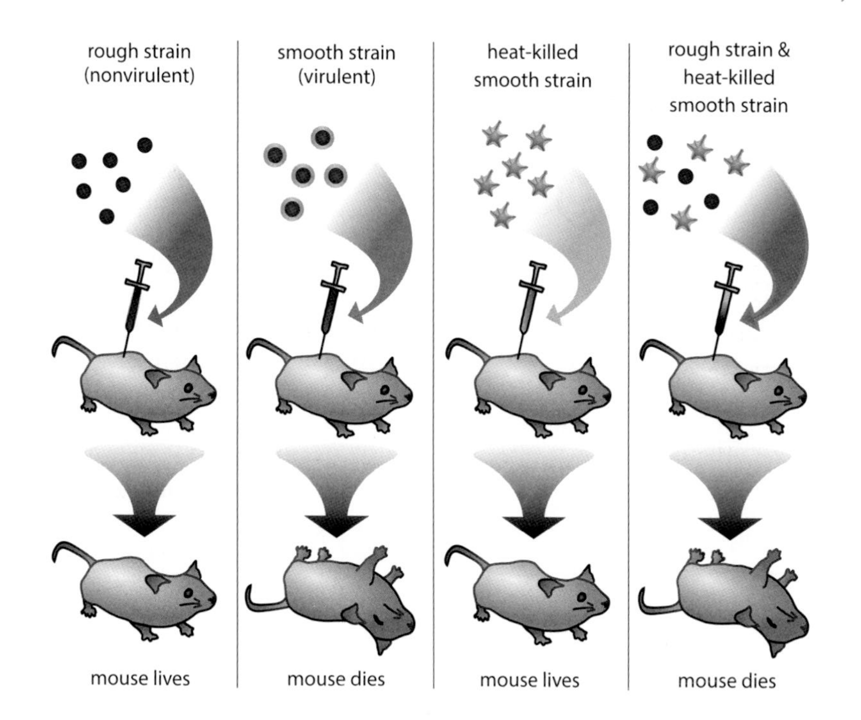

Figure 2.7. Griffith's experiments with rough and smooth bacterial cells. Upon injection in mice, the smooth strain of bacteria killed their hosts, while the recipients of rough strain bacteria lived. When Griffith killed a group of smooth bacteria with heat, they lost their killing ability. When he mixed the dead smooth bacteria with live rough bacteria, and then injected the mixture in mice, the recipients died. It was as if the rough cells became smooth in their virulence.

ments, he could distinguish two major types of cells, rough and smooth. These differed in the shape of the colonies they produced in a petri dish. They also differed in their virulence. Upon injection in mice, the smooth cells killed their hosts, while the recipients of rough cells lived (Figure 2.7).

Remarkably, when Griffith killed a group of smooth cells with heat, and then mixed the dead smooth cells with live rough cells, and then injected the mixture in mice, the recipients died. It was as if the rough cells became smooth in their virulence (Figure 2.7). Griffith never elucidated the reason why.[12]

Before 1944, a group of researchers succeeded in repeating Griffith's *in vivo* results (e.g., in the mouse) strictly *in vitro* (e.g., in the lab rough cells could be transformed into smooth cells without the need for a mouse host). In 1944, at the hospital of the Rockefeller Institute, three investigators (Oswald Avery, Colin MacLeod, and Maclyn McCarty) utilized the *in vitro* system and published an explanation for Griffith's discovery.

They discovered the explanation by purifying the transforming substance from heat-killed smooth cells and then sending it through a battery of tests to determine its identity. For example, Avery and colleagues examined the stability of the purified substance. In pure water, the purified material degraded quickly, while salt solutions preserved it. If the material was kept at high temperatures for an hour, it still transformed rough cells to smooth. When the material was dissolved in a highly acidic solution, the transforming ability disappeared. Consistent with these properties, direct chemical tests for DNA were positive.

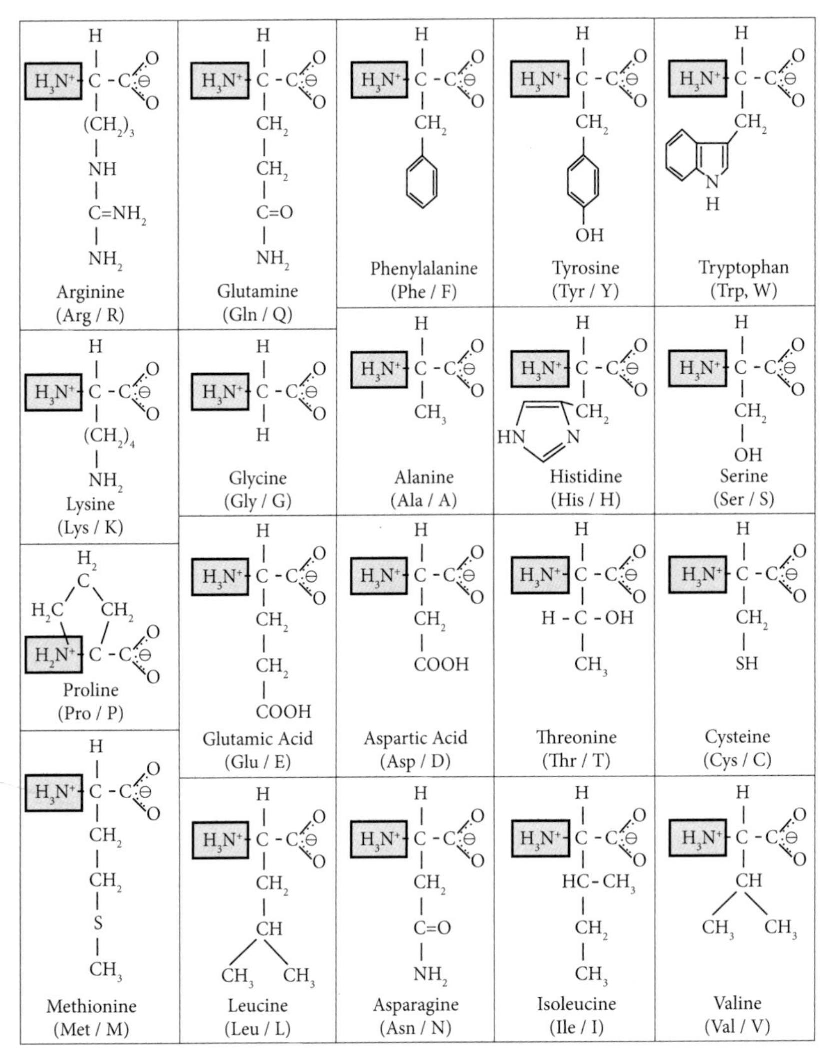

Nitrogen present (phosphorus absent)

Figure 2.8. Distinguishing chemistry of amino acids. Amino acids contain copious amounts of nitrogen (select examples highlighted with black boxes) but, unlike nucleotides, do not contain any phosphorus. In the chemical diagrams themselves, chemical elements are represented with single letter abbreviations: carbon (C), nitrogen (N), oxygen (O), hydrogen (H), sulfur (S). The element phosphorus (P) is absent.

But was the rough-to-smooth transformation due to DNA? Or to some other chemical that contaminated the DNA preparation?

Attempts to remove fats did nothing to diminish the transforming ability. Conversely, the chemical constituents of the purified material were highly suggestive of pure DNA. Plenty of phosphorus was present, relative to nitrogen. Since amino acids do not contain phosphorus, but do contain copious amounts of nitrogen (Figure 2.8), this finding suggested that proteins were largely absent from the purified material. Furthermore, direct enzymatic* elimination of proteins from the purified material did nothing to stop transformation. Enzymatic elimination of another

* Enzymes are molecules (often, but not exclusively, proteins) that catalyze particular types of chemical reactions. In this case, the enzymes in this experiment catalyzed the severing of amino acid chains in proteins.

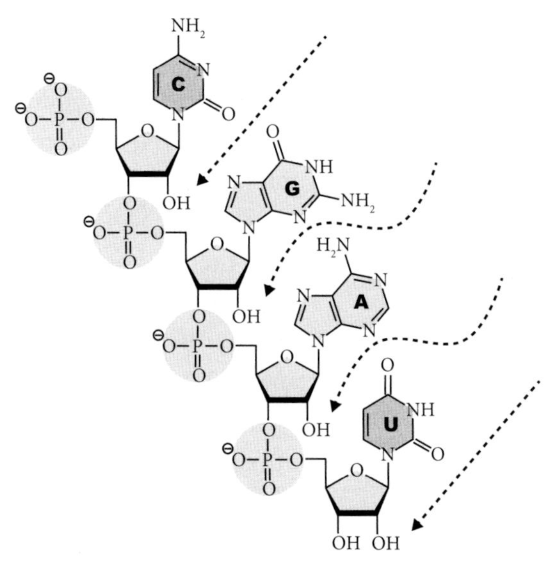

Figure 2.9. Chemical diagram of 4 linked RNA nucleotides. Elements are represented with single letter abbreviations: carbon (C), nitrogen (N), oxygen (O), hydrogen (H), phosphorus (P). Five-sided and six-sided chemical structures consist entirely of carbon and hydrogen, unless otherwise indicated. Chemical bonds indicated by lines (single bond by one line; double bond by two lines). Subscripts denote number of atoms of the adjacent element. The four different nucleotide subunits are designated by the bold, single-letter abbreviations for the defining bases (i.e., the six-sided chemical structures, or the joined five-sided and six-sided chemical structures): Adenine (A), cytosine (C), uracil (U), guanine (G). RNA is distinguished from DNA by the extra "OH" linkage below the five-sided chemical structure (highlighted with dashed arrows).

type of nucleic acid, RNA (Figure 2.9), was equally impotent. Only when enzymes specific for DNA were added to the purified material did the transforming ability vanish.

When the material was examined by other tests for purity, the results consistently showed a concentrated, highly pure substance. Under UV light, the material absorbed in the same wavelengths as DNA. When diluted, even to 1 part in 600,000,000, the material was still effective in transforming rough cells to smooth.

The results of Avery, MacLeod, and McCarty pointed squarely at DNA.[13]

Thus, red hair, chin shape, zebra stripes, the elephant's trunk, the giraffe's neck, the bald eagle's feathers, and the butterfly's wing patterns now appeared to be encoded, not by chains of amino acids, but by chains of nucleotides on chromosomes.

These bacterial transformation results were just the beginning of this shift in scientific opinion. New evidence demonstrated that DNA was, in fact, a very large molecule with much more theoretical capacity to store information. In addition, the chemistry of DNA was reanalyzed. The Austro-Hungarian biochemist Erwin Chargaff found that the different nucleotides did, in fact, vary in at least two ways. First, in species as diverse as bacteria, yeast, and human, the nucleotide ratios were not constant. This was compelling evidence that DNA might, in fact, have the potential to specify the diverse traits in these creatures.

Second, Chargaff found that, within an individual, the amount of the *adenine* nucleotide* ("A") was always approximately the same as the amount of the *thymine* nucleotide ("T"). The amount of the *cytosine* nucleotide ("C") was always approximately the same as the amount of the *guanine* nucleotide ("G"). However, the amount of A or T was not equivalent to the amount

* That is, the amount of the nucleotide containing the adenine nitrogenous base.

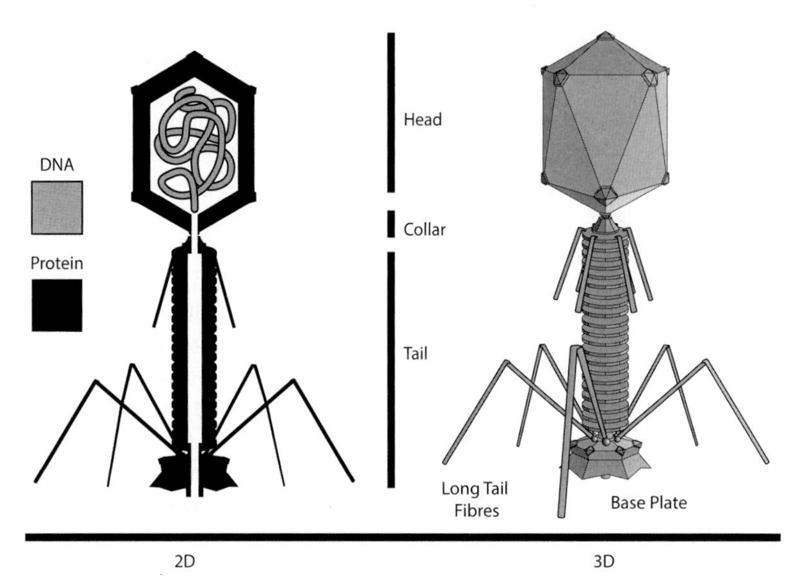

Figure 2.10. Structure of a bacteriophage.

of G or C.[14] Due to poor technology in the early 1900s, the ratios among the four DNA nucleotides were measured in error.

Then in 1952, two American biologists, Alfred Hershey and Martha Chase, reported a stunning find in viruses. Like Avery, MacLeod, and McCarty before them, Hershey and Chase utilized the chemical differences between proteins and DNA to identify which was the heritable material. But unlike Avery, MacLeod, and McCarty, Hershey and Chase utilized a bacterial system in which the primary subject of study was not bacteria. It was viruses — *bacteriophages* — that infect bacteria.

The experimental setup was straightforward. Bacteriophages have a comparatively simple structure. Their viral coats are composed of protein. The internal contents consist of DNA (Figure 2.10). Some of the amino acids in proteins contain sulfur (Figure 2.11), whereas nucleotides do not (Figure 2.12). Conversely, nucleotides contain phosphorus (Figure 2.12), but amino acids do not (Figure 2.11). Thus, the viral coats could be chemically distinguished from the internal contents.

When bacteriophages infect bacteria, they inject a certain material into their hosts, while another part of them remains behind. The injected material induces the bacterial hosts to make more bacteriophages, eventually causing the bacteria to burst open, or *lyse*, and release freshly synthesized bacteriophages. The fresh viruses then go on to infect other cells. For Hershey and Chase, the key question was the identity of the injected material.

To determine which molecule carried the information for making more bacteriophages, Hershey and Chase labeled one experimental group of bacteriophages with radioactive sulfur. Another they labeled with radioactive phosphorus. After letting the bacteriophages infect the bacteria, they isolated the extracellular bacterial contents and the intracellular bacterial contents. Radioactive sulfur showed up outside the bacteria. Radioactive phosphorus appeared in the bacteria cells (Color Plate 13). Therefore, nucleic acids were the heritable material of bacteriophages.[15]

But what about species that have multiple chromosomes? Which chemical carries the information in these creatures? A simple correlation suggested the answer. As we discussed earlier, sperm and egg are formed via the process of meiosis. Since meiosis results in just one member

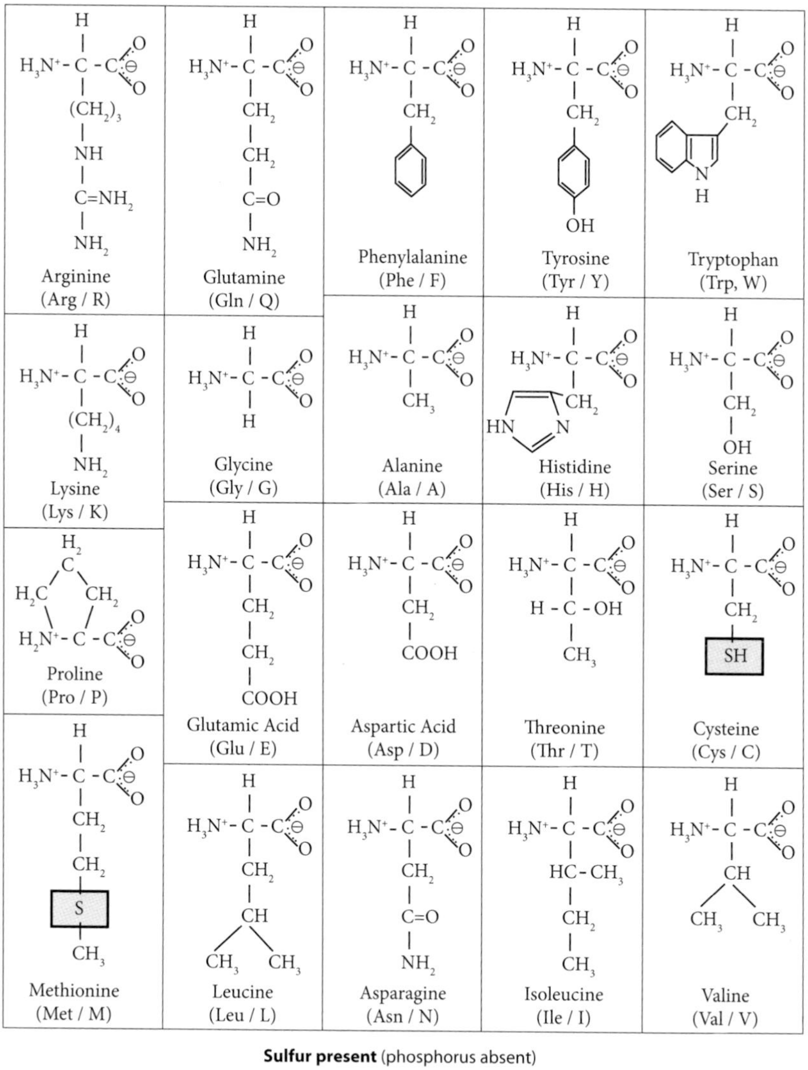

Arginine (Arg / R)
Glutamine (Gln / Q)
Phenylalanine (Phe / F)
Tyrosine (Tyr / Y)
Tryptophan (Trp, W)
Lysine (Lys / K)
Glycine (Gly / G)
Alanine (Ala / A)
Histidine (His / H)
Serine (Ser / S)
Proline (Pro / P)
Glutamic Acid (Glu / E)
Aspartic Acid (Asp / D)
Threonine (Thr / T)
Cysteine (Cys / C)
Methionine (Met / M)
Leucine (Leu / L)
Asparagine (Asn / N)
Isoleucine (Ile / I)
Valine (Val / V)

Sulfur present (phosphorus absent)

Figure 2.11. Additional distinguishing chemistry of amino acids. Unlike nucleotides, amino acids contain sulfur (examples highlighted with black boxes), but do not contain any phosphorus. In the chemical diagrams themselves, chemical elements are represented with single letter abbreviations: carbon (C), nitrogen (N), oxygen (O), hydrogen (H), sulfur (S). The element phosphorus (P) is absent.

of each chromosome pair in sperm or egg, meiosis reduces the genetic material by half in germ cells. Fertilization restores the full complement of genetic material. If it didn't, the chromosome number would increase each generation and eventually balloon to unmanageable amounts. Consequently, if DNA was the heritable material in species with chromosomes, DNA content should track with the pattern predicted by meiosis — twice as much DNA should be in somatic cells as compared to sperm and egg.

DNA matched this prediction. But the amount of protein did not.

In addition, since mitosis keeps the amount of genetic material constant in somatic cells, DNA content should be constant in these cells as well.

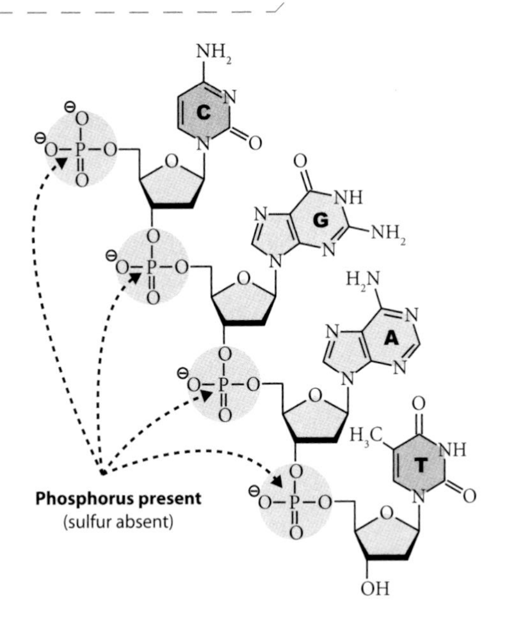

Figure 2.12. Distinguishing chemistry of DNA nucleotides. Elements are represented with single letter abbreviations: carbon (C), nitrogen (N), oxygen (O), hydrogen (H), phosphorus (P). Five-sided and six-sided chemical structures consist entirely of carbon and hydrogen, unless otherwise indicated. Chemical bonds indicated by lines (single bond by one line; double bond by two lines). Subscripts denote number of atoms of the adjacent element. Unlike amino acids, nucleotides possess phosphorus. The element sulfur (S) is absent.

It was.

Combined with the fact that the four nucleotide ratios varied widely among a diversity of species, these observations were highly suggestive of DNA being the physical substance of heredity. Furthermore, in species with chromosomes, DNA was found in the nucleus — the site of hereditary transmission. Proteins were found in both the nucleus and *cytoplasm* (the extra-nuclear space; Figure 2.13). Finally, the fact that UV light was *mutagenic* — UV light altered

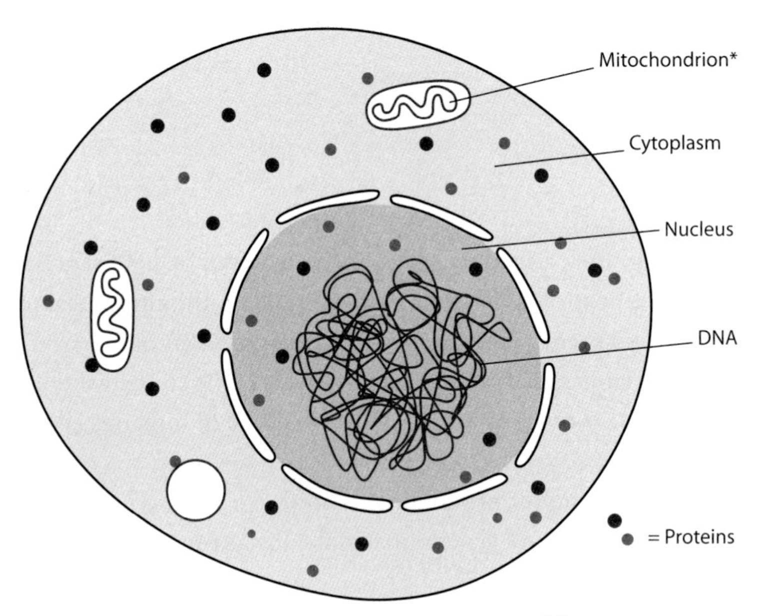

Figure 2.13. Basic elements and partitions of the cell. The two major partitions of the cell are the nucleus and cytoplasm. Proteins are found in both; DNA is found in the nucleus, but not the cytoplasm. *Technically, DNA is also found in another subcellular compartment, the mitochondria — a fact which is discussed at length in later chapters.

genetic material — pointed to DNA. The mutagenic activity was highest in the wavelengths of light in which DNA highly absorbed light, arguing that DNA was the substance of heredity in species with chromosomes.

Thus, traits were traced to the DNA in chromosomes.[16]

All of this evidence was compelling, but not complete. If DNA was linked to traits, how could nucleic acids physically carry the information for red hair, chin shape, zebra stripes, pine needles, and all the other traits that characterize the diversity of species on earth?

The answer to the *how* of DNA function was bound up in the structure of DNA. Many lines of evidence constrained the path to elucidating this structure. First and foremost, any potential structure for DNA would have to demonstrate how it could carry complex hereditary information. If the structure was simply repetitive unchanging units, it would imply a situation much more akin to the discarded tetranucleotide hypothesis than to a bona fide basis for heredity. In addition, any potential structure for DNA would have to satisfy the ratios that Chargaff discovered. In other words, the structure would have to explain why the amount of A and T in a cell were roughly equivalent, and why the amount of G and C in a cell were about the same.

Chemically, a proposed structure for DNA would need to be stable. If DNA was indeed the molecule of heredity, it would need to be stable over many generations. Since elephants produce elephants each generation, and snakes more snakes, and eagles more eagles, the structure of DNA would need to explain this fact. In contrast, a delicate molecule would fail to account for this consistency in traits.

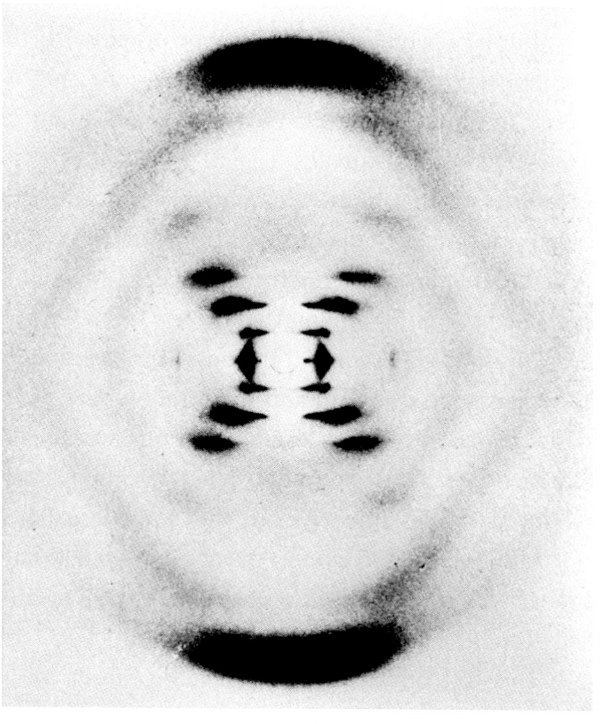

Figure 2.14. X-ray image of DNA.

Finally, the structure of DNA would have to suggest a means by which it could be replicated. Without faithful transmission, hereditary information would be diluted and extinguished in just a few cell divisions.

As science moved into the 1950s, indirect observations on the structure of DNA became available. Because the molecular structure of DNA was too small to be seen with visible light, even

Figure 2.15. Solution to the structure of DNA. James Watson (left) and Francis Crick (right) assembled a physical model of DNA that solved the many constraints on the three-dimensional structure of DNA.

high-powered microscopes could not find the answer. However, x-rays could penetrate the spaces between the parts of the DNA structure. The x-ray image of DNA (Figure 2.14) was critical to orienting the molecule.

The limited components of DNA quickly narrowed the number of possible structures. For example, the carbohydrate and phosphate elements of nucleotides did not possess complexity such that they could carry reams of information. The carbohydrates and phosphates of nucleotides came in only one version. The nitrogenous bases, however, came in four different possibilities (Figure 2.5). Hence, the former suggested themselves as structural components; the latter, as information carriers.

But how would the structure be oriented? If the carbohydrates and phosphates formed the backbone, which way would it face — in or out? Would it be a stiff, straight chain of bases, or would the structure have a twist to it? How many chains would compose the molecule — one? Two? Perhaps even three?

After much toil, in 1953, James Watson and Francis Crick finally put the pieces together — literally. Using cut-outs of the various subunits (Figure 2.15), Watson and Crick put together a structure that arranged the nitrogenous bases in a manner that could carry information and that explained the ratios among them.

In short, DNA formed the now-familiar twisted double helix (Color Plate 14).[17] The nitrogenous bases faced inward and the carbohydrate-phosphate backbone outward. The weak electrostatic attraction between the nitrogenous bases held the structure together. Combined with the fact that the carbohydrate-phosphate backbone was water-soluble, the attraction and backbone produced a very stable structure that was consistent with the known spatial relationships.

However, the structure was not completely stable. Since the attraction between the nitrogenous bases were weak, they could be broken with sufficient force. Put simply, the DNA double helix could be unzipped (Color Plate 14).

This fact immediately suggested a means by which DNA could be replicated during cell division. If DNA was unzipped, each chain could act as a template for synthesis of a new DNA chain. Since A always paired with T, and since G always paired with C, the sequence of

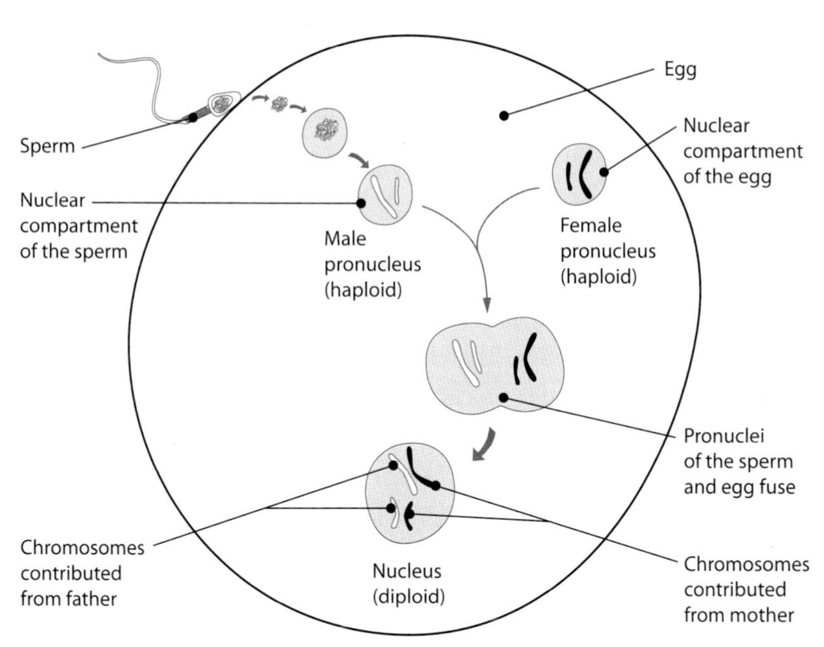

Figure 2.16. Fate of chromosomes during fertilization. When sperm fuses with egg, the sperm chromosomes ("male pronucleus") join the same nuclear compartment as the egg chromosomes (which began in the "female pronucleus"). However, the chromosomes are not melded together. Rather, these paternal and maternal chromosomes exist as individual entities, each carrying information from one of the parents. "Haploid" and "diploid" refer to states in which chromosomes exist as isolated entities or as pairs, respectively.

nitrogenous bases on one strand was sufficient to determine which bases belonged on the other — and vice-versa (Color Plate 14).[18]

The physical basis of heredity — the secret of life — had been solved.

Since the discovery of the double helix, the structural relationship between DNA and chromosomes has been uncovered. DNA doesn't exist in chromosomes as a long, straight stretch of helix. Chromosomes represent densely packed forms of DNA. In fact, to condense the packaging of DNA, cells wrap these helices around proteins in progressively higher levels of compaction to form the familiar chromosome shape (Color Plate 15). Thus, by watching chromosomes during cell division, we can watch the movement of DNA at a very zoomed-out level.

This perspective reveals what happens to DNA each generation. When sperm and egg fuse, the chromosomes of the sperm join the same nuclear compartment as the chromosomes of the egg (Figure 2.16). However, when this happens, the chromosomes are not melded together. The DNA helices along each of these chromosomes do not fuse. Rather, these paternal and maternal chromosomes exist as individual entities, each carrying information from one of the parents. In other words, pairs of chromosomes — and, therefore, of DNA helices — exist. Since both the father and the mother contribute an equal number of chromosomes, both parents make an equal contribution to the features of the newly conceived offspring. Some of the information from each parent might be hidden via the phenomena of dominance and recessiveness. But the information is still physically present.

In summary, the first major insights to mechanisms controlling trait behavior were slowly but steadily uncovered over a span of nearly 100 years. Hidden inside males and females was the microscopic code for visible traits. Males and females pass on this code via sperm and egg. The union of these cells produces the first cell — the *zygote* — which contains the instruction manual, not only for visible traits, but also for everything else that constitutes the individual members of a species.

⁂ ⁂ ⁂ ⁂

We observed at the beginning of this chapter that species are ultimately defined by their traits. We then observed that traits are defined by genetics. Therefore, the origin of species is a fundamentally genetic question. Genetics defines the edge pieces of the puzzle of the origin of species.

Yet the physical basis for heredity — the nature of the code of life — was not uncovered until nearly 100 years after Darwin wrote *On the Origin of Species*.

Consider the significance of this fact, especially in light of the puzzle pieces that Darwin did possess in 1859. Fossils aren't inherited in sperm and egg. A miniature adult is not passed on through germ cells. A geographic location is not the substance of heredity. Instead, a set of instructions (encoded in DNA) is. Darwin tried to assemble a jigsaw puzzle of sorts without any edge pieces to guide his progress.

Without this genetic knowledge, could Darwin have speculated intelligently on the origin of species? If he had no idea how traits were coded and inherited each generation, could he have identified the origin of a particular trait? Before the advent of genetics, would his explanation have had any hope of being accurate?

The history of genetics poses a second set of questions to Darwin. Not only was his question a fundamentally genetic one, but his specific answers to the origin of species were also deeply tied to this field. For example, Darwin proposed that all species had one or a few common ancestors. In other words, he said that the vast diversity of life belongs to one or a few family trees. Genealogical relationships are directly recorded in genetics — and nowhere else.

Furthermore, Darwin claimed that new species arose via the process of survival of the fittest, or *natural selection*. Natural selection is useful to evolution if — and only if — the survivors pass on their superior traits to offspring. In other words, the mechanism of evolution is inextricably tied to inheritance. Inheritance is directly recorded only in genetics.

Finally, Darwin placed the origin of species on a very long timescale. However, the process of inheritance also acts like a timekeeper, independently recording the length of time over which species appeared (a concept we'll explore in detail in later chapters). How could Darwin have written *On the Origin of Species* without any genetic data to test his ideas? Since both his question and his hypotheses were deeply tied to inheritance, what prompted him, not only to pen, but also to vigorously argue for his proposal?

When Darwin wrote his most famous work, he took a scientific risk of massive proportions.

Chapter 3
Cracking the Code

In a jigsaw puzzle, edge pieces constrain the arrangement of the center pieces. Since species are defined by their traits, the origin of traits constrains the puzzle of the origin of species. The origin of leopards and cheetahs depends on an answer to the origin of spots. The origin of toucans relies on the answer to the origin of large, colorful beaks. The origin of the blue whale is bound up in the origin of baleen. The origin of scorpions and the origin of stingers go hand in hand. The answer to the origin of traits represents the edge pieces to the puzzle.

In the previous chapter, we took steps toward understanding the origin of traits — but we did not reach the answer. For example, we observed the behavior of traits at the visible level and traced the control of this behavior to the DNA double helix. Nevertheless, we never uncovered *how* DNA controlled the behavior of traits.

The mystery of the *how* concealed the answers to several critical questions. Could the mechanism by which DNA controlled the behavior of traits be altered? Could it be changed to an entirely different program? Could leopards become whales? Could toucans change into scorpions? Could jellyfish become jaguars? The answers to these questions awaited the discovery of the mechanism by which DNA interfaced with traits.

Standing in the way of the answer were several paradoxes. First, in each generation, all traits are erased — only to be rebuilt again. This fact is a curious phenomenon in its own right. It's one thing to observe red hair appear and disappear on a family tree. It's something entirely different to discover that all traits — hair color, facial features, hands, legs, feet, etc. — are absent when sperm and egg meet, yet eventually appear in the adult.

Second, when this massive amount of change was compared to the theoretical changes involved in the origin of species, the paradox deepened. For example, consider some of the questions we have asked of species. Can a fish spawn a spider? Can elephants give birth to giraffes? Could butterflies sire birds? The differences between these pairs of creatures are striking — fins versus eight legs, trunks versus long necks, scaled wings versus feathered wings. The morphological changes required to transform one of these creatures into another are numerous. But none of these theoretical transformations are as profound as the transformations that occur during the process of development. For example, all of the species transformations listed above are between species with heads (spiders have fused heads), trunks, limbs, respiratory systems, digestive systems, and excretory systems. To go from a single cell to, say, an adult zebra, far more visible change is required. At fertilization, head, trunk, limbs, etc. are absent.

Third, despite this enormous upheaval in morphology, the outcome of the developmental process was extremely consistent. Development of sperm and egg faithfully results in the same traits from generation to generation. In fact, it results in consistent *species-specific* traits. It must — or we would have no concept of species in our vocabulary. If each species could spawn something entirely different every generation, *species* wouldn't be a scientific term.

Now consider again what it would take to form a new species. For a fish to become a spider, significant morphological changes must occur — an endoskeleton must transform into an exoskeleton, fins must become legs, an aquatic form of respiration must transform into a terrestrial form of respiration, etc. In other words, for new species to form, the developmental process would have to become less consistent.

Today, a small measure of developmental inconsistency occurs. Though the process of development produces a very consistent, species-specific outcome each generation, it doesn't produce *identical* individuals each generation. For example, you differ from your parents. All three of you are clearly human, but you are not identical to your parents. As another example, no two zebra individuals are alike. Again, zebra parents and zebra offspring are both members of the same species, but none of the three are identical. A little bit of change happens each generation. But is the accumulation of these little changes over time enough to produce a new species?

The solution to these paradoxes held the key to discovering the origin of traits — the edge pieces to the puzzle — and the origin of species.

Prior to the discovery of the DNA double helix, candidate solutions to these paradoxes began to emerge. One candidate arose from the fields of physics and chemistry. For example, consider the physics of transforming a single cell into a complex adult. This single cell would have to overcome many chemical and physical barriers in order to produce an adult. In other words, left to themselves, physical structures don't spontaneously assemble into creatures with heads and tails. Instead, thanks to the Second Law of Thermodynamics,[1] they degrade — much like an old library collects dust and eventually collapses without maintenance. To make a trunk or a long neck or a wing, cells would have to find a way to put energy into the system to overcome this thermodynamic barrier.

In the decades prior to the 1950s, the fundamental principles of cellular energy management began to be uncovered.[2] The major players were proteins. Unlike human assembly and construction processes, cells don't use electrical outlets or fossil fuels to power their division and growth. Instead, they utilize the sources available to them — nutrients in food stuffs. Once your stomach and intestines break down the meals you supply them (using, among other things, proteins to break down the food), the resultant chemical products are absorbed into the bloodstream (via protein channels, among other mechanisms) and passed on to cells where they are absorbed or transported inside (via proteins). In the womb, the developing and assembling baby receives these nutrients ultimately via the placental bloodstream.

Once inside the cell, more proteins break these tiny nutrients down into even smaller molecules. At the molecular level, sugars, fats, and other nutrients store an enormous amount of energy. The chemical bonds that hold the individual atoms of sugars and fats together can be broken. Doing so is an energetically favorable process — a process that proteins catalyze.

By analogy, food breakdown is like trying to roll a ball down a hill. Once you get the process started, natural forces like gravity takes over, and the ball picks up speed. Foodstuffs are like a ball sitting at the top of a hill.

However, to make the analogy more accurate, we'd need to put a little mound of dirt in front of the ball. With a little shove over the dirt, the ball will continue naturally on its own down the hill. But it won't go down without a little shove. In the cell, chemical bonds in nutrients don't immediately break. Because chemical "mounds of dirt" — in technical terms, *activation energies* — exist, chemical bonds aren't spontaneously severed. Proteins[3] help these processes get over the chemical "mounds of dirt." Once started down this chemically favorable path, the chemical breakdown reactions continue (Figure 3.1).

The reverse process — resynthesizing these bonds — is energetically prohibitive. For the cell to assemble itself into an adult body with a trunk or long neck or wings, it has to make more fats, sugars, and whatever other molecules it needs to make these structures. Self-assembly is a very energetically challenging process. Again, by analogy, it's like trying to roll the ball back up the hill. Going against gravity doesn't happen naturally. It requires energetic input — your muscles, sweat, and hard work (Figure 3.1).

Alternatively, you might be able to get a ball up a hill by rolling it down another. If you give the ball a big enough push at the top of one hill, it might have enough energy to roll down one and then roll up another. Similarly, to become a full-grown adult, our single cell has to find a way

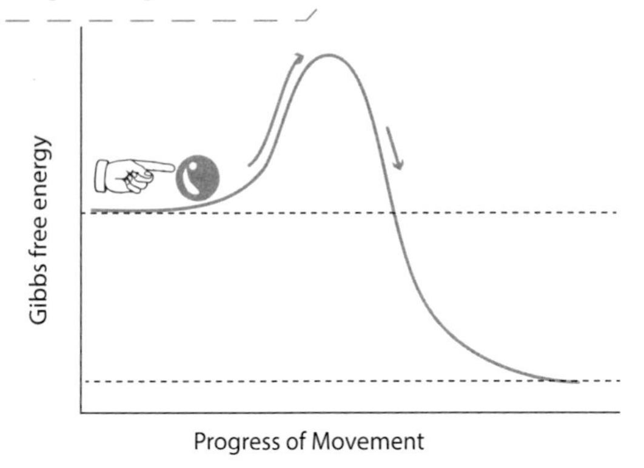

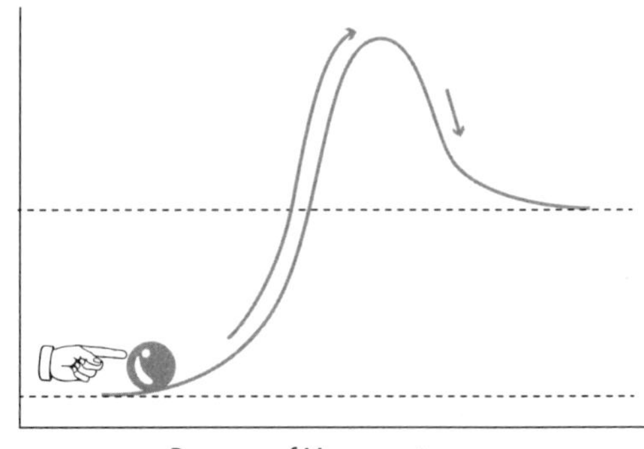

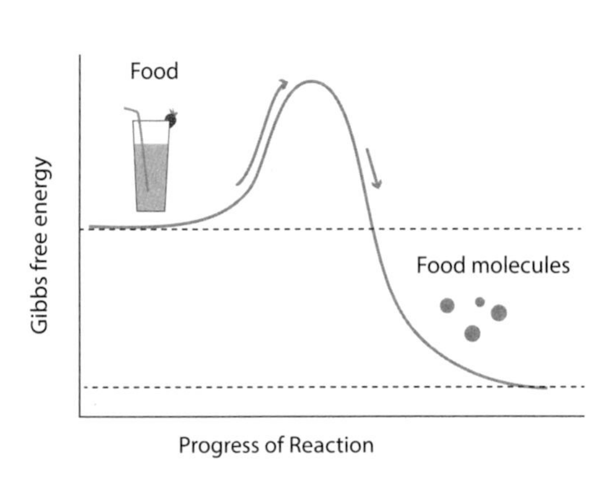

Figure 3.1. Illustration of basic cellular biochemical principles. Favorable chemical reactions in the cell can be compared to the process of rolling a ball down a hill. More realistically, they can be compared to rolling a ball over a small hump and then down a hill. These small humps represent activation energies, and proteins aid in reducing the size of this energetic barrier, allowing chemical reactions to proceed. Unfavorable chemical reactions can be compared to rolling a ball up a hill. Coupling these two rolling processes can allow unfavorable chemical reactions to proceed. In the cell, favorable chemical reactions (i.e., catabolic reactions) are coupled to unfavorable chemical reactions (i.e., anabolic reactions) via energy intermediates.

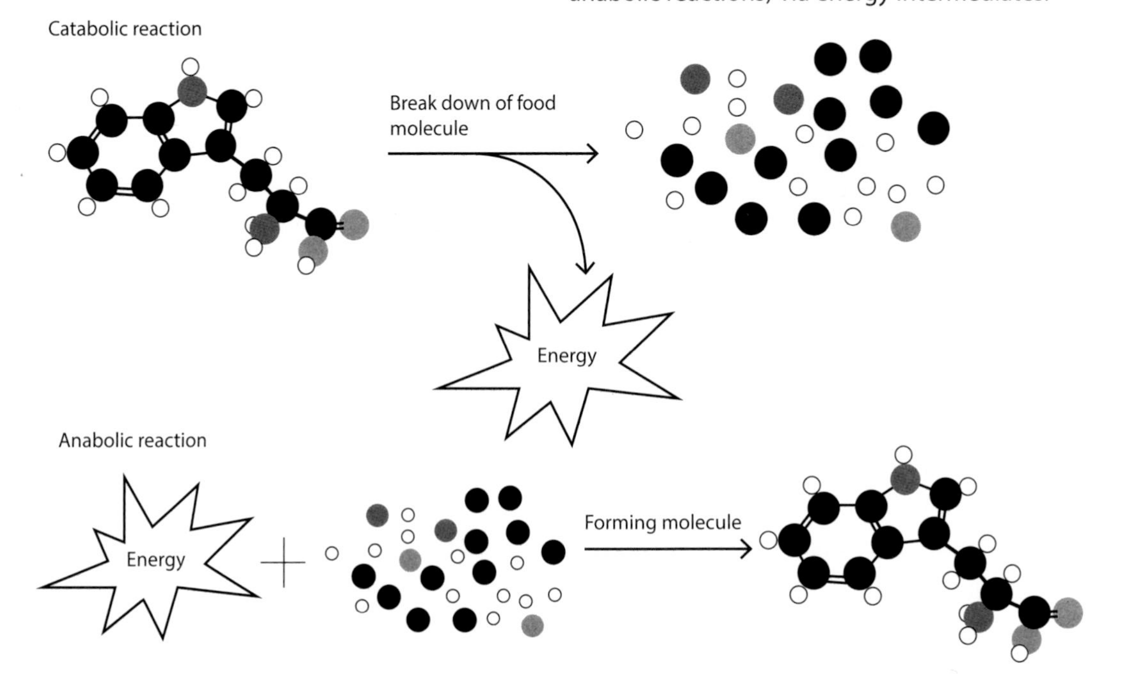

to couple the energetically-favorable breakdown of food stuffs to the energetically-unfavorable synthesis of body parts (Figure 3.1).

Proteins couple food breakdown to molecular synthesis. Not surprisingly, the process is complex. Several different types of nutrients get passed to cells (e.g., carbohydrates, proteins, fats). A mind-boggling number of individual chemical steps break down these various molecules to individual chemical components.

Theoretically, a single protein coupling machine could perform this task. In practice, it would have to be enormous in size — perhaps too enormous with respect to the size of a cell. Similarly, in building construction, humans don't solve the energy coupling problem with a one-size-fits-all machine. We don't have a single, massive structure at construction sites that pumps oil from the ground and then burns it in an engine that synthesizes concrete pillars, steel beams, windows, and dry wall. Instead, we refine the oil to gasoline at an oil refinery. A manufacturing plant assembles vehicles. The gasoline acts as an energy intermediate/currency that a wide variety of individual machines can use. A similar principle holds true for coal or nuclear power plants, except that the energy currency is electricity.

In the cell, division of labor also occurs. For example, many different proteins are involved in food breakdown. The result of this process is a form of energy currency. In the cell, the currency doesn't take the form of gasoline or electrical outlets. Instead, it's primarily in the form of a molecule called *ATP* (Figure 3.2; yes, it's one of the nucleotides we encountered in the previous chapter). Then other proteins couple the breakdown of ATP to the synthesis of cell assembly products.

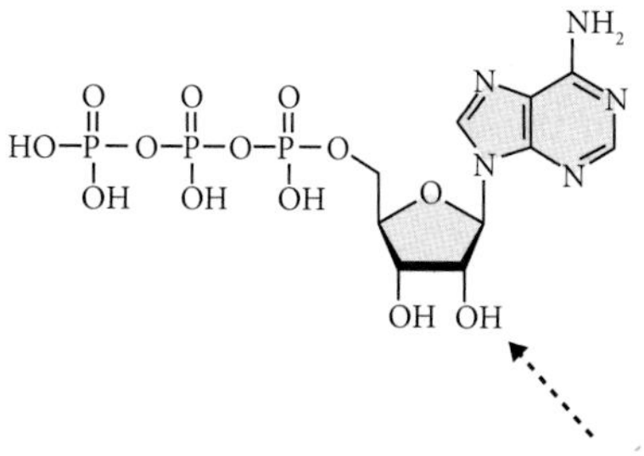

Figure 3.2. Adenosine triphosphate (ATP). Elements are represented with single letter abbreviations: carbon (C), nitrogen (N), oxygen (O), hydrogen (H), phosphorus (P). Ring structures (five sided, or joined five-sided and six-sided) consist entirely of carbon and hydrogen, unless otherwise indicated. Chemical bonds indicated by lines (single bond by one line; double bond by two lines). Subscripts denote number of atoms of the adjacent element. This structure is an RNA molecule (rather than a DNA molecule), as indicated by the extra "OH" linkage below the five-sided ring (highlighted with dashed arrow).

When we compare specific elements of this biological process to specific elements in the process of constructing a building, the cellular importance of proteins grows even larger. Both processes involve a nondescript starting point that transforms into a highly complex final result. Extending the analogy further, both construction projects need a way to transform energy and raw materials to something useful. Both need tools to connect the transformed energy and refined materials toward a desired end. At the building construction site, transformed energy and refined materials are taken for granted. We outsource these functions to power plants and manufacturing sites, respectively. Then we transport the products to the site. Tools are synthesized elsewhere and brought in as well. In contrast, the proteins of the cell act as the power plant, the factory, and the tools. Since the cell compartmentalizes these functions within itself, proteins do the jobs that humans spread out over a wide range.

Given the integral role that proteins had in transforming a non-descript single cell into an adult with diverse visible traits, proteins were a strong candidate for connecting DNA to the traits that define species.[4]

For a single cell to self-assemble into an adult, what we've discussed is a good start to this process. But only a start. Transformed energy, manufactured materials, and powered tools (i.e., powered by ATP) are critical to getting the cell from zygote to birth. Yet, while these components are necessary, they aren't sufficient.

If these things were all you had at a building construction site, you wouldn't see your final product form. Dump transformed energy, manufactured materials, and powered tools in a pile, and a building won't spontaneously assemble. Even if you send numerous construction crews, you can't guarantee the outcome you desire — unless you have a blueprint that the crews agree to follow. Similarly, the cell needs a blueprint by which to control the activity of its powered protein tools.

In one sense, we already know where the blueprint lies. Since the visible appearance of a species is consistent from generation to generation, so also must the blueprint be. If this is true, then the blueprint must be heritable. Since the physical basis of heredity is DNA, we have an obvious candidate for the cellular blueprint.

But how could a linear series of nitrogenous bases in a twisted ladder-like structure specify a three-dimensional animal with a trunk, long necks, wings, or red hair? How could DNA organize the protein activity of the cell?

The first answers took years of searching to find. The way in which DNA functions as the blueprint wasn't discovered until over a decade after the structure of DNA was solved. Remarkably, despite requiring years of experimental investigation, the code was cracked before the DNA sequences of any creature were elucidated.[5]

However, the first hints to the relationship between DNA and protein long predated the discoveries of Watson and Crick. In 1902, the British physician Archibald Garrod documented several cases of a human metabolic disorder — one that turns the urine dark. The distribution of the cases hinted at Mendel's principles, suggesting a genetic origin. Since metabolism is controlled by proteins, Garrod's publication intimated a link between genetics and proteins.[6]

In 1941, two Stanford investigators, Beadle and Tatum, documented a similar metabolism-Mendel link in fungi. They x-rayed the fungi to induce defects, and they found a mutant fungus that was unable to synthesize a key cellular vitamin. When they crossed mutant fungi to normal, the offspring showed ratios consistent with Mendelian genetics. Metabolism of vitamins — and, by implication, the proteins controlling it — were connected to genetics.[7]

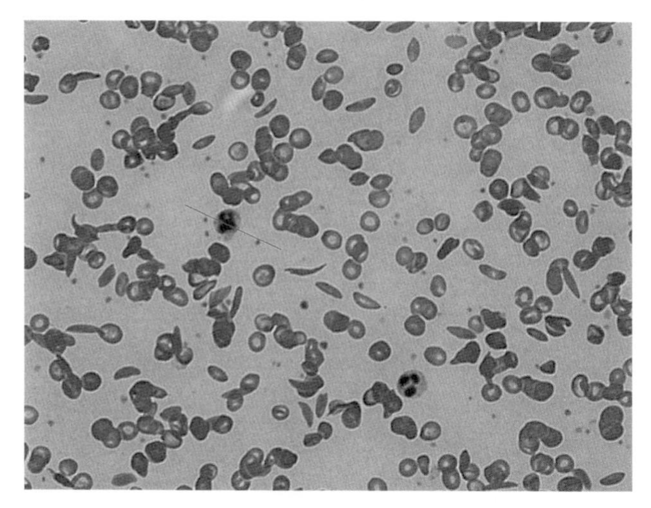

Figure 3.3. Sickle cell anemia. Among many normal blood cells, an abnormally shaped (sickle shaped) red blood cell is visible near the center of the image; the thin diagonal line points toward it.

But how? In humans, a red blood cell disease termed *sickle cell anemia* is inherited in a Mendelian fashion. Under the microscope, the primary manifestation of this disease is a switch in red blood cell shape from a biconcave disk to the sickle shape from which the disease gets its name (Figure 3.3). In 1949, a group of investigators at Caltech showed that, at its root, sickle cell anemia was due to a change in a protein. In the late 1940s and early 1950s, the British biochemist Fredrick Sanger was establishing that proteins have a specific sequence of amino acids.* [8] By 1956, German-born Vernon Ingram in Cambridge, England, demonstrated that the sickle cell protein change was in the specific amino acid sequence of the protein hemoglobin.[9]

Changes in genetics altered the amino acid sequences of proteins.

Naturally, these findings suggested that DNA *encoded* protein sequences. The major question was how.

In theory, something about the structure of DNA could have directly encoded the amino acid sequence. Perhaps proteins controlling red hair and zebra stripe formation were synthesized directly on DNA via some sort of physiochemical complementarity. Evidence soon accumulated that rejected this hypothesis.

As an alternative, Francis Crick (of DNA double helix fame) proposed an adaptor molecule between DNA and protein. Crick's idea was novel. Perhaps too novel — no adaptors had ever been discovered. Crick's hypothesis remained speculative.

Independent of the question of adaptors were basic questions about the nature of the DNA-to-protein code. Presumably, something about the linear sequence of nitrogenous bases must encode the linear sequence of amino acids in the protein. But was there a one-to-one correspondence? Would one nitrogenous base code for one amino acid?

Simple math ruled out this possibility. With only 4 possible nitrogenous bases, all 20 amino acids could not possibly find a unique signature in a single base pair** code. If 2 base pairs coded for a single amino acid, then only 16 of the 20 amino acids could theoretically have a DNA basis (e.g., 4

* This discovery would garner him one of his two Nobel Prizes.

** "Base pair" is the term used when referring to nitrogenous bases in a DNA *double* helix (see Color Plate 14).

possibilities at the first base pair * 4 possibilities at the second base pair = 16 possible base pair combinations). The code must be a three-to-one, base-pair-to-amino-acid ratio (e.g., 4 possibilities at the first base pair * 4 possibilities at the second base pair * 4 possibilities at the third base pair = 64 possible base pair combinations). Or higher.

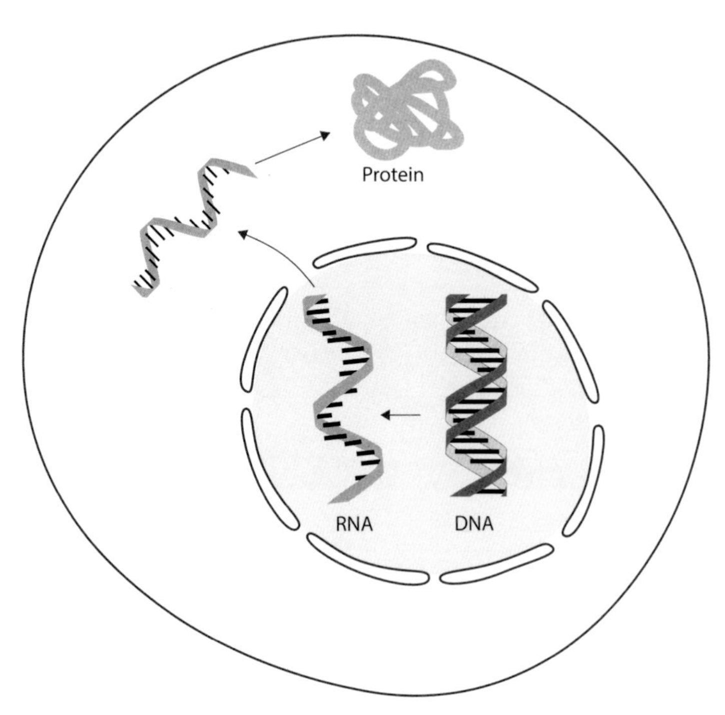

Figure 3.4. Discovery of a chemical messenger. Information in DNA is copied into a chemically analogous form — RNA — which is translated into proteins.

In the years following Watson and Crick's 1953 publication, several independent lines of evidence began to accumulate about the nature of the DNA-protein relationship. Odd, inexplicable links between protein and *RNA* appeared. Eventually, the conclusion was unavoidable. The link between DNA and proteins was indirect and mediated via a molecule that was chemically very similar to DNA — *RNA*. The information for the amino acid sequence in proteins was copied from DNA to RNA, then RNA carried the message to a different part of the cell where it was read and translated into the amino acid sequence of a protein (Figure 3.4).

Shortly after these discoveries, synthetic biochemical experiments in the lab cracked the secrets of this code. It was triplet in nature. Three DNA base pairs were required to code for a single amino acid. Theoretically, with 64 possible DNA base pair combinations, some amino acids would be encoded by more than one DNA base pair triplet. In practice, this prediction held true. Though some of the DNA triplets contained instructions for stopping the RNA translation and protein synthesis process, the vast majority encoded one of the amino acids[10] (Table 3.1).

In the years following these breakthroughs, the flow of information I just described was immortalized as the *central dogma*. Information in DNA is transcribed (copied) into RNA form, and the RNA is translated into an amino acid sequence for protein (Figure 3.5).[11] *Molecular biology* (i.e., the study of life at the molecular level — at the level of chemicals and molecules) revolves around this central dogma.

In the decades following the discovery of the central dogma, specific details have been added to it. For example, tiny cellular machines synthesize a complementary copy of RNA on the DNA

		First nucleotide									
		U	*amino acid*	C	*amino acid*	A	*amino acid*	G	*amino acid*		
Second nucleotide	U	UUU	*Phe*	CUU	*Leu*	AUU	*Ile*	GUU	*Val*	U	Third nucleotide
		UUC		CUC		AUC		GUC		C	
		UUA	*Leu*	CUA		AUA		GUA		A	
		UUG		CUG		AUG	*Met*	GUG		G	
	C	UCU	*Ser*	CCU	*Pro*	ACU	*Thr*	GCU	*Ala*	U	
		UCC		CCC		ACC		GCC		C	
		UCA		CCA		ACA		GCA		A	
		UCG		CCG		ACG		GCG		G	
	A	UAU	*Tyr*	CAU	*His*	AAU	*Asn*	GAU	*Asp*	U	
		UAC		CAC		AAC		GAC		C	
		UAA	*Stop*	CAA	*Gln*	AAA	*Lys*	GAA	*Glu*	A	
		UAG		CAG		AAG		GAG		G	
	G	UGU	*Cys*	CGU	*Arg*	AGU	*Ser*	GGU	*Gly*	U	
		UGC		CGC		AGC		GGC		C	
		UGA	*Stop*	CGA		AGA	*Arg*	GGA		A	
		UGG	*Trp*	CGG		AGG		GGG		G	

Table 3.1. Triplet codon table.

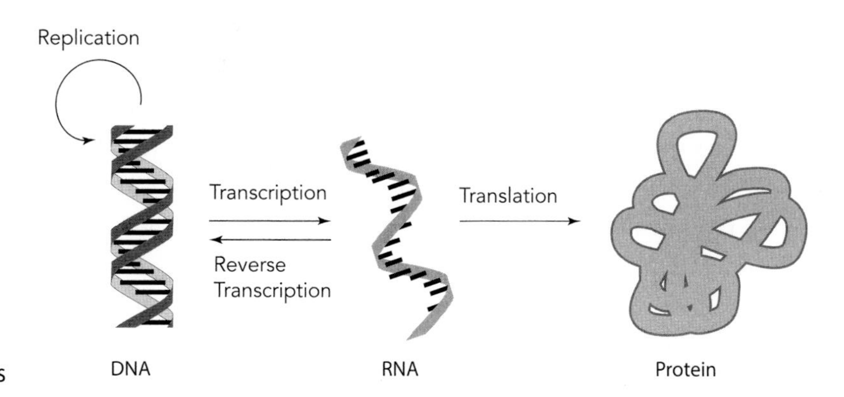

Figure 3.5. The "central dogma" of molecular biology. DNA is replicated during cell division. Information in DNA can also be transcribed into RNA, and, in some cases, the reverse can occur — information in RNA can be reverse transcribed into DNA. The information in RNA is translated into protein.

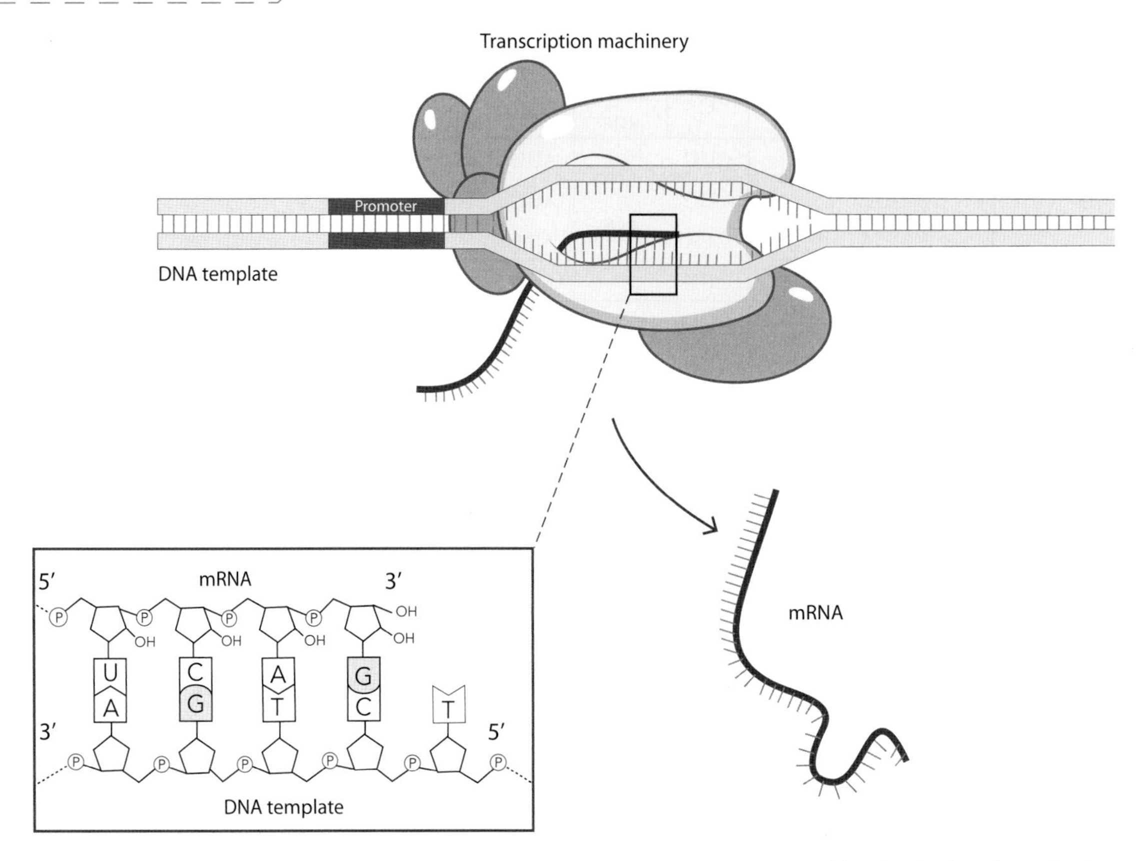

Figure 3.6. The process of transcription. Tiny cellular machines (symbolized by shaded ovals and other oblong shapes) recognize specific sequences in DNA (for example, a section of DNA, here labeled as a "promoter"). These machines then synthesize a complementary copy of RNA on the DNA template. Base pairing between DNA and RNA controls the sequence of the RNA molecule during RNA synthesis.

template. These machines recognize specific sequences in DNA, directing them where to start synthesizing the RNA (Figure 3.6). Base pairing between DNA and RNA controls the sequence of the RNA molecule during RNA synthesis. Since RNA is virtually identical to DNA, except for an extra oxygen atom, the same chemical principles holding the DNA double helix together are what makes the DNA-RNA pairing possible. Thus, once the DNA is unzipped, RNA base pairs with DNA as well as — if not better than — another DNA molecule does (Figure 3.6).

As the tiny reading and copying machine finishes synthesizing the protein-coding RNA, the DNA is re-zipped. The *messenger RNA* (RNA that codes for protein sequences) moves to another part of the cell where it encounters a translation machine called the *ribosome* (Figure 3.7). RNA-based adaptors (Crick's hypothesized intermediates were finally discovered) translate the message into an amino acid sequence (Figure 3.7). Again, base pairing between the messenger RNA and the adaptor RNA (called *transfer* RNA or *tRNA*) makes the translation process possible (Figure 3.7).

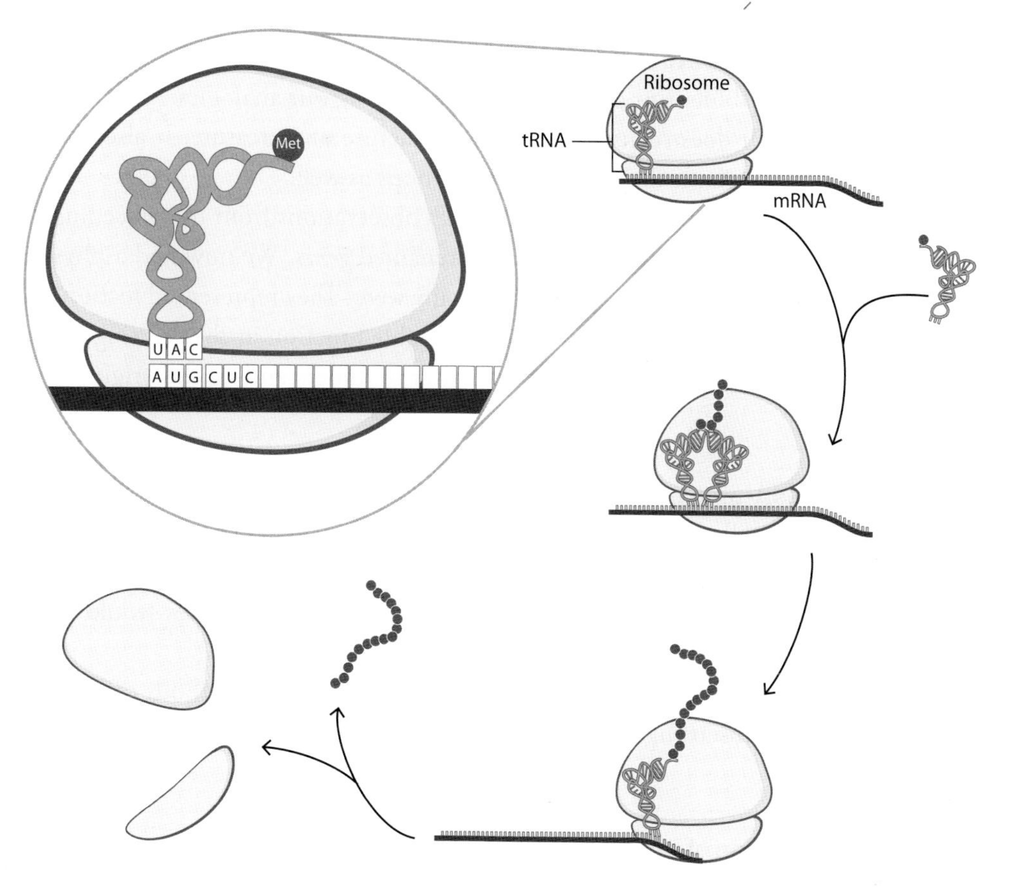

Figure 3.7. The process of translation. RNA that codes for protein sequences encounters a translation machine (the ribosome). RNA-based adaptors (tRNAs) translate the message into an amino acid sequence via base pairing between the messenger RNA and the tRNA. Each adaptor is chemically attached to a single amino acid (symbolized as a small circle). Because the adaptors match three base pairs at a time, the RNA code for protein is triplet in nature. As the ribosome reads through the messenger RNA molecule, the amino acids on each adaptor are taken off and attached to the growing protein chain, one amino acid at a time (because of space constraints, this diagram skips several of the individual amino acid attachment steps). Once the final amino acid is attached, the ribosome structure dissociates, and the protein chain breaks free.

Each adaptor is chemically attached to a single amino acid (Figure 3.7). Because the adaptors match three base pairs at a time, the RNA code for protein is triplet in nature (Figure 3.7). As the ribosome reads through the messenger RNA molecule, the amino acids on each adaptor are taken off and attached to the growing protein chain (Figure 3.7). Once the final amino acid is attached, the ribosome structure dissociates, and the protein chain breaks free (Figure 3.7).[12]

Together, these discoveries uncovered an unprecedented role for DNA in the cell. By our analogy to the construction of a building, we've discovered that the encoded products of DNA, proteins, are multi-taskers. They transform energy, manufacture parts from raw materials, and act as the tools of the cell. By virtue of the fact that DNA codes for protein, we can see that DNA acts, at least in part, as the blueprint of the cell. Yet, extending our analogies even further, DNA isn't a normal blueprint. It doesn't just code for the structure of a creature. It also contains the information for the power plants, factories, and tool shops as well.

These discoveries also shed new light on Mendel's observations from the preceding century. Prior to the 1950s, Mendel's unit factors had been renamed *genes*. We now had a physical basis for understanding what unit factors — genes — actually were. They represented sections of DNA that coded for a protein, which performed some detectable function in the cell (Figure 3.8).

The origin of traits now seemed to be just a matter of understanding the origin of genes.

The solution to the DNA-protein code took the scientific community one step closer to understanding how DNA controlled development. At the same time, other experimental results seemingly nullified this progress. Because the central dogma of molecular biology was cleverly solved without the knowledge of DNA sequences, the discovery of these sequences would represent a curious test of what was commonly accepted.

Even without actual DNA sequences, in the early 1950s indirect methods of DNA quantitation were disclosing an unsettling find. In these experiments, an investigator isolated cells from

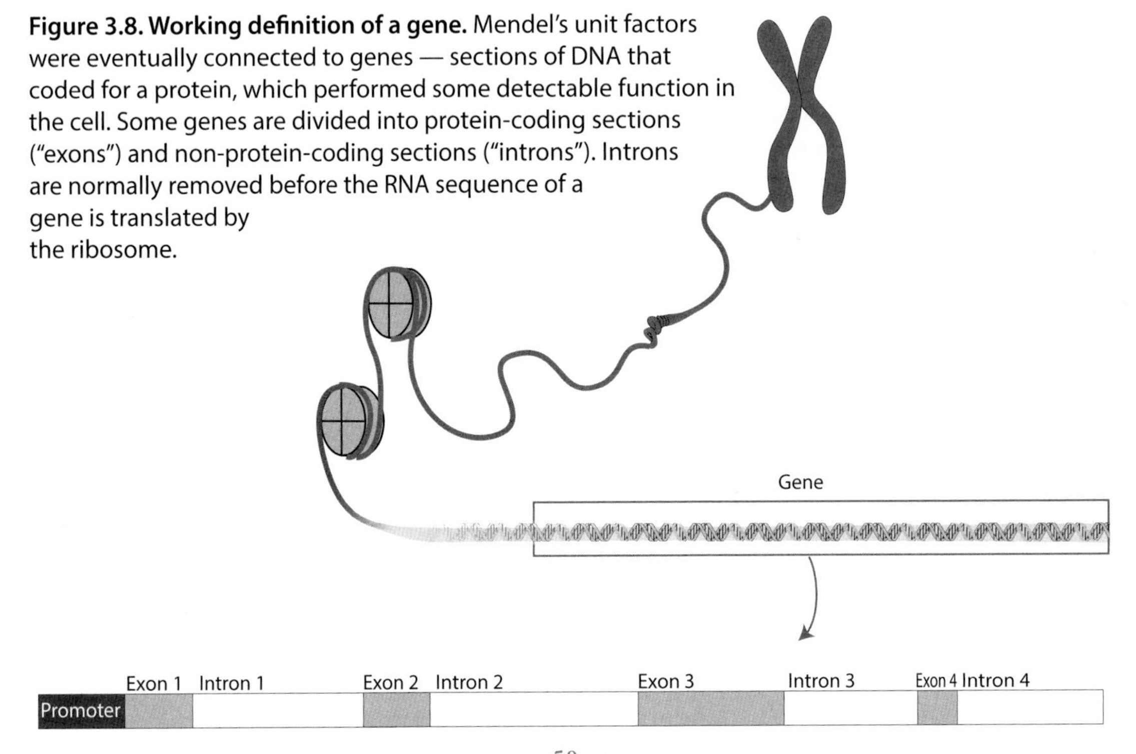

Figure 3.8. Working definition of a gene. Mendel's unit factors were eventually connected to genes — sections of DNA that coded for a protein, which performed some detectable function in the cell. Some genes are divided into protein-coding sections ("exons") and non-protein-coding sections ("introns"). Introns are normally removed before the RNA sequence of a gene is translated by the ribosome.

a species, estimated the number of cells in the sample under a microscope, and then chemically quantified the amount of DNA present. Dividing the amount of DNA by the number of cells yielded an estimate of the DNA per cell in a species.

In these experiments, you might expect simple creatures to have less DNA than more complex ones. Yet humans had the same amount of DNA per cell as a rat — and only slightly more than turtles and snakes. Among species without a backbone (e.g., *invertebrates*), sea urchins had roughly 6 times less DNA than humans, and snails only about 10 times less. Going the other direction on the DNA scale, some amphibians had 10 times *more* DNA than humans. Lungfish had nearly 20 times as much![13] If DNA was the blueprint of life, the amount of blueprint didn't have a clear relationship with the amount and complexity of the creature to be built.

By the late 1960s and 1970s, the first complete DNA sequences — the *genome* — from various species were trickling in.[14] The smallest genomes were solved first — the genomes of viruses. As might be expected from the central dogma, most of the DNA sequence in these genomes coded for proteins.

The first human sequences were from an unusual location. As we alluded to in a previous chapter, most of our DNA is contained in 46 chromosomes. But not all of it. In terms of physical location, these chromosomes reside in the subcellular compartment termed the *nucleus* (Figure 3.9). In the 1960s, DNA was detected elsewhere — in a different subcellular compartment termed the *mitochondria*.

With respect to the energy transformation functions we discussed earlier, mitochondria are the major sites (Figure 3.9). They contain numerous proteins involved in breaking down nutrients and coupling this to synthesis of ATP. Perhaps not surprisingly, then, when human mitochondrial DNA was sequenced in 1981,[15] investigators discovered that it was chock-full of DNA sequences that coded for energy transformation proteins. It didn't code for all the proteins that reside in mitochondria. But it was tightly packed with genes.

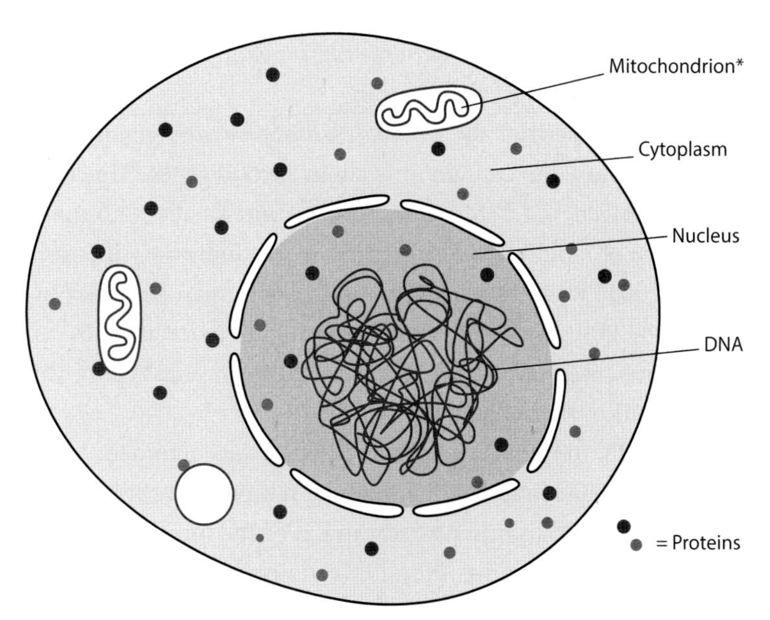

Figure 3.9. Location of DNA in the cell. The primary repository of DNA in the cell is the nucleus. *However, DNA is also found in another subcellular compartment, the mitochondria.

For the next several years, the main genomes that were cracked were from more viruses. The size of the sequenced genomes increased with time — from the 5,000 bases of the bacteriophage phi X174 genome in 1977 to the 237,000 base pair genome of the human cytomegalovirus in 1991. The evidence supporting the general expectations of the central dogma increased as well. The human cytomegalovirus genome was full of genes.

By now, the DNA sequences of the large animal genomes appeared within reach. In fact, in the late 1980s, discussions had already begun on whether the human genome should be formally tackled. Wisely, investigators planned a step-wise approach, choosing first to sequence genomes from key research organisms of increasing genome size. By targeting a bacterium, a yeast, a worm, and a fly, this approach would elucidate the genetics of a broad sampling of life. It would also allow DNA sequencing protocols to be optimized for the monumental task of sequencing the human genome.

Ironically, the first of these sequences to be published was not the smallest. In 1996, the 12 million base pair sequence of the single-celled baker's yeast (*Saccharomyces cerevisiae*) appeared.[16] The 4.6 million base pair genome of the common bacterium *Escherichia coli* (*E. coli*) followed in 1997.[17] Nevertheless, both of these genomes still generally followed the expectations of the central dogma — they were packed with protein-coding genes.

By 1998, the first animal genome sequence appeared. At 97 million base pairs, the genome of the roundworm, *Caenorhabditis elegans*, represented one of the largest genome sequences uncovered to date. However, unlike the genomes of viruses and single-celled creatures, only around 27% of the *Caenorhabditis elegans* genome coded for proteins.[18]

What was the other 73% doing?

In 2000, the fruit fly (*Drosophila melanogaster*) genome of 120 million base pairs was published. It appeared to contain only about 13,600 genes.[19] In contrast, the *E. coli* genome contained 3-fold fewer genes (4,288) — but was 26 times smaller (4.6 million base pairs). Fruit flies followed the pattern of rapidly increasing genome size and slowly increasing gene number.

In 2001, when the 3 *billion* base pair human genome[20] was finally elucidated, only about 20,000–30,000 genes were found — roughly twice as many as the fruit fly. Furthermore, these genes represented only 1.5% of the human genome. Again, vast stretches of DNA didn't fit the expectations of the central dogma.

In 2002, the mouse genome sequence was obtained. Like the human sequence, it was billions of base pairs long. Yet it contained few genes for its size — roughly the same number as the human genome.[21] In other words, much of the mouse genome did *not* code for protein.

What was this non-coding DNA doing?

In the modern era of molecular biology, one way to test the function of a DNA sequence is by removing it — or, in molecular biology parlance, knocking it out. The initial experiments were done on a small scale. For example, *gene deserts* are long stretches of DNA without gene sequences. Over 2 million base pairs (i.e., a small fraction of the mouse genome) of a gene desert region were experimentally removed from the mouse genome, yet, in the parameters that the investigators measured, the mice were normal.[22]

If DNA was the blueprint for a creature, why did most of the blueprint in these multicellular creatures seem to *not* represent instructions for tools, power plants, and factories? In other words, if the code for a species' morphology was hidden in DNA, why did the overwhelming majority of the sequence do something other than encode proteins? What was all this non-protein-coding DNA doing?

Genome sequences revealed additional puzzles — at the level of genes themselves. If the relationship between genes and traits was as simple as the central dogma implied, then the results of gene mutation experiments would have been unremarkable and predictable. For example, each gene knockout would have resulted in an altered trait. In contrast, laboratory tests revealed a diversity of results — that spanned a wide spectrum of outcomes.

For instance, mutant fruit flies had been familiar to geneticists for a century. Long before DNA was established as the physical basis for heredity, heritable changes to the appearance of fruit flies had been documented and studied. One of the most dramatic mutants results in flies with legs protruding from the place where antenna normally attach (Figure 3.10). In other words, these mutant *hox* genes appeared to control the development of multiple traits.

Figure 3.10. *Antennapedia* mutation in fruit flies. In contrast to normal flies, *Antennapedia* mutant flies possess legs where antenna normally occur.

At the DNA level, when the sequence for several of these types of mutants was obtained, it differed from the sequence in non-mutant flies.[23] The mutant DNA affected the activity of a protein that bound to DNA. This protein controlled the transcription, not of single genes, but whole batteries of genes. In general, it's as if these *hox* genes sit near the top of a molecular circuit. When the molecular switch is flipped — when the Hox proteins bind DNA — they regulate the expression of an entire program of development.[24] For example, the program for making a leg or an antenna.

The relationship between genes and traits was not as simple as it first appeared.

Other mutant genes had much less dramatic effects. For example, unlike fruit flies, mice often have, not one, but several copies of a gene in their genome. Where fruit flies might have one particular type of DNA binding gene, mice have several different versions of it (Figure 3.11 — *hox* genes).

This asymmetric relationship uncovered a perplexing pattern in mammals. Take the cell division cycle genes as a representative example. Since progression from a single cell to an adult mouse involves massive amounts of cell division, organisms must carefully regulate the process. (Cancer is an example of the process gone awry.) Not surprisingly, the mouse genome contains genes involved in controlling the cell division cycle. One particular type of regulatory protein is encoded by the *cyclin D* gene. Consistent with the pattern we observed with *hox* genes, mice possess three different versions of the *cyclin D* gene in their genomes—*cyclin D1, cyclin D2*, and *cyclin D3*.

Knocking out the *cyclin D1, cyclin D2*, or *cyclin D3* genes individually produces little effect.[25] If mice lack one of these individual genes, only a few tissues are effected; in other words, knocking out each of these does not result in embryonic lethality. For such a fundamental process as cell

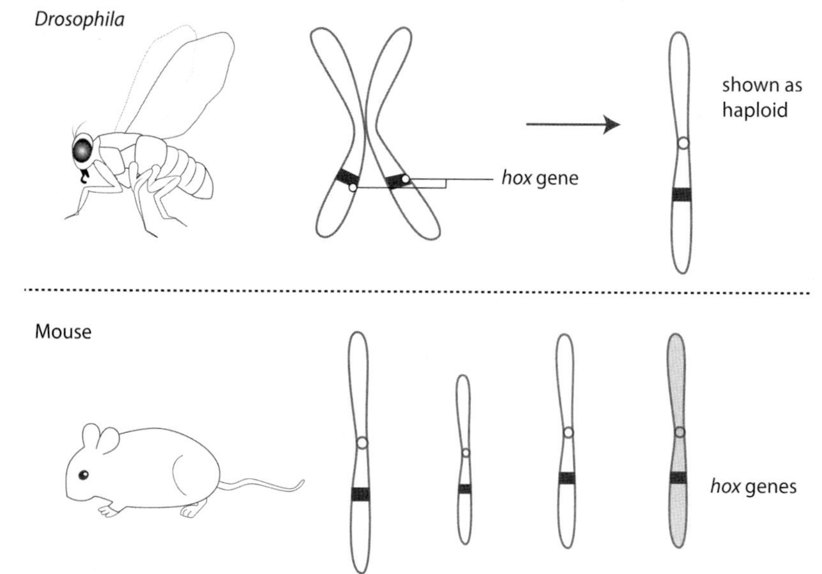

Figure 3.11. Apparent genetic redundancy in more complex species. Similar genes can be found in fruit flies and mice. However, where fruit flies might have only one version of a particular gene — in this case a *hox* gene — mice might have four versions. Technically, because chromosomes come in pairs, fruit flies have two copies (one on each chromosome) of a gene, and mice have eight. But a 1:4 relationship (a "haploid" state) is shown for simplicity.

division, this result is very surprising. You might predict each of these proteins to be essential for life. Instead, only the simultaneous removal of all three *cyclin D* genes from the mouse genome results in embryonic lethality.[26]

In other words, in our fruit fly example, some genes appeared to control multiple traits (*hox* genes). In contrast, in mice, some genes appeared to control very little — some genes appeared to be redundant.

When hundreds of genes were examined individually in mice, similar discoveries were made. Of mouse genes that have been knocked out individually, only around 40% are essential for mouse life.[27]

In other words, even when we restrict our focus to genes, the majority of mouse genes behave like non-protein-coding DNA — they sit in the genome, but appear to have no function.

These conclusions were not limited to mammals. Because of the smaller size and faster reproductive times in yeast and roundworms, more comprehensive tests of gene function were performed in these species. In yeast, over 6,000 genes were knocked out or inhibited — nearly the entire gene set.[28] Eighty-six percent of roundworm genes have been tested in a similar way.[29] In both cases, a small fraction of the total gene set was required for life. Thus, across animal and fungal *kingdoms* (i.e., one of the highest categories of biological classification), gene sequences behaved in odd and inexplicable ways.

✤ ✤ ✤ ✤

The resolution of these paradoxes arose from several cleverly designed experiments. For example, in all of the experiments on genome function that we discussed so far, the setting was the uniform conditions of the laboratory. In the wild, species face a diversity of conditions that are not present in the lab. This fact raised the possibility that the initial findings on genome function were an artifact of the laboratory setting.

In yeast, this hypothesis can be tested fairly easily. Since yeast are single-celled organisms, they can be grown in a wide variety of conditions, yet still be contained in a small physical space. One research team created over 1,000 different conditions. Then they repeated the screen for yeast gene function. Under these new experimental conditions, nearly every gene knockout resulted in defects under at least one condition.[30]

Another artifact of typical laboratory experiments is the narrow outcomes that investigators typically score. For example, in the initial comprehensive tests of roundworm gene function, the researchers recorded only whether the worms lived or died. Viability and lethality are very dramatic outcomes to test. (They also represent the simplest outcomes to test.) In contrast, in the wild, creatures fulfill many more functions than just survival. They must eat, grow, and reproduce, among other things. Conversely, a group of investigators repeated tests of gene function, but looked for more subtle, non-lethal effects of interfering with gene function. The majority of experimentally inhibited roundworm genes resulted in non-lethal defects.[31]

In light of these results in two very different creatures — creatures in different *kingdoms*, no less — we can revisit our initial conclusions in mice. With multiple organ systems and cell types, and with much longer lifespans than either yeast or roundworms, mice present a world of experimental conditions to test and a world of outcomes to score. Current methodologies barely scratch the surface in testing all of these. In short, in mice, the necessary, comprehensive, and labor-intensive experiments on genome function have not yet been performed. Until they are, it would be inappropriate to conclude that most genes — even most DNA — have little function.

In the meantime, one way to find a preliminary answer on the function of DNA sequences is via biochemical testing (i.e., tests that look for signatures of biochemical activity). These types of experiments are much easier to perform than genetic knock out experiments. Perhaps not surprisingly then, biochemical analyses have been done for nearly every base pair in the fruit fly,[32] roundworm,[33] mouse,[34] and human[35] genomes.

The trajectory of these experiments points toward pervasive, genome-wide function, including in the gene desert regions. In humans, preliminary biochemical evidence for function has been found for 80% of the base pairs in the human genome. Since protein-coding genes represent less than 2% of the total genome, a significant chunk of the 80% must represent gene desert regions.

But is this biochemical evidence relevant? Does a biochemical signature in a laboratory experiment have any bearing on how DNA might function in the wild? The history of these experiments suggests an answer.

For example, the first attempts to test the function of human DNA sequences in the laboratory were small. Only 1% of the total human DNA sequence was tested initially.[36] Had the results been largely negative, you might have predicted the end of this project, termed the *ENCODE project*. Instead, the results were so promising that they prompted a test of the remaining 99% of the human genome.

In 2012, the results of the genome-wide study were published.[37] These studies were the ones claiming evidence for function in around 80% of the human genome.[38] One of the ENCODE project researchers speculated that, eventually, evidence would accumulate for function in nearly 100% of the human genome.[39]

This expectation is plausible for at least two reasons. First, when DNA sequences are plotted against organismal complexity, only a subsection of the genome shows a good positive correlation. Surprisingly, the protein-coding DNA is *not* a good predictor of biological complexity. Instead, non-protein-coding DNA tracks much better.[40] This correlation implies that organismal complexity is encoded in the non-protein-coding section of the genome — which implies that this part of the genome is functional.

Second, despite the genomic comprehensiveness of the ENCODE project, it was biologically shallow. Consider: DNA is the instruction manual for building an organism, especially the traits that define each species. In creatures like mammals, these traits are generally absent at conception but present at birth. Therefore, from a purely theoretical perspective, much of the genome is likely used between conception and birth. Then, after birth, it might never be called upon again. The ENCODE project sampled hardly any of these conception-to-birth windows. I wouldn't be surprised if the evidence for function sharply increases as investigators sample a greater diversity of cell types, of tissues, and of temporal windows of biological development.

Thus, when examined historically, the evidence for genome-wide function is gaining strength.

Together, these experiments suggested that the majority — if not the vast majority — of gene and non-genic DNA sequences were functional. The function might not be essential for life. But the genome appears to contain enormous amounts of information that act in ways yet to be fully explored.

What *specifically* might non-protein-coding DNA be doing during development? Our analogy to construction sites for buildings suggests an answer. Consider the elements of the construction process that we've discussed: We've explored the need for transformed energy, manufactured materials, powered tools, and a blueprint. Now consider what's missing: On their own, these four elements won't automatically produce a skyscraper — or even a woodshed.

Why not?

Let's derive the answer by comparing the construction site for a skyscraper to that for a woodshed. The major difference between these sites is not at the level of tools, power sources, or manufactured parts. To be sure, the former would require a slightly different set of tools than the latter. Also, steel beams and air conditioning units would be required in the former, and not the latter. But hammers, saws, screwdrivers, and other common tools would be present at both. Wood, nails, screws, and other common materials would be found in both places. In other words, the biggest differences between the sites would not be in the types of tools and materials present. It would be in the manner in which these tools and materials are used — in the timing and location of their application.

This difference is not borne out exclusively in the blueprint. A piece of paper doesn't automatically result in the correct timing and location of the tools and materials. Rather, for these critical parts of the process to result in meaningful activity, a foreman must interpret the blueprint, coordinate the activity of worker crews, and ensure that the blueprint is correctly followed and enforced.

In our cellular construction project, what acts as the analogy for the foreman and human workers? In a sense, we're asking what takes the place of the human mind. It's one thing for the DNA to contain the instructions for how the pieces of the body are supposed to be put together. Enforcing and coordinating the necessary steps to make sure the pieces are

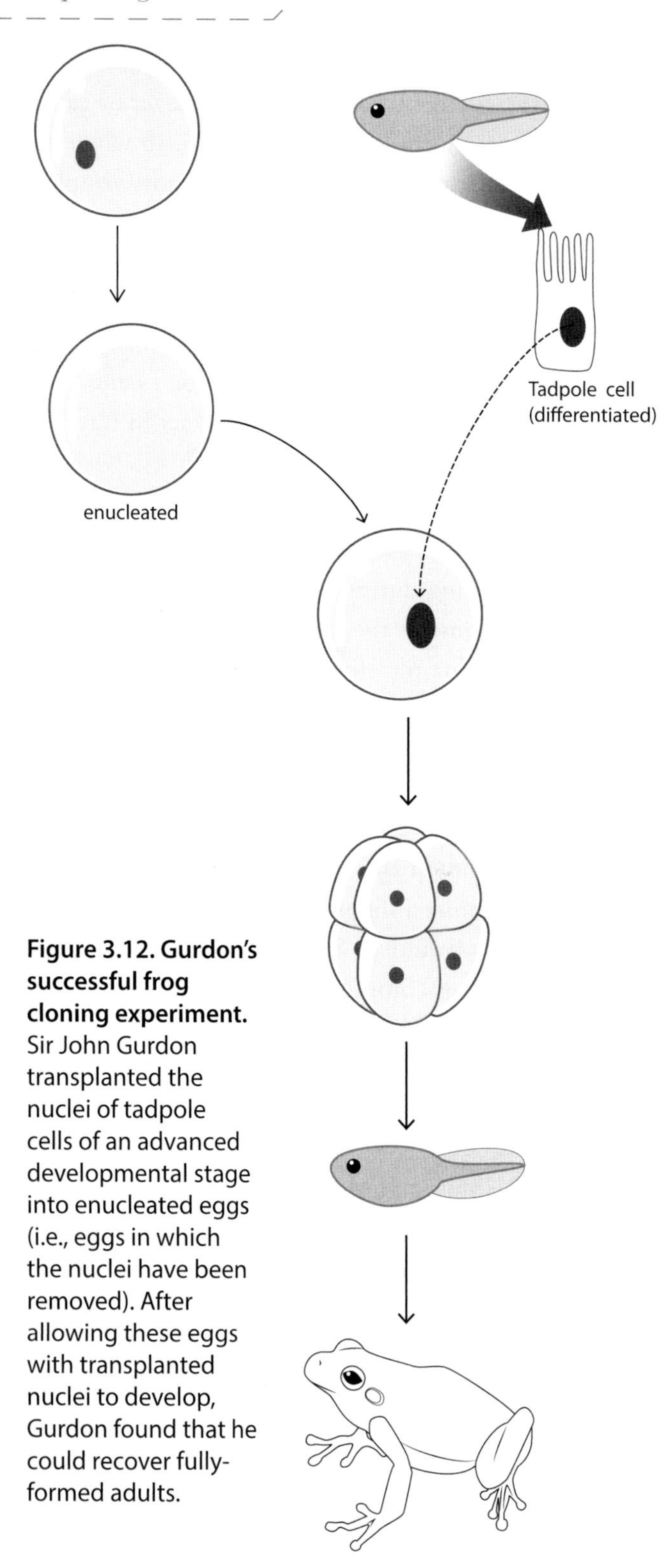

Figure 3.12. Gurdon's successful frog cloning experiment. Sir John Gurdon transplanted the nuclei of tadpole cells of an advanced developmental stage into enucleated eggs (i.e., eggs in which the nuclei have been removed). After allowing these eggs with transplanted nuclei to develop, Gurdon found that he could recover fully-formed adults.

put together in the correct temporal and spatial order is an entirely different task. In the cell, something must coordinate the temporal and spatial activity of the proteins — or a monster will result. What performs this task? And how does it do it? Does DNA get broken up and distributed among various cells?

Experiments by Sir John Gurdon in the 1950s and 1960s ruled out the latter. Gurdon was one of the first to perform cloning experiments. Other investigators before him had transplanted the nuclei of very early stage frog embryos into eggs in which the nuclei had been removed (i.e., *enucleated* eggs). Gurdon did the experiment with tadpole cells of an advanced developmental stage as sources of donor nuclei. After transplanting the nuclei into enucleated eggs and giving them a chance to develop, Gurdon found that he could recover fully-formed adults (Figure 3.12).[41]

In the context of what we've already uncovered about DNA, Gurdon's result suggested a curious conclusion. Since Gurdon used nuclei from highly developed cells as donors, he showed that development was reversible (at least, reversible at the level of single cells). In other words, since DNA is the stuff of heredity, the process of development appeared to *not* destroy the developmental potential of DNA. Furthermore, since nuclei — DNA — from advanced developmental stages still retained the capacity to recapitulate the developmental process, Gurdon's results argued that the DNA sequence was the same in both embryonic and more developed cells.[42] If development entailed shunting parts of the total DNA sequence — sections of each helix — into various cells, Gurdon's results would never have occurred.

In the decades following, other experiments extended Gurdon's conclusions. Cloning was successfully performed in mammals, and cloning was also achieved with fully developed cells from an adult.[43] In addition, DNA sequencing studies confirmed the preservation of DNA during development. Despite obvious visible differences under the microscope among various cells of the body (Color Plate 16), the DNA in these cells was the same.*

These results raised, again, the question of what substances in the cell substituted for a human foreman and workers. Clearly, since messenger RNA and protein levels differ among cells,[44] something must be coordinating the timing and use of these substances. But what?

In the decades following the discovery of the central dogma, numerous additional discoveries have uncovered the means by which protein synthesis is *regulated*. The most obvious way to regulate the synthesis is by controlling when and where the RNA transcription machinery operates. These machines are, themselves, proteins. However, they bind to sequences that do not, themselves, code for protein.[45]

Sometimes the transcription process results in RNA that doesn't get translated to protein. In fact, the DNA encoding these RNAs doesn't look like protein-coding sequence. Instead, the sequence falls in the category of non-protein-coding DNA. The transcription of this non-protein-coding DNA produces RNAs that bind to DNA and affect the binding of transcription proteins.[46]

After transcription of messenger RNAs, regulation acts at each subsequent step of the process leading to protein synthesis. Eventually, messenger RNAs get degraded by the cell. Some RNAs are degraded quickly; others have longer lifespans.

If the messenger RNA survives long enough to reach the ribosome, further regulation modulates protein synthesis. Some messenger RNAs are translated quickly; others, more slowly.[47] Like processes in the nucleus, some RNAs are synthesized, not for the purpose of being translated into protein, but for the purpose of binding to messenger RNAs and regulating the process of translation. In recent years, the number and types of newly discovered RNA molecules is expanding far beyond what anyone imagined.[48]

Together, the interactions among DNA, RNA, and protein coordinate the enforcement of the blueprint contained in DNA (Figure 3.13). In mind-numbing ways that we're just beginning to uncover, the newly fertilized egg of a creature reads the instructions in DNA on how

* Naturally, this blanket rule has nuance and exceptions. For example, as we discovered in the previous chapter, gametes have half the DNA of somatic cells; some cells fuse and double their DNA content; some immune cells alter their DNA in the process of making antibodies; etc. But the general principle holds true--development does not occur by the fractionation and partitioning of subsets of DNA sequence.

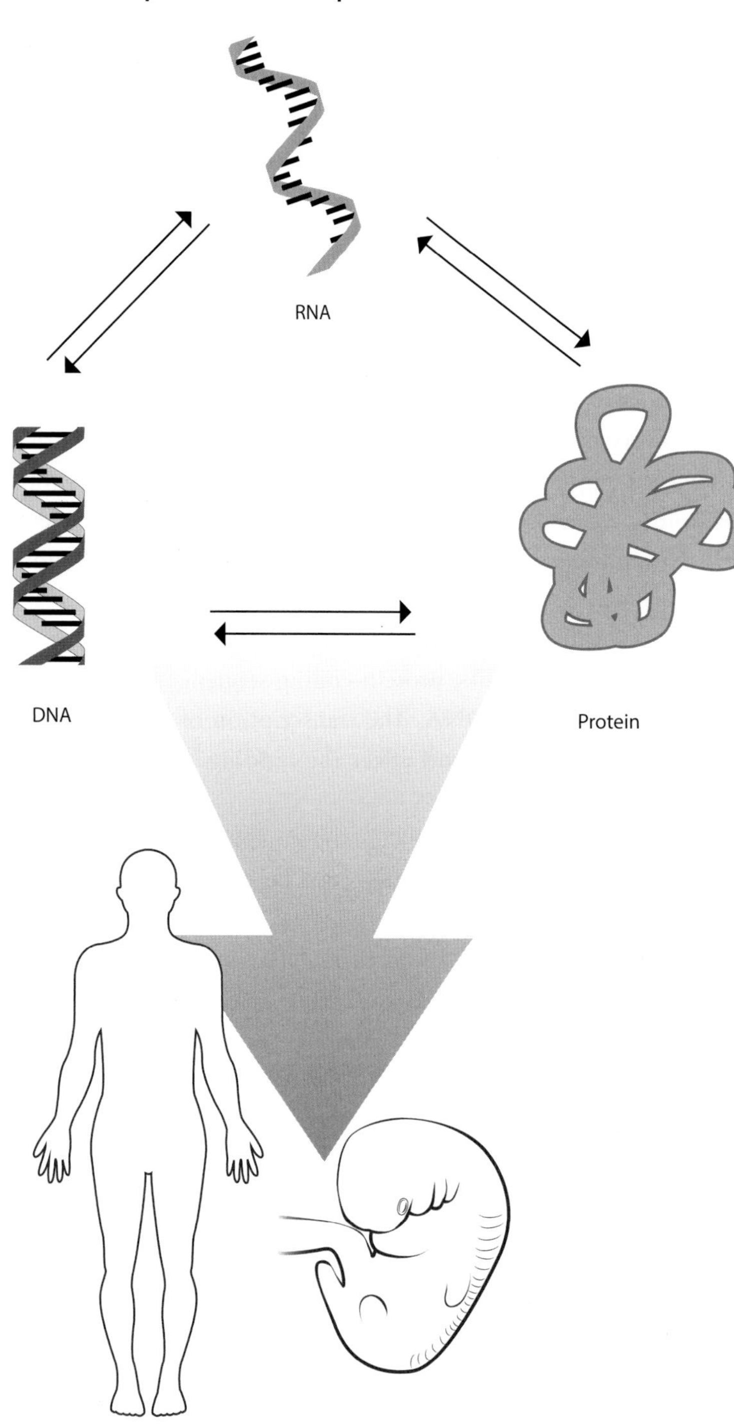

Figure 3.13. The interactions among DNA, RNA, and protein control the process of development.

to transform energy, manufacture materials, connect them via powered tools in the right ways at the right locations and the right times, and then uses these instructions to assemble itself into a three-dimensional creature of enormous complexity. And it does so with extreme consistency and precision each time. As we have observed, the non-protein-coding DNA is an integral part of this process.[49]

Despite the limitations of our current knowledge of function in the genome, the findings of molecular biology over the last several decades have uncovered a satisfying outline for the process of development. I used the word *outline* very deliberately. Even though the discoveries of the past several decades have uncovered a universe of activity inside cells, the process of development is one of the most baffling, mysterious, and unsolved puzzles in all of biology. Though great strides have been made since 1953, no one has discovered the step-by-step process by which an animal is built from a single cell. In fact, no one individual might ever possess this instruction manual. For one person to wrap their mind around all of the biological processes, steps, molecules, interactions, and developmental programs that are involved in development, they would need a lifetime of study — if not more. Consequently, possessing an outline of the process is a significant achievement.

With this outline in hand, we can begin to sketch plausible hypotheses on how species acquire their traits each generation. For example, we can speculate on how zebras get their stripes. At the most basic level, proteins must surely be involved in the laying down of stripes in the zebra embryo. Since many proteins catalyze chemical synthesis and degradation steps, stripe production might be effected by proteins that catalyze the steps of the synthesis of a dark- or light-colored pigment.

At a deeper level, we can begin to formulate ideas on how this synthesis is regulated. If the pigment-synthesizing protein is allowed to perform catalysis in any and every cell all the time, the individual will be a solid color throughout. Conversely, if the protein is inhibited in any and every cell all the time, the individual will be a solid color throughout — but likely a different color than the individual with universally uninhibited protein activity.* Instead, if the creature prevents the protein (and, therefore, the pigment) from being made except in certain patches of skin cells, the creature will have a very distinct splotchy or patchy coat color.

You can imagine the result if activity is limited to certain stripes on the skin.[50]

In other creatures, speculating on the development of certain structures is more difficult. For example, unlike zebra stripes, the giraffe's neck is much more than a surface decoration on an otherwise common anatomical pattern. Anatomically, like most mammals, giraffes have seven neck vertebrae.[51] Yet their necks contain much more than bone; they also harbor muscles, blood vessels, and nerves. Consequently, producing longer necks requires changes, not only to bone, but also to muscle, blood vessel, and nervous system development. At the molecular level, major changes to the standard developmental pathway for vertebrae would be required to produce the giraffe's signature structure.

In theory, these changes could take one of two forms. On the one hand, the standard developmental pathway for neck vertebrae production might involve changes to multiple proteins. Regulation might be altered for multiple individual proteins involved in blood vessel production, skeletal muscle production, and innervation. On the other hand, if master regulators of these developmental pathways exist, the regulation of these regulatory proteins (or, possibly, regulatory RNAs) might be changed.

A similar scenario exists in the case of the elephant's trunk. Unlike the faces of so many

---

* In some pigment synthesis pathways, other pigments represent intermediate steps in the process of producing the final pigment. Thus, if the final step of the process is inhibited, the pigment produced in the immediately preceding step might accumulate.

other vertebrates, the development of the trunk involves the production of tens of thousands of additional muscles. Again, at the molecular level, major changes would be required to a standard developmental pathway — this time, the pathway for the development of the face. The regulation of multiple proteins might be each altered individually, or the regulation of a master regulator might be changed.

Thus, from relatively simple tasks like stripe production to comparatively challenging tasks like vertebrae elongation and trunk production, modern genetics is beginning to outline the steps by which these structures are built.

Together, the observations of the last two chapters reveal how traits arise each generation. Despite the size and shape differences between sperm and egg, each carries the same number of chromosomes. Along these chromosomes, DNA double helices are tightly compacted. Along each double helix, the sequence of base pairs codes for RNA and proteins, and the interaction among these three molecules executes the developmental plan for each species.

The specific DNA differences among species explain specific aspects of the developmental process. For example, though visible traits are erased each generation, they are rebuilt in an extremely consistent manner. Consistent with this visible phenomenon, the vast majority of base pairs among individuals within a species are the same. For example, individual humans differ from one another at 0.1% to 0.6% of their base pairs.[52] Similarly low percentages hold true in horses,[53] donkeys,[54] and dogs.[55] Since 99% of the inherited DNA is the same each generation, it's no wonder that the developmental process produces a consistent output each time.

This less-than-one-percent difference also explains the other side of the developmental coin. For example, though the output of the developmental process is extremely consistent each generation, offspring are not carbon copies of their parents. In other words, a small amount of change still happens each generation. Since DNA codes for traits, this change must be due to underlying DNA differences between offspring and parents. A less than 1% DNA difference fits this pattern of small changes each generation.

Extrapolating these processes backwards in time, we reach the answer to the bigger origins question. What we've observed thus far explains the origin of traits — but only over each generation. If we want to understand the origin of species, we must uncover the origin of the first traits. Since traits are ultimately encoded by DNA, the origin of species is a question of the origin of DNA differences within and between species. The answer to this question reveals whether a fish can spawn a spider — and whether it ever did.

In other words, by analogy to a jigsaw puzzle, the DNA differences among species represent the edge pieces to the puzzle — they set the hard limits and constraints on a potential explanation for the origin of species.

When Darwin wrote *On the Origin of Species*, he had no knowledge of the genetic processes that we explored in this chapter. DNA wasn't recognized as the physical basis of heredity. No one had any idea how many DNA differences divided species. In fact, the DNA sequence of our own species wasn't solved until 2001. Since our species is the best-studied multicellular species on the planet, you can appreciate how recently the genetics of other species have been elucidated.

In other words, the answer to the origin of species can be uncovered for the first time right now.

Color Plate Section

Color Plate 1. Polar bear (*Ursus maritimus*).

Color Plate 2. Arctic fox (*Vulpes lagopus*).

Color Plate 3. Arctic hare (*Lepus arcticus*).

Color Plate 4. Black bear (*Ursus americanus*).

Color Plate 5. Eurasian red fox (*Vulpes vulpes*).

Color Plate 6. Antelope jackrabbit (*Lepus alleni*).

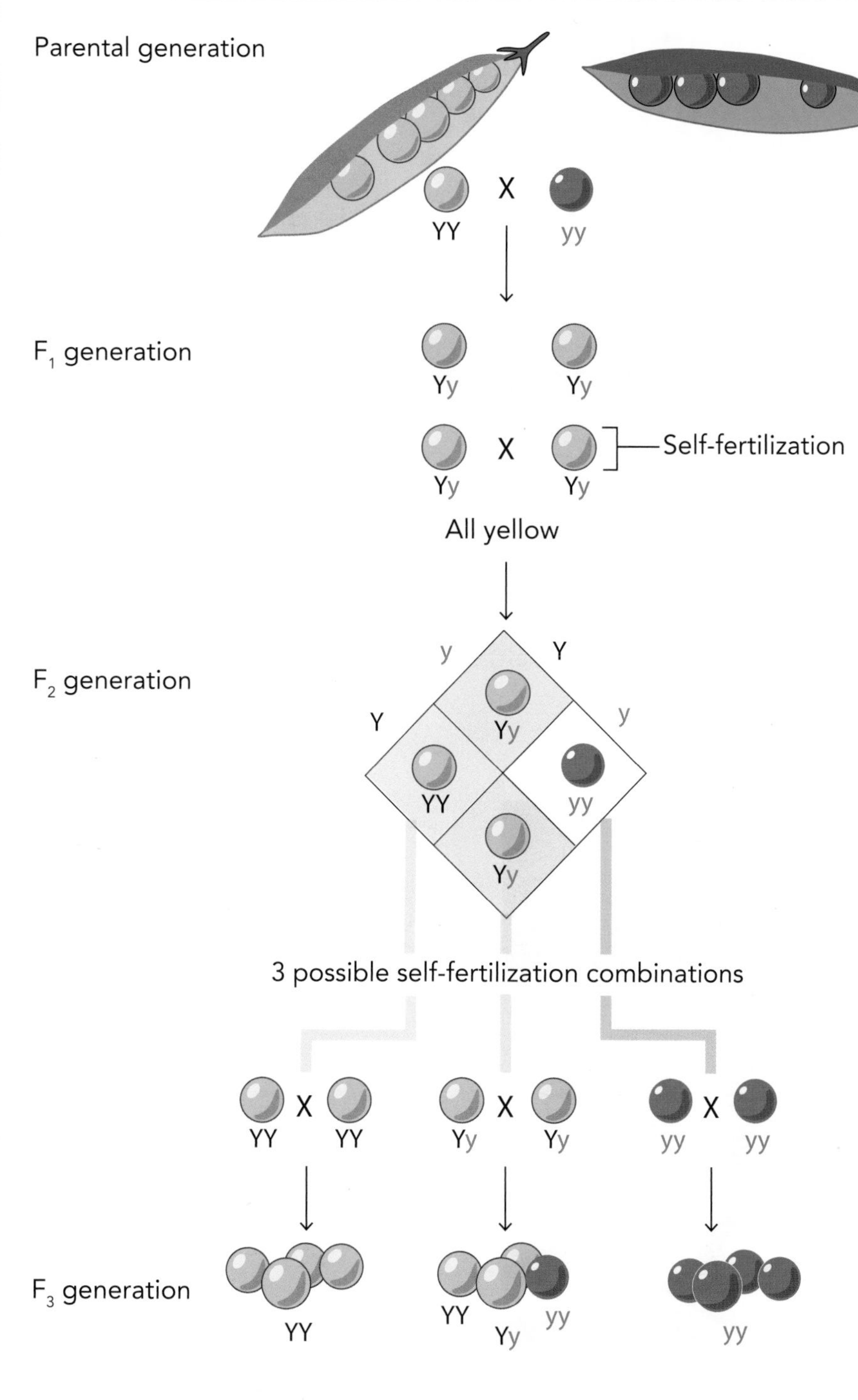

Color Plate 7. Mendel's single-trait crosses. Mendel crossed pea plants that were pure-breeding for yellow seeds with pea plants that were pure-breeding for green seeds ("Parental generation"). All the offspring were yellow ("F_1 generation"). When the F_1 generation plants were self-fertilized, the offspring ("F_2 generation") bore yellow and green seeds in a 3:1 ratio, respectively. When the F_2 generation plants were self-fertilized, three types of outcomes ensued. The F_2 generation plants with green seeds continued to bear offspring with only green seeds. About one-third of the F_2 generation plants with yellow seeds bore offspring with only yellow seeds. The other two-thirds of the F_2 generation plants with yellow seeds bore offspring with yellow seeds and green seeds in a 3:1 ratio, respectively. Mendel inferred that genetic information for seed color occurred in discrete unit factors, symbolized by "Y" and "y." By postulating that the differing versions of these unit factors segregated from one another each generation, Mendel was able to make sense of his results.

Color Plate 8. Hybrid offspring of horse-zebra cross (left) and donkey-zebra cross (right).

Color Plate 9. Partially striped hybrid offspring of donkey-zebra cross.

Parental generation

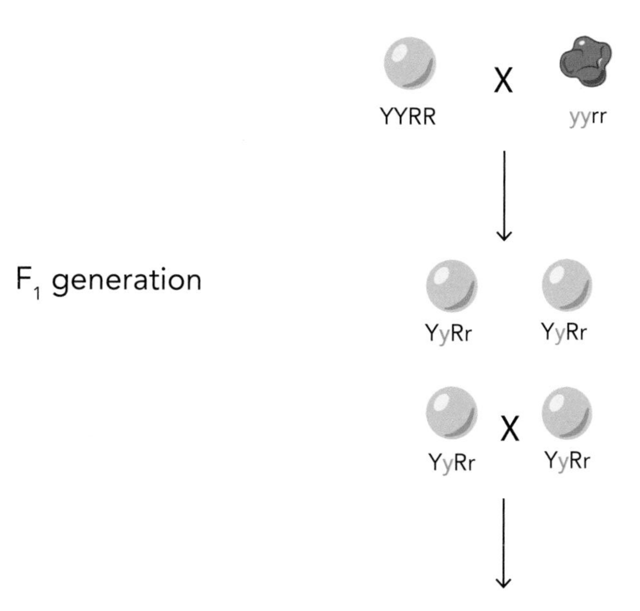

F_2 generation

	YR	yR	Yr	yr
YR	YYRR	YyRR	YYRr	YyRr
yR	YyRR	yyRR	YyRr	yyRr
Yr	YYRr	YyRr	YYrr	Yyrr
yr	YyRr	yyRr	Yyrr	yyrr

Color Plate 10. Mendel's double-trait crosses. Mendel crossed pea plants that were pure-breeding for smooth yellow seeds with pea plants that were pure-breeding for wrinkled green seeds ("Parental generation"). All the offspring were smooth and yellow ("F_1 generation"). When the F_1 generation plants were self-fertilized, the offspring ("F_2 generation") bore a diversity of trait combinations—smooth yellow seeds, wrinkled yellow seeds, smooth green seeds, and wrinkled green seeds in a 9:3:3:1 ratio, respectively. This ratio can be predicted if we treat unit factors for each trait as coming in two forms (e.g., symbolized by "Y" and "y"; or "R" and "r") that sort independently of one another. In the F_2 generation, these predictions take the form of a Punnett square. The Punnett square predictions can be derived by the intersection of each row and column. In other words, at the intersection of a particular row and column, the predicted outcome can be derived by combining the Y/y/R/r notation at the top of the particular column with the Y/y/R/r notation at the beginning of the particular row.

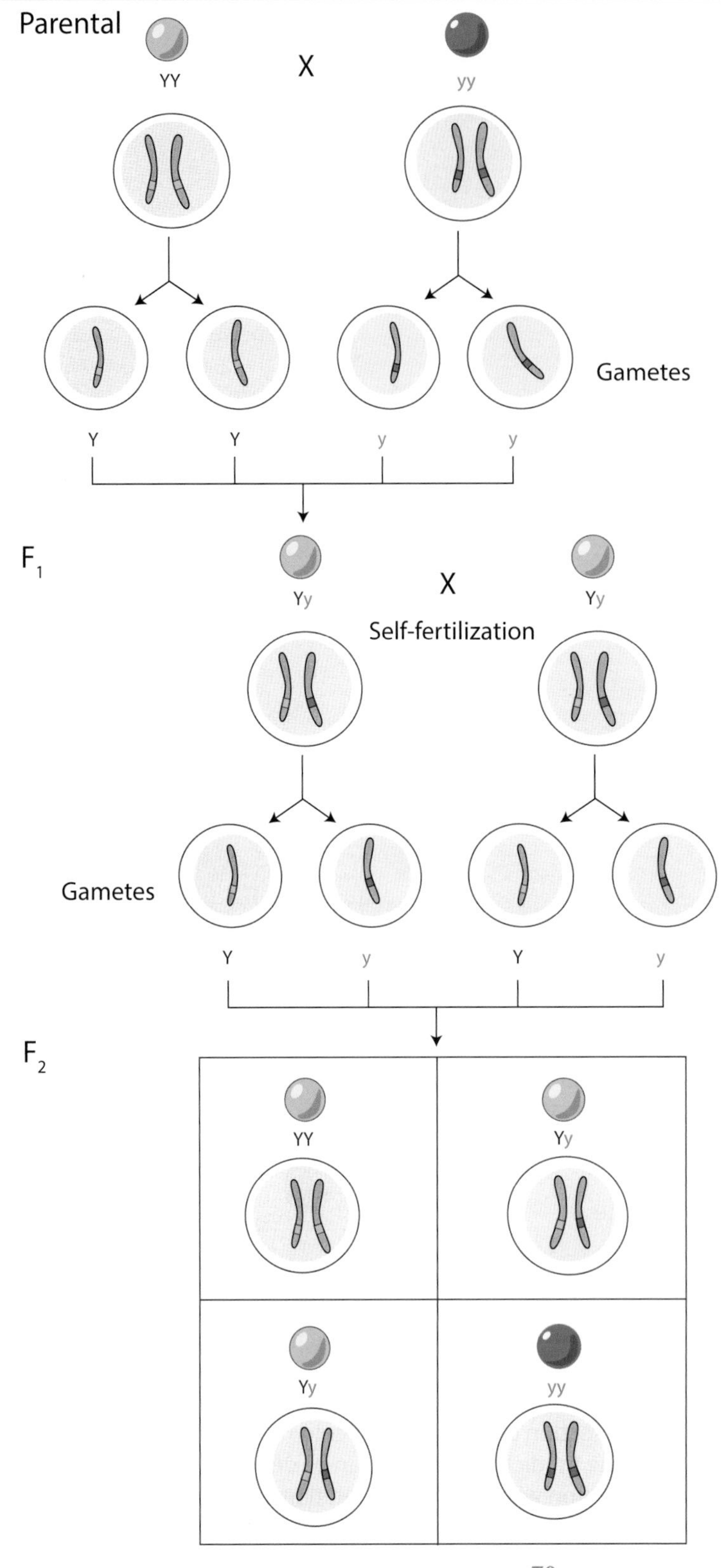

Color Plate 11. Correlation between chromosome behavior and unit factor behavior for a single trait cross. During meiosis, the behavior of chromosomes correlated with the behavior of Mendel's unit factors. In other words, the pairing of chromosomes in adults (i.e., the "parental" generation); the reduction of chromosome pairings to individual chromosome units in gametes; the segregation of these individual chromosomes in these gametes; and the re-pairing of these chromosomes in fused gametes in the next generation (i.e., the "F_1" generation) predicted the behavior of Mendel's unit factors. This chromosomal behavior — and the consequent predictions that it made for Mendel's unit factors — held true even in subsequent generations: for example, the F_2 generation.

In this diagram, the correlation between chromosomes and unit factors is visualized with color-coded bars on each chromosome. The bars represent the potential physical location for the unit factors for seed color that each chromosome encodes; the individual letters (i.e., "Y" and "y") are shown to make the correlation even more clear.

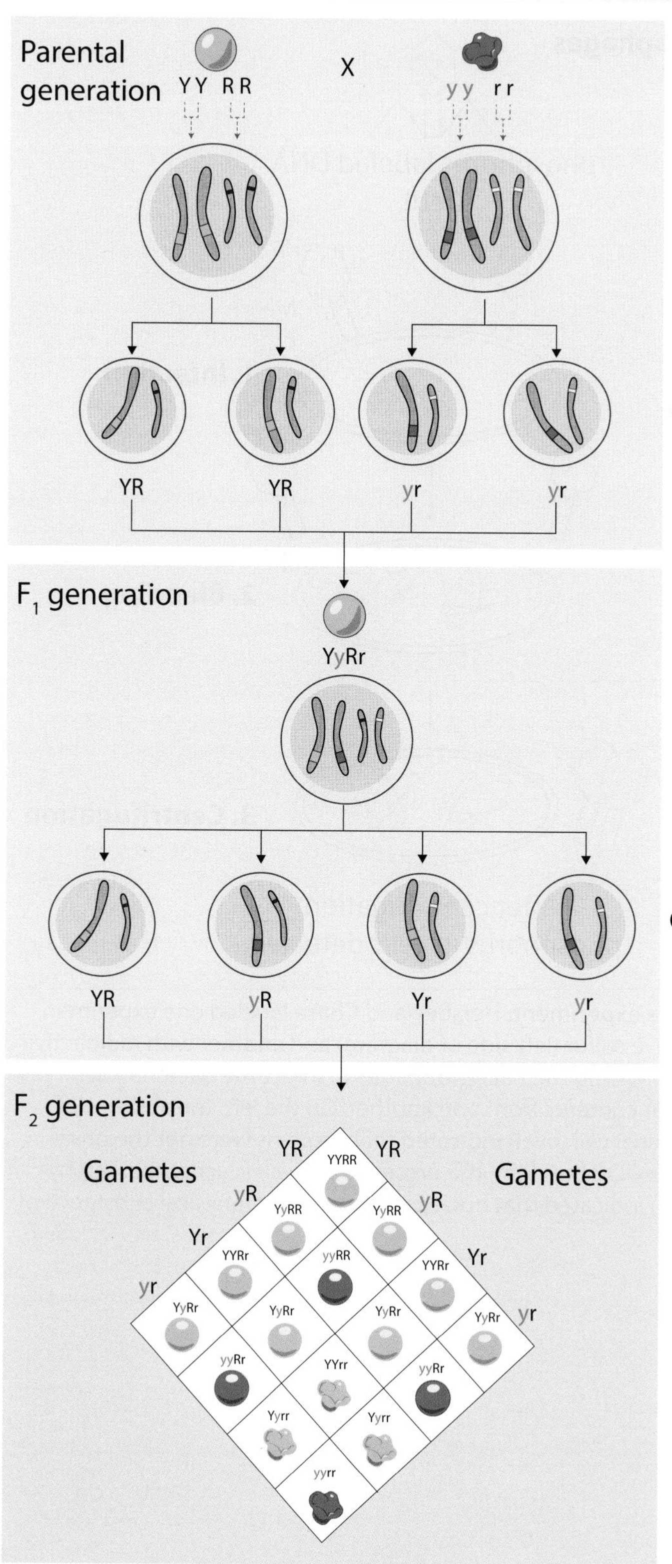

Color Plate 12. Correlation between chromosome behavior and unit factor behavior for a double trait cross. During meiosis, the behavior of chromosomes correlated with the behavior of Mendel's unit factors, even when multiple traits were observed. In other words, the multiple sets of chromosome pairs in adults (i.e., the "parental generation"); the reduction of each of these chromosome pairings to individual sets of chromosome units in gametes; the segregation of individual chromosomes in a chromosome pair in these gametes; the independent assortment of individual chromosomes from separate chromosome pairs in these gametes; and the re-pairing of these chromosomes in fused gametes in the next generation (i.e., the "F_1 generation") predicted the behavior of Mendel's unit factors. This chromosomal behavior — and the consequent predictions that it made for Mendel's unit factors — held true even in subsequent generations: for example, the F_2 generation.

In this diagram, the correlation between chromosomes and unit factors is visualized with color-coded bars on each chromosome. The bars represent the potential physical location for the unit factors for seed color or for seed shape that each chromosome encodes; the individual letters (i.e., "Y" and "y"; "R" and "r") are shown to make the correlation even more clear. Chromosomes are not depicted in the F_2 generation due to space constraints.

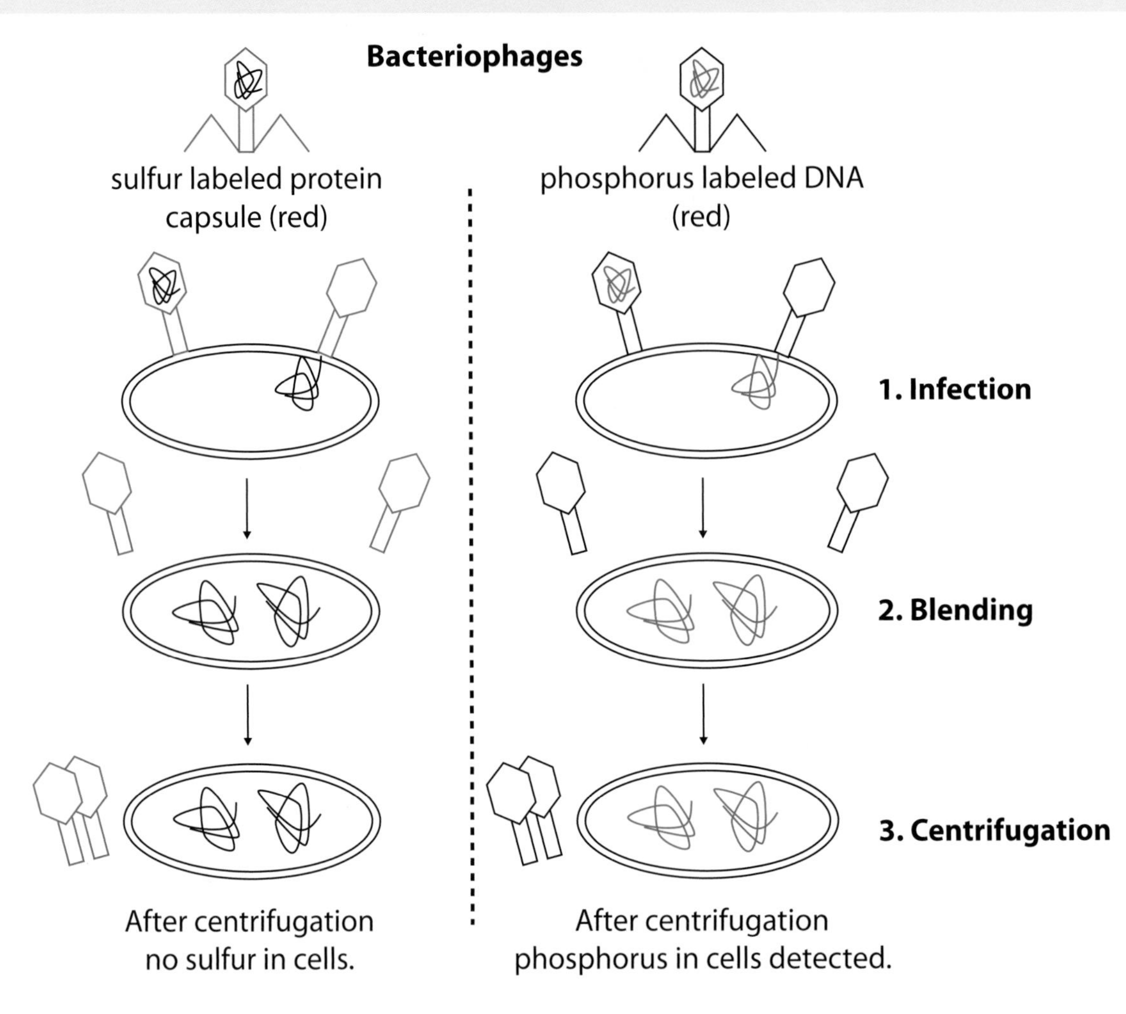

Color Plate 13. Diagram of Hershey-Chase experiment. Hershey and Chase labeled one experimental group of bacteriophages with radioactive sulfur (left side of diagram) and another with radioactive phosphorus (right side of diagram). After allowing the bacteriophages to infect the bacteria, they separated the intracellular and extracellular contents from one another. On the left, the absence of protein (marked by sulfur) within the bacterial cell (oval) indicated that proteins were not the physical substance of heredity in bacteriophages. On the right, the presence of nucleic acids (marked by phosphorus) within the bacterial cell (oval) indicated that nucleic acids were the physical substance of heredity in bacteriophages.

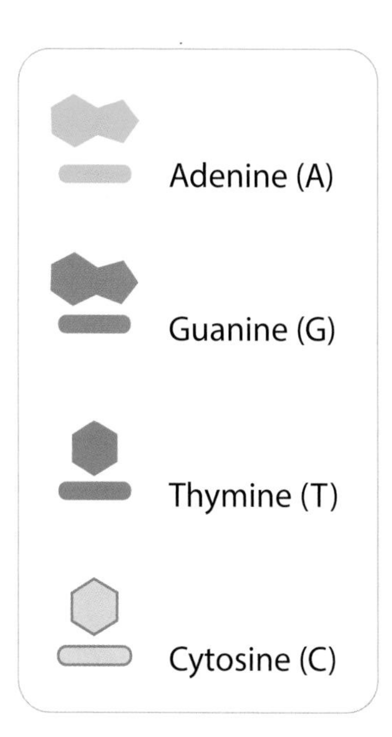

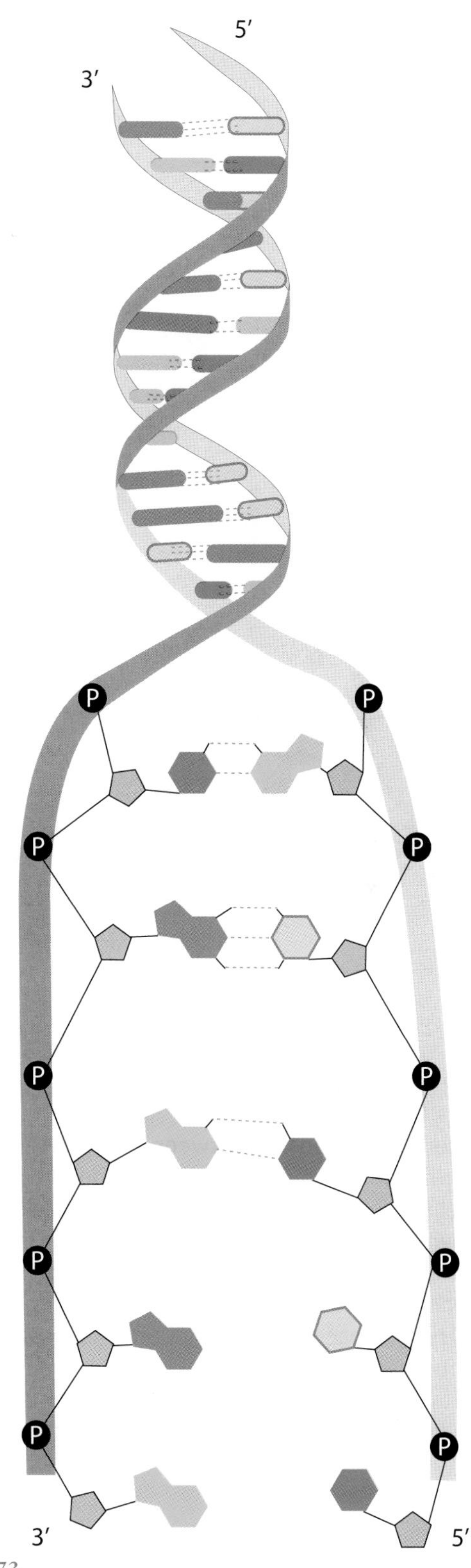

Color Plate 14. DNA double helix. The pairs of nitrogenous bases (adenine (A), guanine (G), thymine (T), and cytosine (C)) face inward, and the carbohydrate-phosphate backbone (gray ribbon, with phosphorus indicated with light "P" on dark circle) outward. The weak electrostatic attraction between the nitrogenous bases (show as dotted blue lines) holds the structure together. However, these attractions can be broken, which allows the DNA double helix to be unzipped (lower half of figure). Since A always pairs with T, and since G always pairs with C, the sequence of nitrogenous bases on one strand is sufficient to determine which bases belonged on the other—and vice-versa.

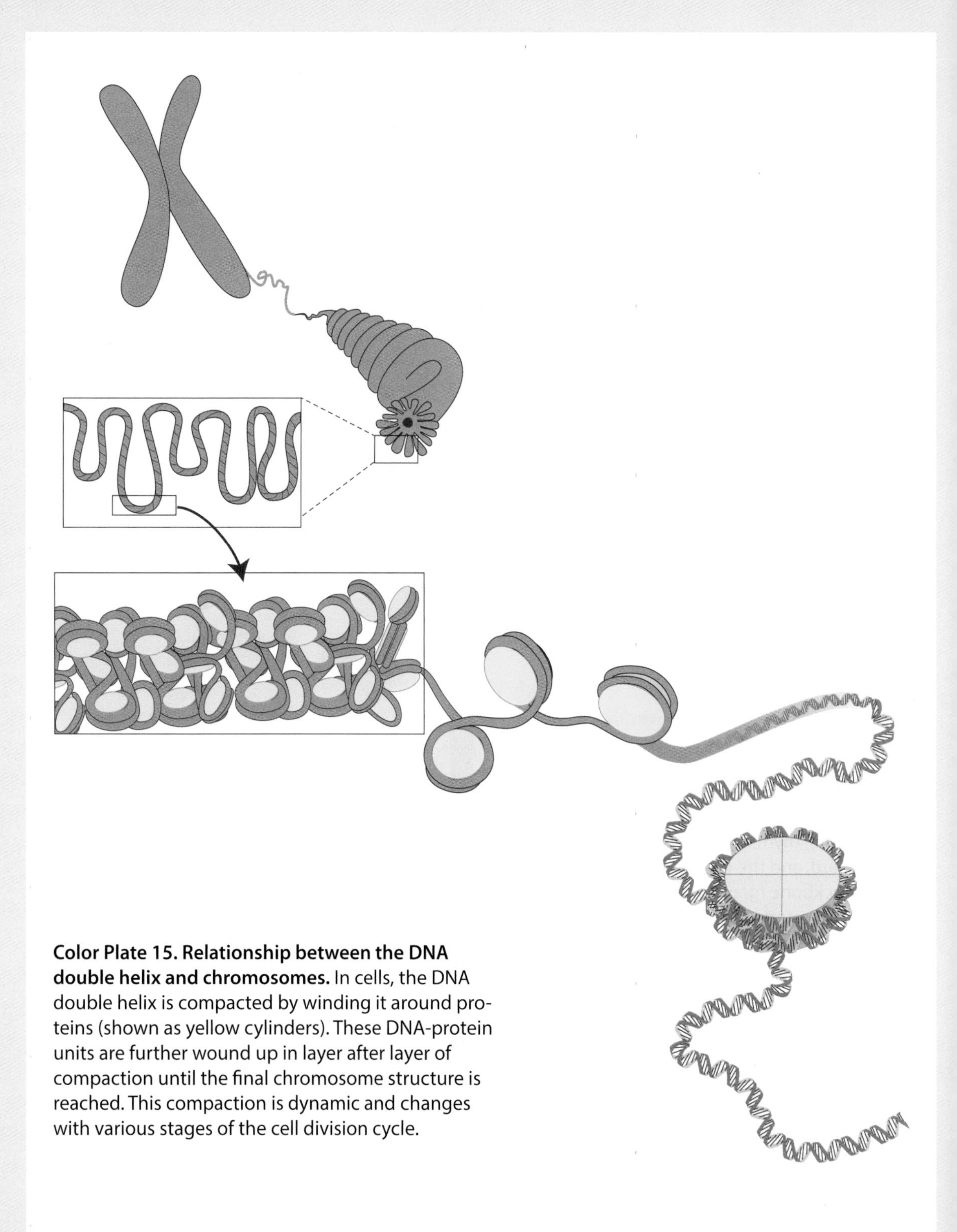

Color Plate 15. Relationship between the DNA double helix and chromosomes. In cells, the DNA double helix is compacted by winding it around proteins (shown as yellow cylinders). These DNA-protein units are further wound up in layer after layer of compaction until the final chromosome structure is reached. This compaction is dynamic and changes with various stages of the cell division cycle.

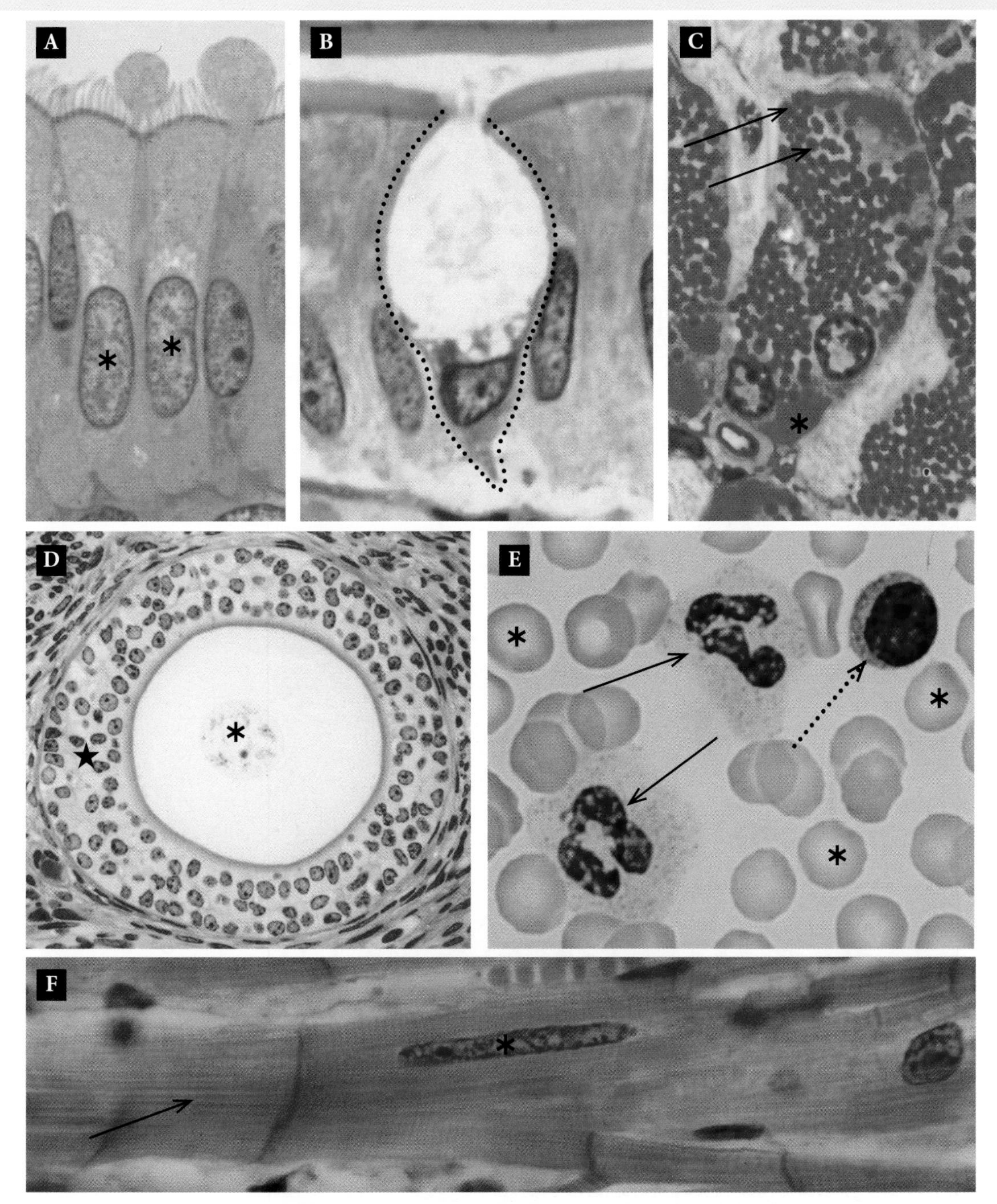

Color Plate 16. Cell diversity in mammals. (A) Cells from the lining of the oviduct (columnar epithelial cells), nuclei marked with asterisks; **(B)** goblet cell of the digestive tract outlined with dotted line; **(C)** pancreatic cell (acinar cell) marked by asterisk, with intracellular granules highlighted with arrows; **(D)** nucleus of egg cell (oocyte) marked with asterisk, follicular cells marked with star; **(E)** blood smear with red blood cells (marked with asterisks) and white blood cells (neutrophils marked by solid arrow, lymphocyte with dotted arrow); **(F)** nucleus of cardiac muscle cell marked with asterisk, with muscle striations highlighted by arrow.

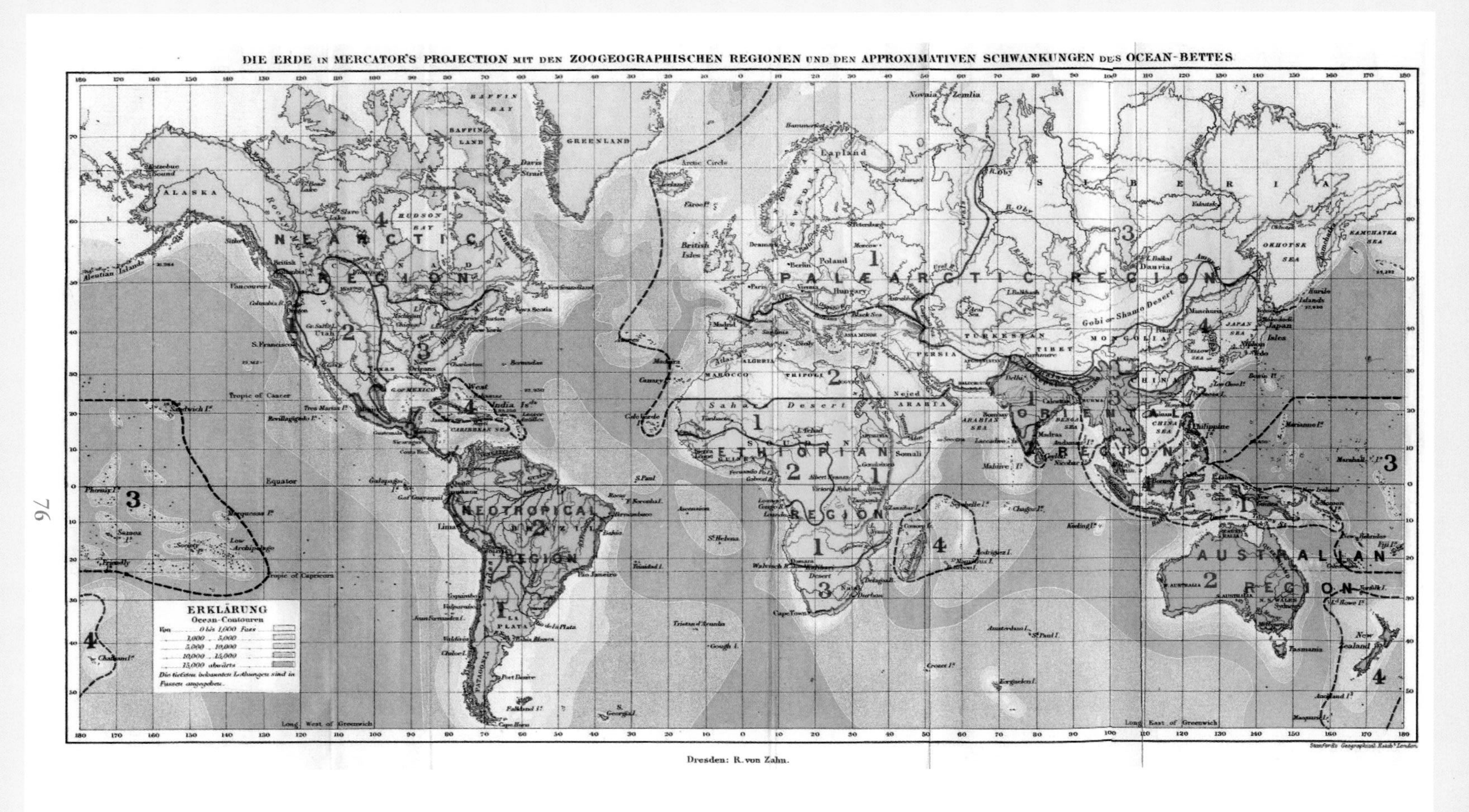

Color Plate 17. Wallace's biogeographic zones of the world.

Color Plate 18. Plains zebra (*Equus quagga*).

Color Plate 19. Mountain zebra (*Equus zebra*).

Color Plate 20. Imperial zebra (*Equus grevyi*).

Color Plate 21. Common ostrich (*Struthio camelus*).

Color Plate 22. Somali ostrich (*Struthio molybdophanes*).

Color Plate 23. Darwin's rhea/lesser rhea (*Rhea pennata*).

Color Plate 24. Greater rhea (*Rhea americana*).

Color Plate 25. Sumatran orangutan (*Pongo abelii*).

Color Plate 26. Bornean orangutan (*Pongo pygmaeus*).

Color Plate 27. Western gorilla (*Gorilla gorilla*).

Color Plate 28. Eastern gorilla (*Gorilla beringei*).

Color Plate 29. Przewalski's horse (*Equus przewalskii*).

Color Plate 30. African wild ass (*Equus africanus*).

Color Plate 31. Onager (*Equus hemionus*).

Color Plate 32. Kiang (*Equus kiang*).

Color Plate 33. Black rhinoceros (*Diceros bicornis*).

Color Plate 34. White rhinoceros (*Ceratotherium simum*).

Color Plate 35. Indian rhinoceros (*Rhinoceros unicornis*).

Color Plate 36. Sumatran rhinoceros (*Dicerorhinus sumatrensis*).

Color Plate 37. Javan rhinoceros (*Rhinoceros sondaicus*).

Color Plate 38. Baird's tapir (*Tapirus bairdii*).

Color Plate 39. Brazilian tapir (*Tapirus terrestris*).

Color Plate 41. Mountain tapir (*Tapirus pinchaque*).

Color Plate 40. Malayan tapir (*Tapirus indicus*).

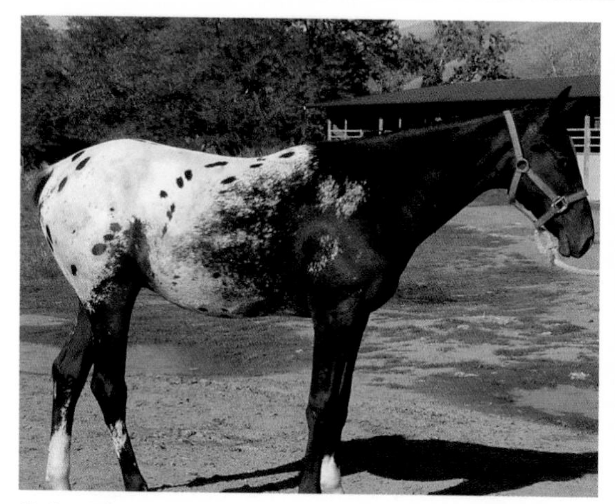
Color Plate 42. Horse breed (*Equus caballus*).

Color Plate 43. Horse breed (*Equus caballus*).

Color Plate 44. Horse breed (*Equus caballus*).

Color Plate 45. Horse breed (*Equus caballus*).

Color Plate 46. Horse breed (*Equus caballus*).

Color Plate 47. Horse breed (*Equus caballus*).

Color Plate 48. Horse breed (*Equus caballus*).

Color Plate 49. Horse breed (*Equus caballus*).

Color Plate 50. Horse breed (*Equus caballus*).

Color Plate 51. Horse breed (*Equus caballus*).

Color Plate 52. Horse breed (*Equus caballus*).

Color Plate 53. Horse breed (*Equus caballus*).

Color Plate 54. Horse breed (*Equus caballus*).

Color Plate 55. Horse breed (*Equus caballus*).

Color Plate 56. Horse breed (*Equus caballus*).

Color Plate 57. Horse breed (*Equus caballus*).

Color Plate 58. Donkey breed (*Equus asinus*).

Color Plate 59. Donkey breed (*Equus asinus*).

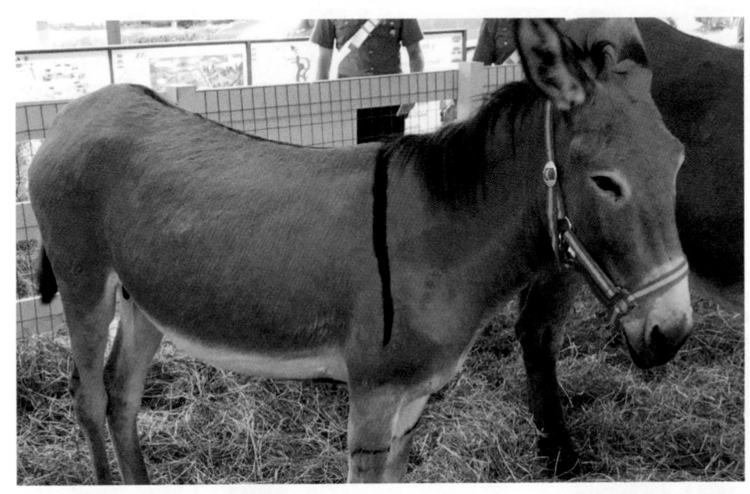

Color Plate 60. Donkey breed (*Equus asinus*).

Color Plate 61. Donkey breed (*Equus asinus*).

Color Plate 62. Donkey breed (*Equus asinus*).

Color Plate 63. Plains zebras have been bred for minimal striping.

Color Plate 64. Plains zebras have been bred for minimal striping.

Color Plate 65. African Cape buffalo (*Syncerus caffer caffer*).

Color Plate 66. African forest buffalo (*Syncerus caffer nanus*).

Color Plate 67. Reticulated giraffe (now *Giraffa reticulata*).

Color Plate 68. Rothschild's giraffe (now *Giraffa camelopardalis*).

Color Plate 69. Masai giraffe (now *Giraffa tippelskirchi*).

Color Plate 70. Southern giraffe (now *Giraffa giraffa*).

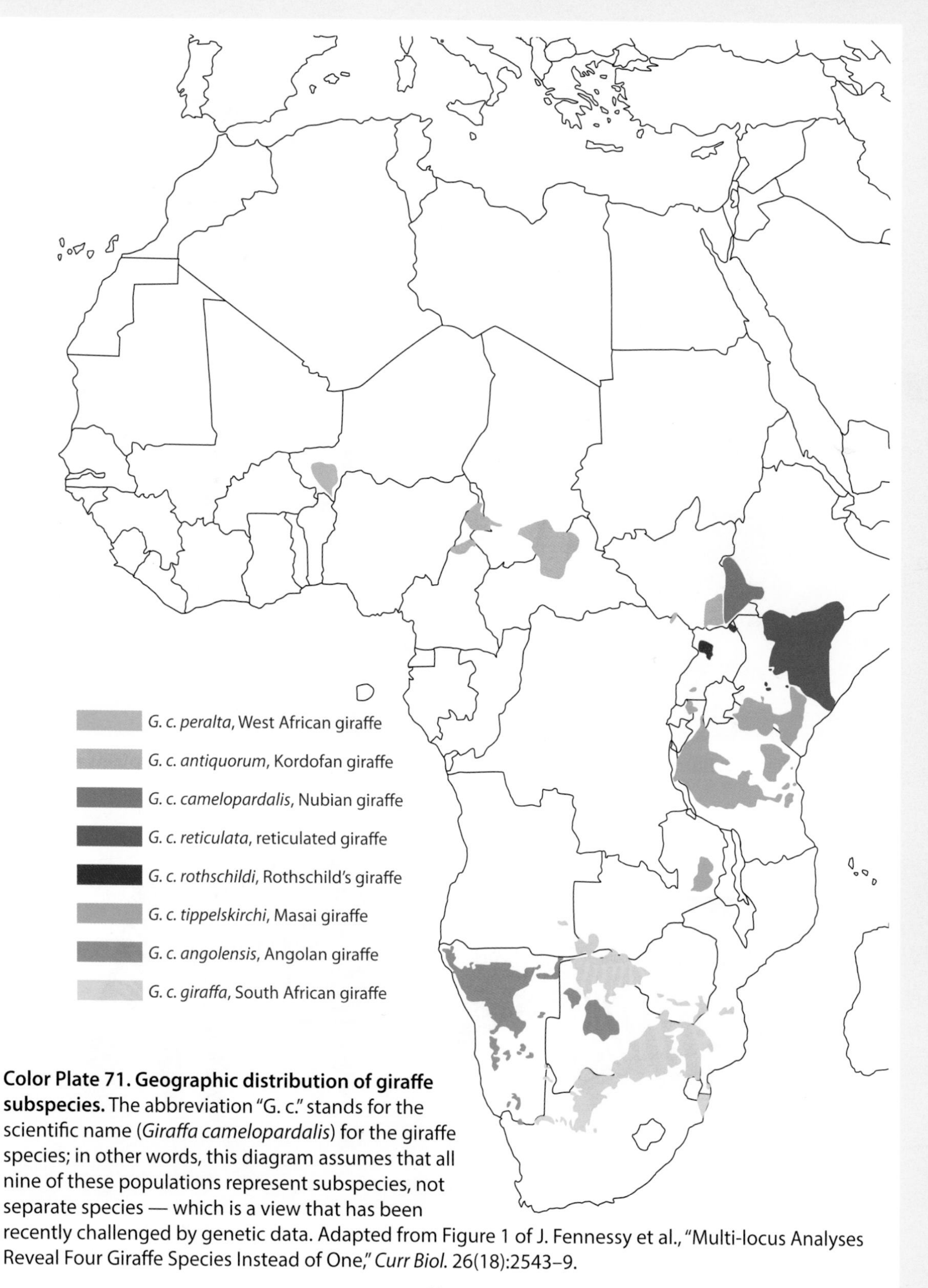

Color Plate 71. Geographic distribution of giraffe subspecies. The abbreviation "G. c." stands for the scientific name (*Giraffa camelopardalis*) for the giraffe species; in other words, this diagram assumes that all nine of these populations represent subspecies, not separate species — which is a view that has been recently challenged by genetic data. Adapted from Figure 1 of J. Fennessy et al., "Multi-locus Analyses Reveal Four Giraffe Species Instead of One," *Curr Biol.* 26(18):2543–9.

Color Plate 72. Burchell's zebra (*Equus quagga burchellii*) — a subspecies of the plains zebra.

Color Plate 73. Maneless zebra (*Equus quagga borensis*) — a subspecies of the plains zebra.

Color Plate 74. Crawshay's zebra (*Equus quagga crawshaii*) — a subspecies of the plains zebra.

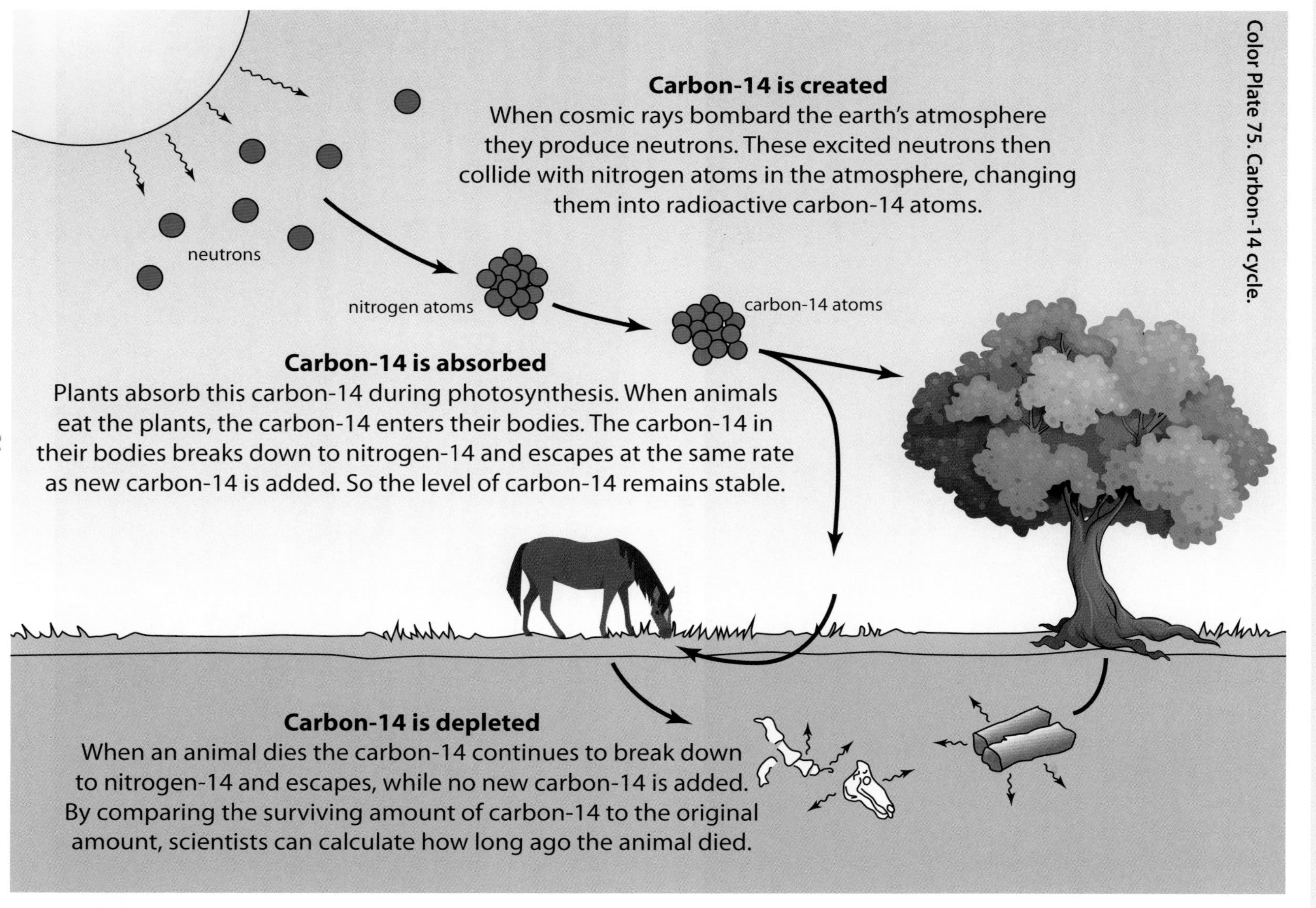

Color Plate 75. Carbon-14 cycle.

Color Plate 76. Canyon below Mount St. Helens.

Color Plate 77. Finely laminated layers from Mount St. Helens eruptions.

Color Plate 78. Grand Canyon.

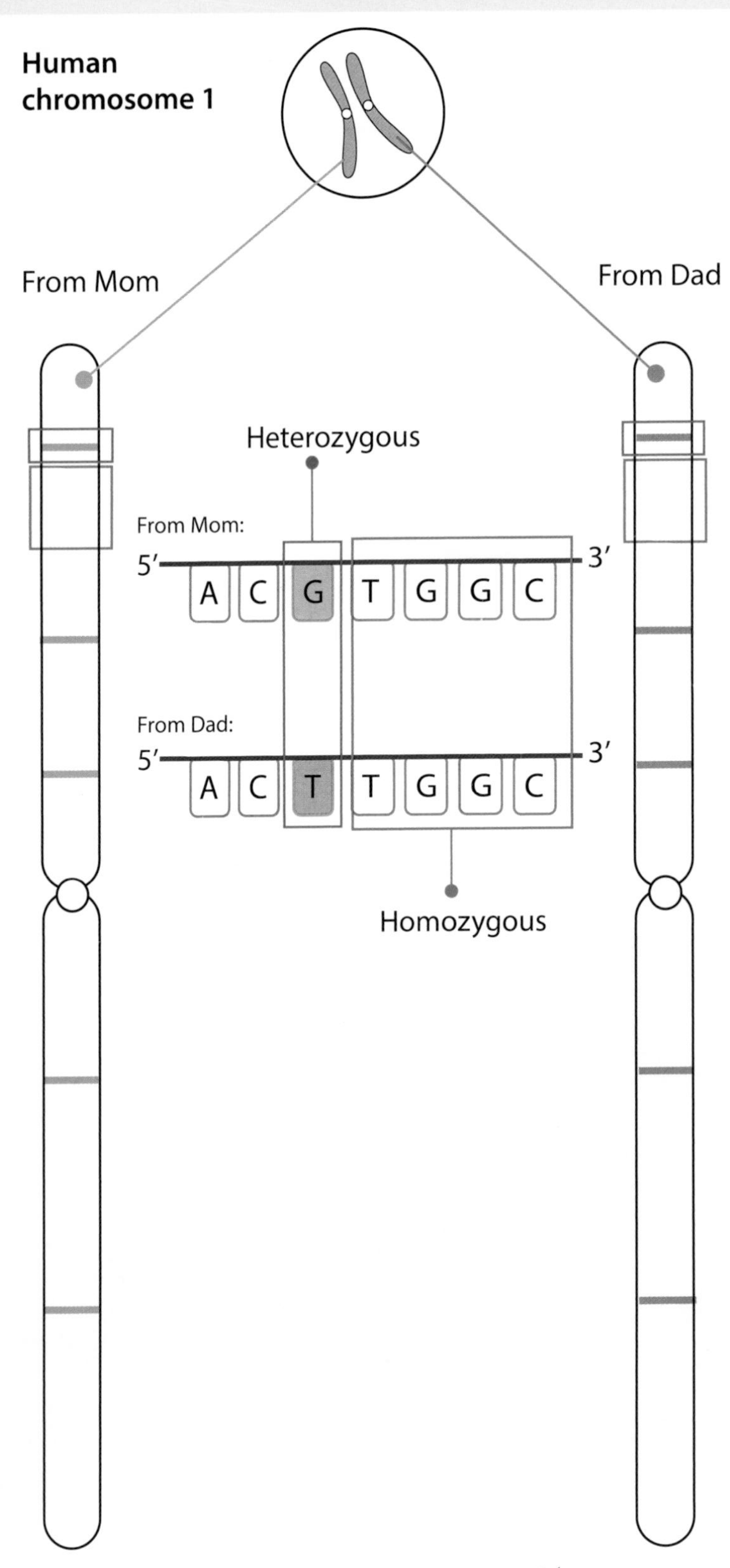

Color Plate 79. Illustration of the fact of heterozygosity. In a normal human chromosome pair, one member of each pair is inherited from each parent. Because the parents have different DNA sequences, the offspring have DNA differences between the members of each chromosome pair — a condition known as heterozygosity. These inter-chromosomal DNA differences are symbolized by colored lines on each chromosome; color represents parent of origin, not a specific nucleotide. Conversely, since the absolute amount of heterozygosity in any pair of human chromosomes is low, these chromosomes have been depicted with few lines/with large blocks of white, which represent DNA sequences that are identical — a condition known as homozygosity. Note that this diagram is not to scale. On average, tens of thousands of heterozygous nucleotides exist on each chromosome, yet they represent less than 0.1% of the total sequence on a chromosome. A few heterozygous sites are shown simply for illustrative purposes, to represent the fact that heterozygosity exists, and that heterozygous sites represent a small fraction of the total sites.

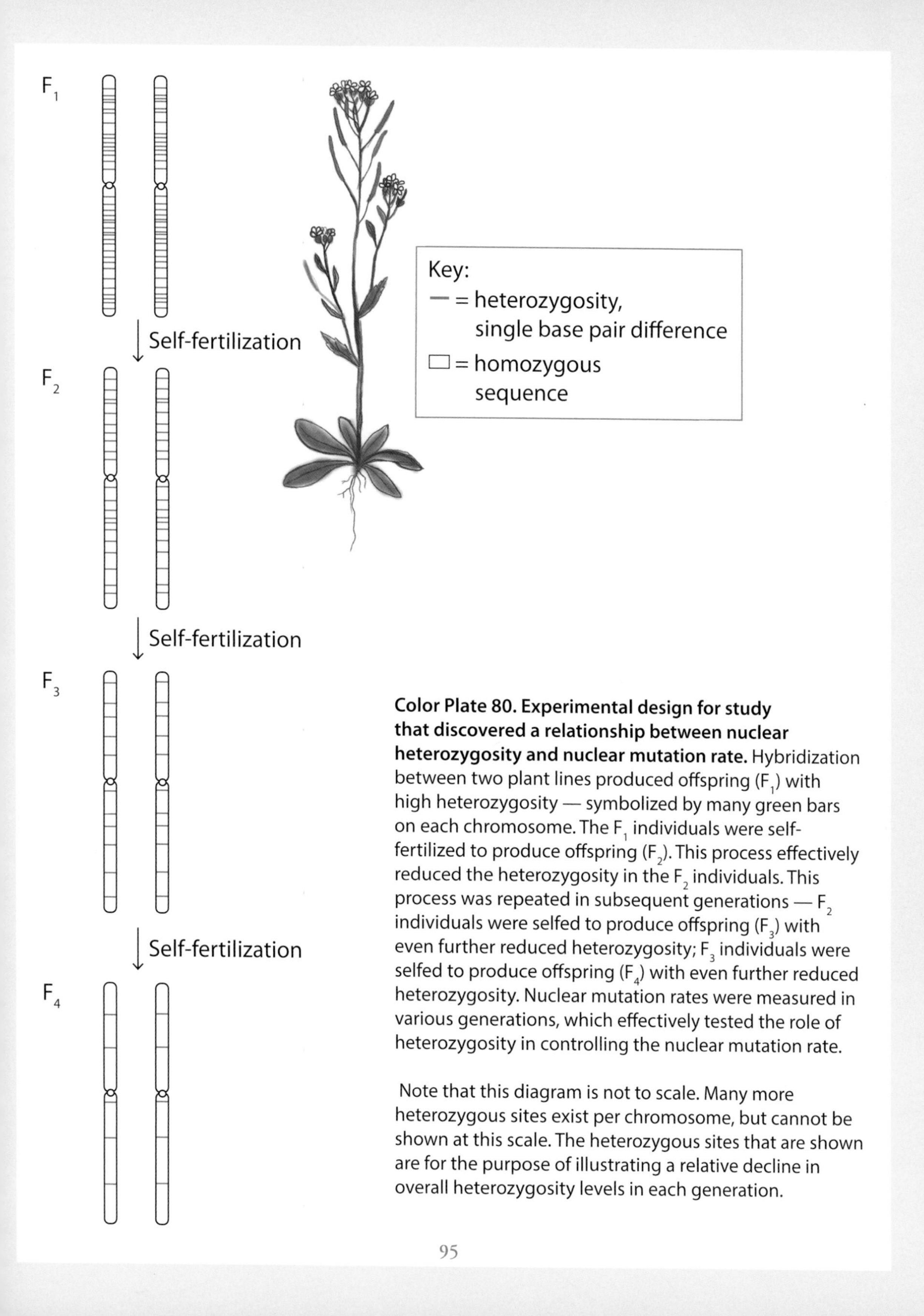

Color Plate 80. Experimental design for study that discovered a relationship between nuclear heterozygosity and nuclear mutation rate. Hybridization between two plant lines produced offspring (F_1) with high heterozygosity — symbolized by many green bars on each chromosome. The F_1 individuals were self-fertilized to produce offspring (F_2). This process effectively reduced the heterozygosity in the F_2 individuals. This process was repeated in subsequent generations — F_2 individuals were selfed to produce offspring (F_3) with even further reduced heterozygosity; F_3 individuals were selfed to produce offspring (F_4) with even further reduced heterozygosity. Nuclear mutation rates were measured in various generations, which effectively tested the role of heterozygosity in controlling the nuclear mutation rate.

Note that this diagram is not to scale. Many more heterozygous sites exist per chromosome, but cannot be shown at this scale. The heterozygous sites that are shown are for the purpose of illustrating a relative decline in overall heterozygosity levels in each generation.

Color Plate 81 (on the following three pages). Nuclear DNA differences can be used to determine the relative hierarchy of speciation or population splitting events.

(i) Under the creationist hypothesis of created nuclear heterozygosity, each individual in the original population has high levels of heterozygous DNA.
To score DNA differences between individuals, each of the two copies of DNA in each individual is compared to every other copy of DNA in every other individual. Each position is scored for the presence or absence of mismatches. As long as one of the comparisons at a particular DNA position (i.e., row of comparisons) is a mismatch (even if all the others are matches), the position is scored as a mismatch. In this population, every DNA position (i.e., row of comparisons) contains at least one mismatch. Thus, these two individuals would be scored as different at each of the nine DNA positions.

(ii) Under the model of population splitting that I've outlined in this book, splitting events involve shifts from heterozygosity to homozygosity. However, due to the criteria by which mismatches are scored, this initial shift towards a more homozygous state results in the same number of DNA positions with a mismatch (i.e., all nine positions/rows). In other words, Population A and Population B would be scored as different at all nine positions.

(iii) When a population splits from Population B, shifts toward homozygosity are involved again, and nuclear DNA differences can reveal the order of splitting events (see iv to vi).

(iv) Again, among all three populations, each copy of DNA would be scored against every other copy of DNA. In the first of these comparisons, Population A is compared to Population B. As we observed in (ii), all of the nine DNA positions have at least one mismatch.

(v) The second pairwise nuclear DNA comparison is shown — Population A versus population C. Again, all of the nine DNA positions have at least one mismatch.

(vi) The third pairwise nuclear DNA comparison uncovers the critical differences —Population B versus Population C. Because Population B was already homozygous at DNA positions 3, 6, and 8, Population C could not have shifted from a heterozygous to a more homozygous state. Instead, it would have inherited the identical DNA base pairs at these three positions. Thus, when these two populations are scored at these three positions, each one of these comparisons shows matches.

(vii) Consequently, Population A would be different from both Population B and Population C at nine DNA positions; in other words, it would be the same genetic distance from both populations. In contrast, Population B and Population C would be closer to one another than either would be to Population A. These differences can be visualized in a graphical tree format, which allows for comparisons to be made to other DNA trees, such as mitochondrial DNA-based trees and Y chromosome-based trees. Though the branch lengths would not necessarily correspond to absolute time (i.e., would not represent a strict DNA clock), the relative timing of the branching events could be compared to the relative timing of branching events as inferred from mitochondrial DNA comparisons or Y chromosome comparisons.

(i)

Population A possesses significant level of heterozygosity in nuclear DNA

(only heterozygous DNA positions shown)

Heterozygous individual 1		Heterozygous individual 2	
A	C	A	C
T	G	T	G
G	T	G	T
C	A	C	A
C	A	C	A
C	A	C	A
C	A	C	A
T	G	T	G
G	A	G	A

Population A

DNA position (row number)	Pairwise comparison 1	Pairwise comparison 2	Pairwise comparison 3	Pairwise comparison 4
1	A — A	A — C	C — A	C — C
2	T — T	T — G	G — T	G — G
3	G — G	G — T	T — G	T — T
4	C — C	C — A	A — C	A — A
5	C — C	C — A	A — C	A — A
6	C — C	C — A	A — C	A — A
7	C — C	C — A	A — C	A — A
8	T — T	T — G	G — T	G — G
9	G — G	G — A	A — G	A — A

Observation: Each of the 9 DNA positions has a mismatch in at least one of the four pairwise comparisons

(ii)

Heterozygous individual	
A	C
T	G
G	T
C	A
C	A
C	A
C	A
T	G
G	A

Formation of Population B from Population A via shifts from heterozygosity to homozygosity

Population A ⇨ Population B

Less heterozygous individual	
A	C
T	G
T	T
C	A
C	A
A	A
C	A
G	G
G	A

DNA position (row number)	Pairwise comparison 1	Pairwise comparison 2	Pairwise comparison 3	Pairwise comparison 4
1	A — A	A — C	C — A	C — C
2	T — T	T — G	G — T	G — G
3	G — T	G — T	T — T	T — T
4	C — C	C — A	A — C	A — A
5	C — C	C — A	A — C	A — A
6	C — A	C — A	A — A	A — A
7	C — C	C — A	A — C	A — A
8	T — G	T — G	G — G	G — G
9	G — G	G — A	A — G	A — A

Observation: Each of the 9 DNA positions has a mismatch in at least one of the four pairwise comparisons

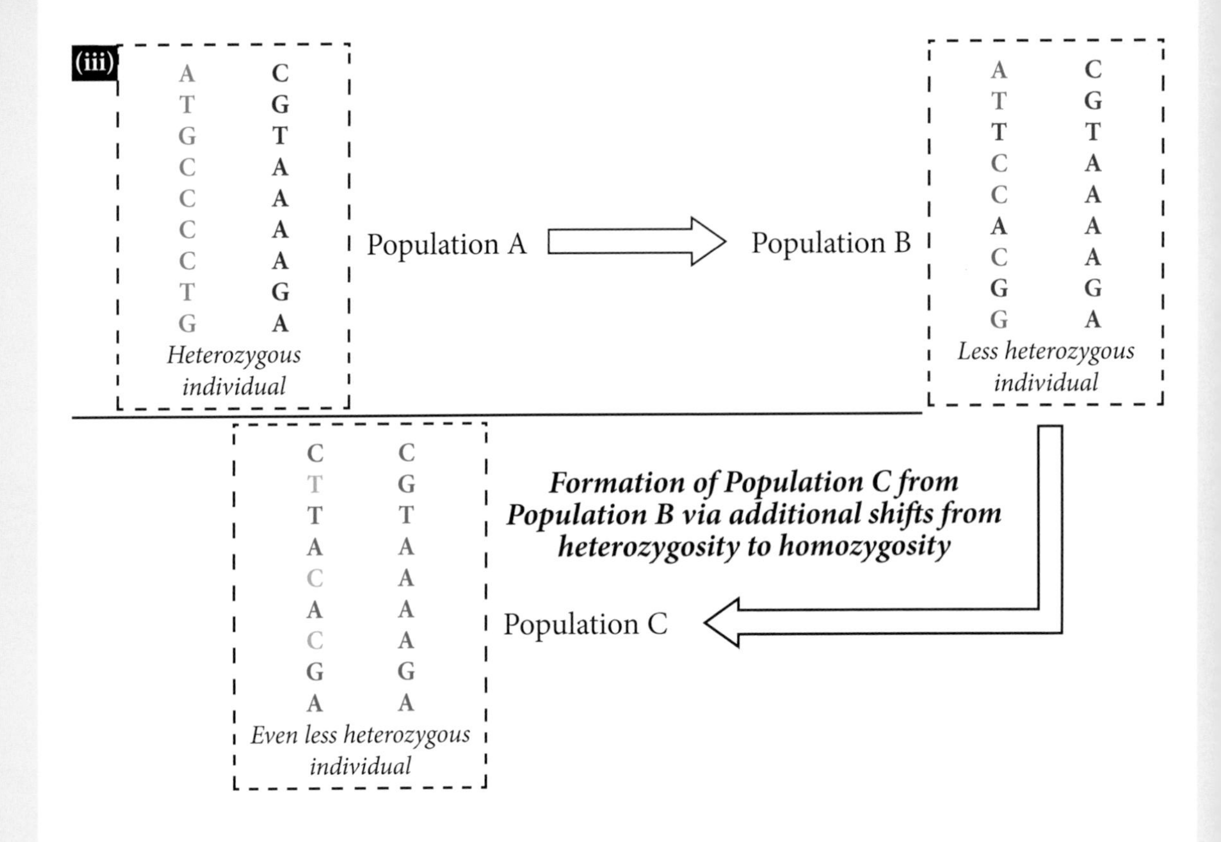

(iv)

Population A versus Population B:

Each of the 9 DNA positions has a mismatch in at least one of the four pairwise comparisons

DNA position (row number)	Pairwise comparison 1	Pairwise comparison 2	Pairwise comparison 3	Pairwise comparison 4
1	A — A	A — C	C — A	C — C
2	T — T	T — G	G — T	G — G
3	G — T	G — T	T — T	T — T
4	C — C	C — A	A — C	A — A
5	C — C	C — A	A — C	A — A
6	C — A	C — A	A — A	A — A
7	C — C	C — A	A — C	A — A
8	T — G	T — G	G — G	G — G
9	G — G	G — A	A — G	A — A

(v)

Population A versus Population C:

Each of the 9 DNA positions has a mismatch in at least one of the four pairwise comparisons

DNA position (row number)	Pairwise comparison 1	Pairwise comparison 2	Pairwise comparison 3	Pairwise comparison 4
1	A — C	A — C	C — C	C — C
2	T — G	T — T	G — G	G — T
3	G — T	G — T	T — T	T — T
4	C — A	C — A	A — A	A — A
5	C — A	C — C	A — A	A — C
6	C — A	C — A	A — A	A — A
7	C — A	C — C	A — A	A — C
8	T — G	T — G	G — G	G — G
9	G — A	G — A	A — A	A — A

(vi)

Population B versus Population C:

Now 3 of the 9 DNA positions has a perfect match in all four of the pairwise comparisons

DNA position (row number)	Pairwise comparison 1	Pairwise comparison 2	Pairwise comparison 3	Pairwise comparison 4
1	A — C	A — C	C — C	C — C
2	T — T	T — G	G — T	G — G
3	T — T	T — T	T — T	T — T
4	C — A	C — A	A — A	A — A
5	C — C	C — A	A — C	A — A
6	A — A	A — A	A — A	A — A
7	C — C	C — A	A — C	A — A
8	G — G	G — G	G — G	G — G
9	G — A	G — A	A — A	A — A

(vii)

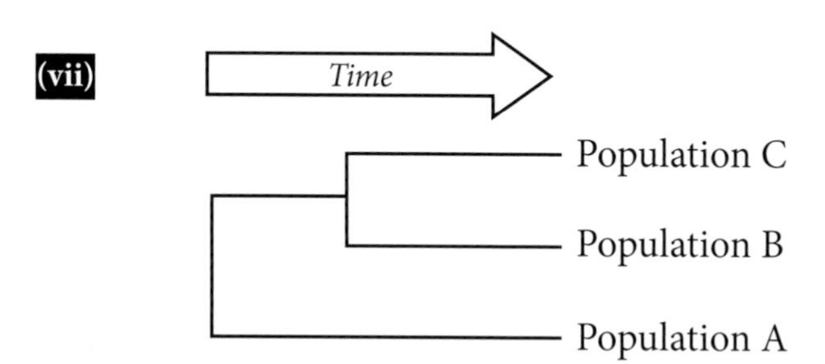

Relative number of nuclear DNA differences among populations reflects the relative order of divergence events

Color Plate 82 (on following page). In the absence of chromosome diversification mechanisms, limited chromosomal variety ensues. If no mechanisms exist by which to diversify chromosome combinations, then a set of heterozygous parents (heterozygous sites shown as colored lines on each chromosome) in part (a) and part (b) give rise to just four possible chromosome pairs in their offspring. If the offspring continue to reproduce, then only four additional chromosome pairings — for a total of ten possible chromosome pairings (two original pairings, four offspring pairs, four additional offspring pairings) — are possible in any subsequent generation. Today, chromosomes around the globe appear quite scrambled (symbolized at the bottom of this chart) — as if some mechanism for chromosomal diversification exists. Again, actual number of heterozygous sites not shown to scale, due to limits of this display.

a) Chromosome #1 pair of the father

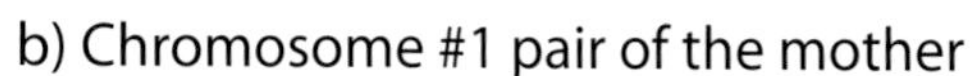

b) Chromosome #1 pair of the mother

In offspring, four possible pair combinations of chromosome #1:

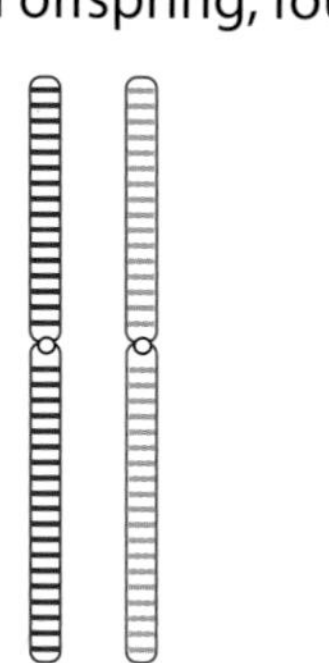

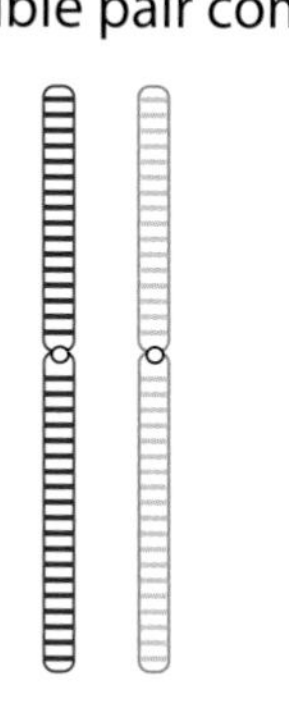

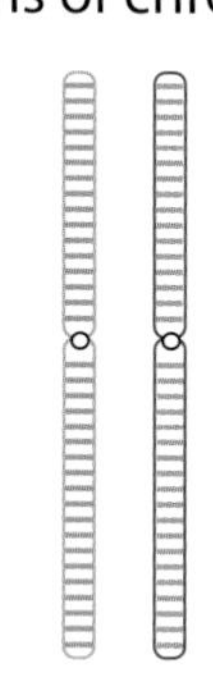

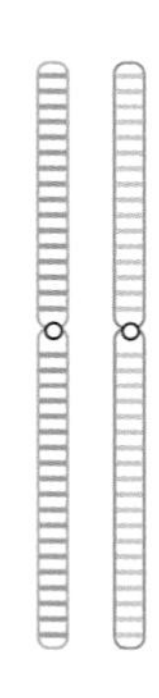

Subsequent additional combinations:

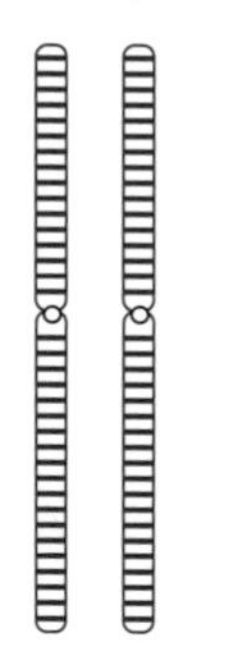

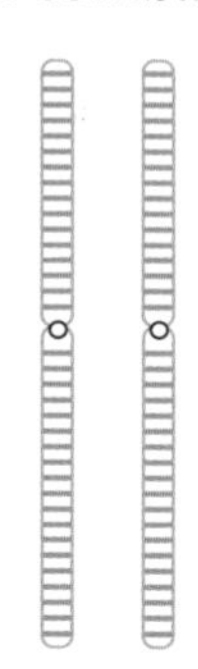

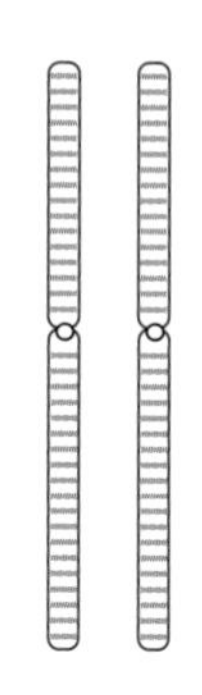

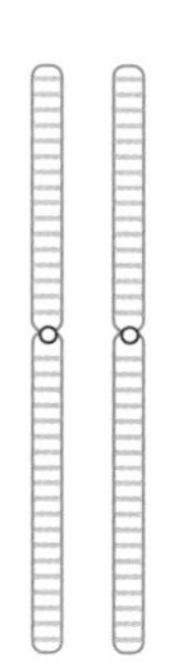

Example of scrambled chromosome #1 pair:

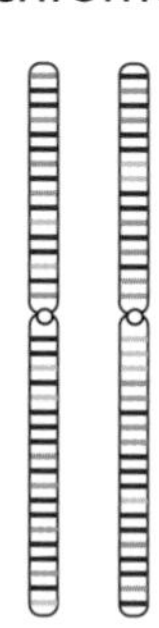

Color Plate 83 (on following page). Recombination and gene conversion produce a dizzying array of chromosomal combinations. If we start with a set of heterozygous parents (heterozygous sites shown as colored lines on each chromosome) in part (a) and part (b), the process of recombination swaps chunks of chromosomes in a chromosome pair, leading to diverse chromosome sequences in gametes (i.e., sperm and egg). Gene conversion adds diversity by swapping tiny sections of each chromosome — as small as a single heterozygous site. On average, each chromosome will experience one recombination event, while every other chromosome (12 / 23 = 0.5) will experience a gene conversion event. This process repeats itself in subsequent generations (only one subsequent generation is shown in this diagram), which compounds the chromosome diversity. Again, actual number of heterozygous sites not shown to scale, due to limits of this display.

Representative gametes (i.e., sperm or egg),
after recombination, gene conversion:

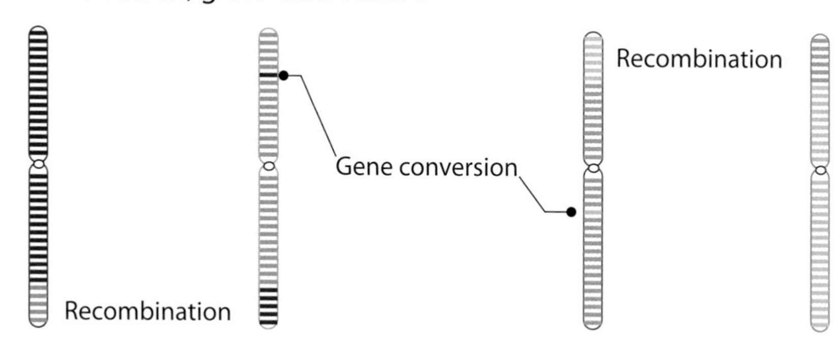

Offspring, representative chromosome pairs

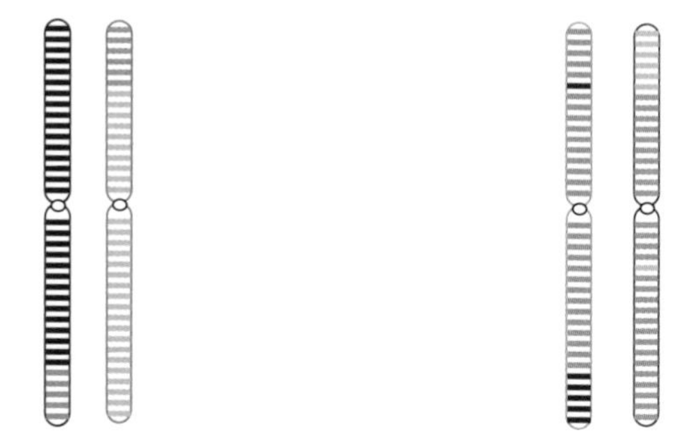

Offspring, representative gametes (i.e., sperm or egg),
after recombination, gene conversion:

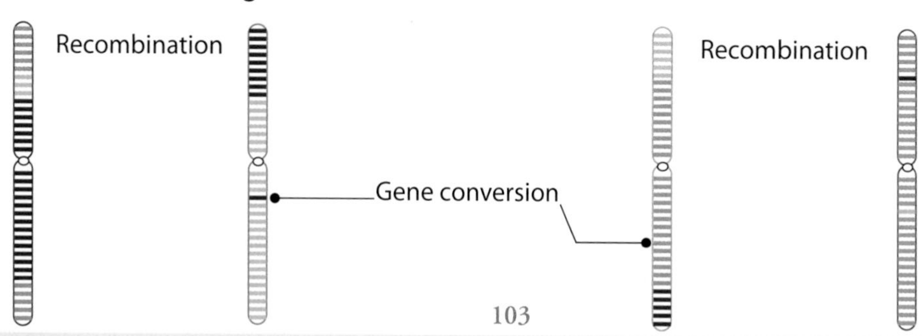

1) Speciation Question

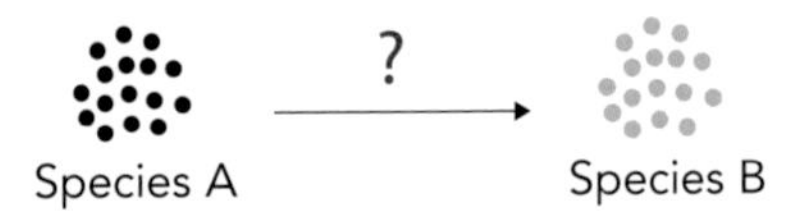

2) Evolutionary Answer

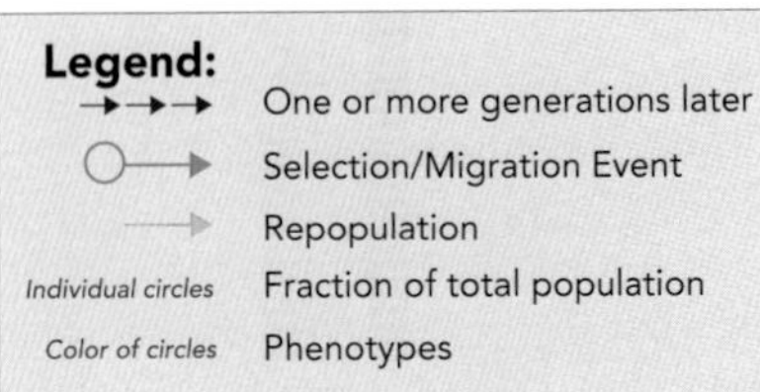

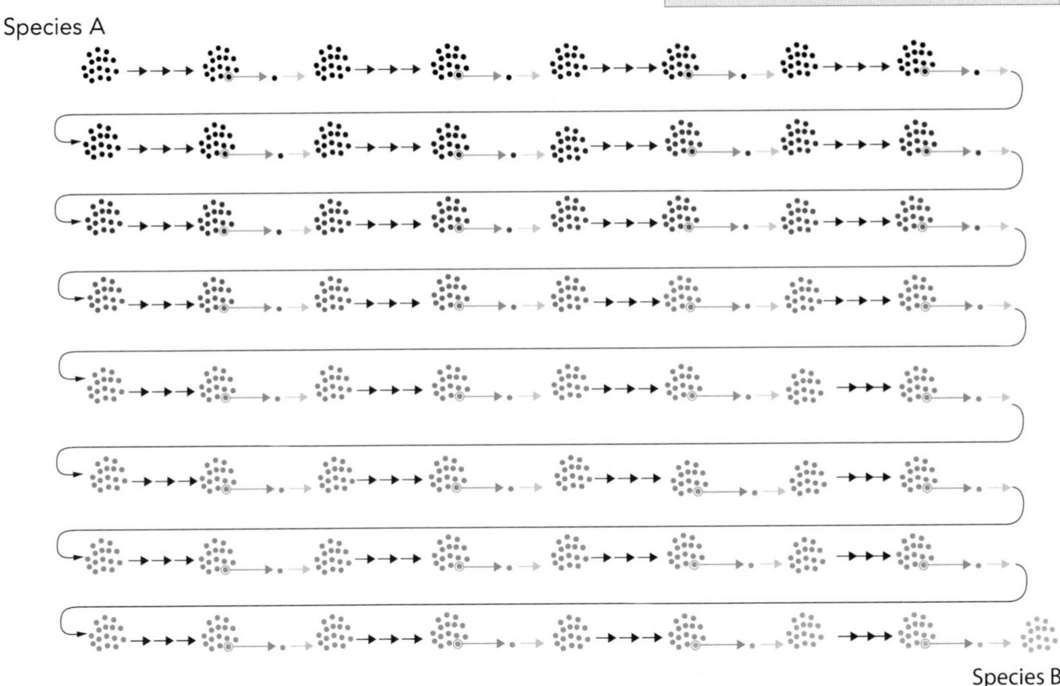

3) Creationist Answer

Species A → Species B

Color Plate 84. Contrasting models on the origin of species. 1) The formation of new species involves the isolation of distinct individuals and their regrowth into a new population. For simplicity, the process of speciation is here depicted as the formation of distinct phenotypes or morphologies. **2)** Under the evolutionary model, the process of speciation is slow. Because evolutionists ultimately explain all genetic differences as the result of mutation, and because very few mutations occur each generation, genetic differences accumulate very slowly, leading to very slow morphological change. In contrast, the isolation and regrowth of these phenotypically distinct individuals can happen comparatively quickly. But since these steps must be repeated multiple times, the overall process is slow. **3)** Under the creationist model of preexisting genetic differences, massive genetic differences arise in a single generation, which hold the potential for immediately resulting in morphological differences. As under the evolutionary model, isolation and population regrowth under the creationist model can happen quickly.

Part Two

All Things (re) *Considered*

Chapter 4
The Riddle of Geography

If Darwin had no knowledge of genetics, how could he write a book on the origin of species? If genetic data were absent from his thesis, then how could he have made any semblance of a scientific argument for the origin of species? Furthermore, why did his arguments gain such traction in the scientific community?

Consider the weakness of the data available to him. Fossils don't directly record genealogical relationships. As we discussed in chapter 2, fossils inform ancestry only after a model of genetics is assumed. Thus, fossil insights to the question of ancestry must follow the discovery of genetics; they cannot precede it. Since Darwin wrote *On the Origin of Species* before genetics was even a scientific field, fossils could not have rigorously informed the ancestry question.

Similarly, geography doesn't directly record inheritance. As an illustration, I have a closer genealogical relationship with people several thousand miles away (my maternal relatives live in Germany) than I do with my next-door neighbors.

Finally, anatomy and physiology have a nonlinear relationship with DNA sequences (see chapter 3). Therefore, they are poor markers of DNA change. What could Darwin have possibly used to make his case?

Darwin's success derived from an idiosyncrasy of the scientific method. Actually, what I'm about to describe is not an idiosyncrasy. It's a fundamental property of the scientific method. But it's unfamiliar to many — hence, the appearance of peculiarity.

This idiosyncrasy played more than one role in Darwin's fate. It didn't just result in his early successes. In the years following Darwin's death, it dramatically recast his conclusions.

❧❧ ❧❧ ❧❧ ❧❧

The peculiarity of the scientific method is a consequence of its philosophical foundations. Logically, the scientific method is an inductive method of reasoning. As such, it requires certain rules to be followed for it to be useful.

The rules are not intuitive. Consequently, they can cause much confusion. Hence, they are much easier to appreciate in the context of an illustration. Imagine for a moment that a young man and his high school sweetheart marry shortly after graduation. For the first several months of their married life, the newlyweds find their time filled with bliss as they exult in all the joys that accompany the union of two separate lives.

But as the months pass, the wife watches her husband develop a depressing pattern. He begins arriving home from work later and later in the evening.

Disappointed by the loss of attention and interest from her husband, she confronts him about his tardiness. He quickly evades her probing with vague explanations, and then he changes the subject.

Over the next several weeks, her suspicions grow. She begins tracking not only his work hours but also his expenses. To her consternation, she finds numerous receipts for flowers, chocolates, and even jewelry — all items that she has not received from him.

The next time he arrives home late, she accosts him with the evidence. Again, he avoids any direct answers and hurries to bed.

As the husband readies himself for work the next morning, she notices that he smells strongly of cologne. As soon as he drives out of sight, she resolves to catch him in the act.

By 4:30 in the afternoon, she is sitting in a rental car in the parking lot of her husband's work, with a direct line of sight to the building's exit. Sure enough, just as she suspected, her husband leaves promptly at five o'clock. (She never thought his late evenings were spent on the job.)

As he drives out of the parking lot, she follows behind him at a safe distance. Several turns later, she notices that her husband appears to be following a familiar route. Just then, he pulls into the driveway — not of some secret lover, but of their own house!

Then she remembers. Today is their anniversary.

Suddenly the pieces fall into place. Her husband's long hours were for the purpose of earning extra cash to buy gifts for his bride — not for pursuing an illicit tryst with a mistress.

Flush with embarrassment, she returns the rental car and arrives home in shame. How could this have happened?

Her failure to correctly interpret her husband's actions illustrates one of the most important elements of the process of inductive reasoning.

Inductive reasoning is the opposite of *deductive* reasoning. In the latter, conclusions follow from a premise via a series of logically coherent steps. A premise might be, "When my stomach growls, I'm hungry." If it's close to noon and my stomach is making noise, I can logically deduce that I'm hungry. As long as the premise is true and the reasoning logically sound, the conclusions are true.

In contrast, inductive reasoning starts with observations and tries to work back to a premise or general principle. In the case of the newly married wife in our example, she started with the observation that her husband was acting suspiciously. Then she remembered that husbands who cheat on their wives act suspiciously. Not surprisingly, she concluded that her husband must be cheating on her.

The problem with inductive reasoning is that there may be multiple explanations for an observation. Again, in the case of our suspicious wife, husbands who plan to surprise their wives on their anniversary also act suspiciously. She failed to eliminate this explanation before reaching her conclusion. Consequently, when acting on this erroneous conclusion, she plunged herself into a highly embarrassing situation.

The same danger holds true in science. Whether the goal is medical research or species research, the same principles apply. For example, let's say you're trying to find a cure for cancer. You might grind some dandelion leaves, extract the juices, and then add them to a Petri dish containing cancer cells (Figure 4.1). Then you might let the cells sit with the extract overnight in an incubator. On the next day you might observe the dish and discover that the cancer cells have all died. Before you jump to the conclusion that you've found a cure, you must apply inductive reasoning extremely carefully.

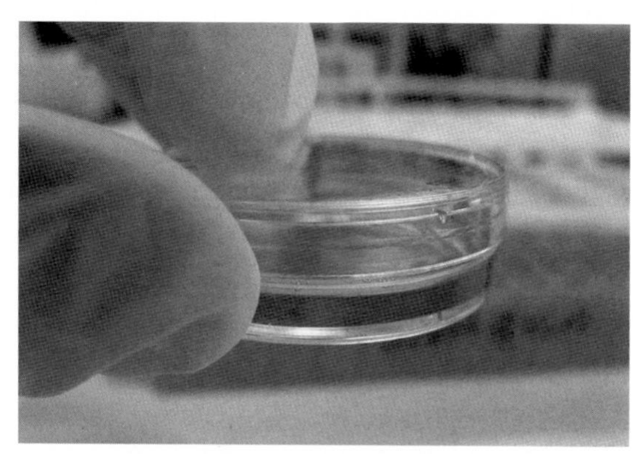

Figure 4.1. Petri dish

Your first step should be an investigation of what else could explain the death of the cancer cells. Perhaps you used a contaminated pipet tip (Figure 4.2) when you added the dandelion extract. Maybe someone used a pipet to transfer toxic chemicals into various storage tubes, and then accidentally added this tip back to the box that you used. If so, then the explanation for the death of the cancer cells is likely this toxic chemical, not your dandelion extract.

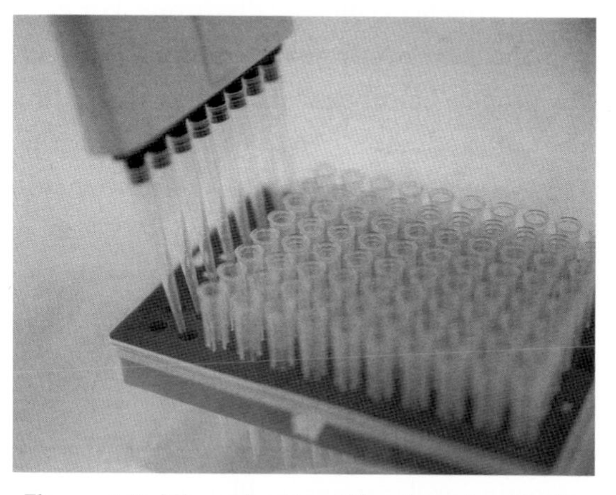

Figure 4.2. Micropipette with micropipette tips

Conversely, if you dissolved the dandelion extract in some chemical, the dissolving chemical (the *solvent*) may have killed the cells instead of the extract. This would be critical to know before trying the extract on human patients. An alternative preparation of the extract might fail to cure the patient — and simultaneously prevent effective treatments from being administered, thus effectively killing him by neglect.

Finally, some chance, inexplicable accident might have killed the cancer cells. While this explanation might initially seem *ad hoc*, it is all too familiar to laboratory researchers. For example, when I was an undergraduate, we explored the molecular biology of plants. When I obtained my first promising experimental result, I was very excited. Yet, in the weeks and months following the initial experiment, I was never able to replicate the first result. Eventually, much to our disappointment, we had to dismiss the initial findings as some chance fluke and move on to other experiments. We had no other option. Chance is always a competing explanation for the results of scientific experiments.

The fact that so many alternative explanations exist does not hinder the scientific method. Many of these explanations can be explored and eliminated in the laboratory with the appropriate tests. For instance, in our cancer example, several easy experiments have the potential to eliminate several of them. By repeating the experiment, by using a control set of cells in which you add just the solvent without the extract, and by double checking the pipet tips, you could test the alternative explanations we just discussed.

If you take all of these steps and discover that, by the next morning, the cells with the extract are still dead while the control set of cells are still alive, you can be more confident that you've found something with potential relevance to the cancer therapy. To translate this discovery into the clinic, many more experiments would be required. Ultimately, if your candidate chemical passed numerous laboratory tests, it would then need to be tested on patients as part of a clinical trial.

But long before you can dream of winning the Nobel Prize, all your observations would have to be obtained with a sound application of the method of inductive reasoning. Without it, the consequences could be disastrous. For example, if you brought your candidate chemical to a clinical trial, it would be critical to have a control group that did not receive the chemical. Otherwise, if your cancer patients were all cured, how would you know that their condition wasn't the result of spontaneous remission? Without the control group, you might erroneously conclude efficacy for an impotent drug. If you then administered your impotent chemical to all sorts of sick patients, you could kill them by neglect — your cocktail might stall the administration of actual anti-cancer drugs that could cure them.

Theory aside, in the previous chapters we already observed the method of inductive reasoning in practice. When Avery, MacLeod, and McCarty tried to uncover the chemical behind bacterial transformation, they rigorously applied the method of inductive reasoning.[1] To arrive at the conclusion that DNA was responsible for bacterial transformation, they had to eliminate several competing explanations — like proteins, RNA, etc. The experiments of Hershey and Chase had to as well.[2] After bacteriophage infection, it would not have been enough for them to identify bacteriophage DNA inside bacterial cells. They also needed to demonstrate that bacteriophage proteins were *not*.

Inductive reasoning is the foundation of the scientific method.

In *On the Origin of Species*, Darwin carefully employed the method of inductive reasoning. As we observed in chapter 1, Darwin's ideas represented a paradigm shift from the century prior. In the 100 years leading up to 1859, competing explanations — creationist ideas — for the origin of species dominated scientific thought. Consequently, Darwin took great pains in his seminal work, not only to undergird his own ideas with evidence, but also to reveal the shortcomings of creationist theories.

Their elimination was critical to Darwin's success.

The match between species and their environment has long fascinated scientists and lay people alike. Wherever you travel in the world, characteristic species exist, and these species seem to be well-suited to their environments. In Darwin's day, this match led creationists to speculate that species have always been in their native habitats. A consequence of this view is that species do not share ancestry with any other species. In other words, the creationists of 1859 believed that both the ancestry and geography of a species was fixed.

But does this hypothesis fit the facts? Darwin's 150-year-old reasoning is easily transferrable to modern examples.[3] For instance, from South Africa all the way up to the Horn of Africa, zebras are as iconic as any African mammal (Figure 4.3).[4] As wild species, they do not exist anywhere else in the world. Why?

At first pass, ecology hints at the answer to this riddle. Among the ecosystems present in the zebra's native habitats, stripes seem to serve a helpful function. Though still a matter of professional scientific dispute,[5] their stripes seem to camouflage them against predators (Figure 4.4). Whatever their function, it's not hard to see that zebras flourish on the savanna. If this were all we knew, we might conclude that zebras were created in Africa.

However, Africa isn't the only location on earth with the ecological potential to harbor zebra species. The Terai-Duar in India, the Brazilian *Cerrado*, and a significant portion of northern Australia (Figure 4.5) are all savannas. Yet no zebras are native to any of these locations.

Zebras aren't the only species that fit this curious pattern. Today, the wild camel (*Camelus ferus*) exists in the deserts of Central Asia — in China and Mongolia (Figure 4.6).* With a double set of eyelashes, broad hoofs that expand in the sand, and long-term energy reservoirs in their humps, camels thrive in a desert environment. Yet deserts exist, not only in Central Asia, but also in Africa, North America, South America, and Australia (Figure 4.7). Wild camels do not.

* The dromedary in North Africa and Australia is a domesticated version, not wild; humans artificially introduced camels in these locations.

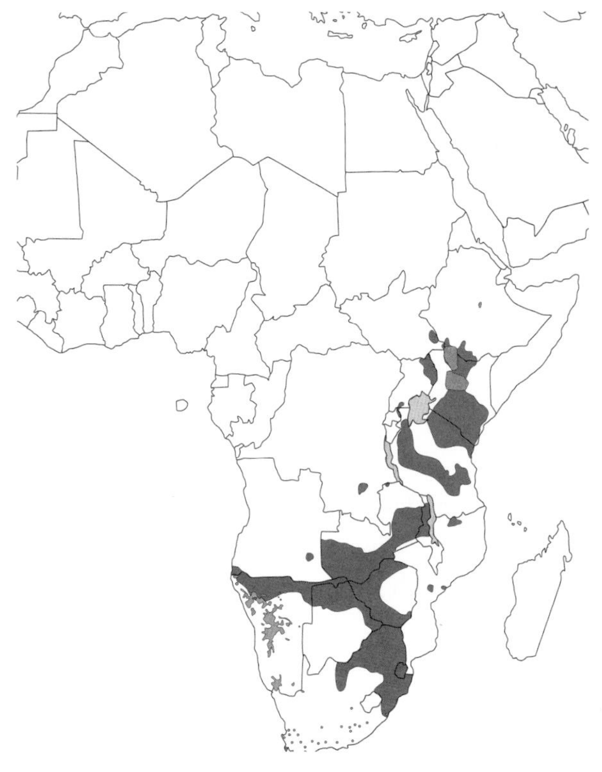

Figure 4.3. Geographic distribution of zebra species. The geography of the three zebra species is shown by gray shading. The darkest gray represents the distribution of the plains zebra; the lighter gray, the distribution of the imperial zebra (north) and mountain zebra (south). The lightest gray shading depicts inland bodies of water. Adapted from http://maps.iucnredlist.org/map.html?id=7950, http://maps.iucnredlist.org/map.html?id=41013, http://maps.iucnredlist.org/map.html?id=7960.

Figure 4.4. A well-camouflaged pair of zebras.

Species like orangutans and gibbons thrive in the tropics and rainforests of Southeast Asia (Figure 4.8). With long arms and feet that look like hands, they have easy access to the rich plant life in the trees.[6] Tropical climates and rainforests also exist in Africa and in South and Central America (Figure 4.9). But no orangutans or gibbons are found there.

In the savannas of Africa,* lions also roam the zebra's native habitat. With a wealth of diverse and readily available mammalian prey,[7] lions don't just roam but rule the savanna. Plenty of mammalian prey exists in the savannas of Australia and Brazil.[8] Yet lions are absent.

In short, like so many other species, zebras fit their native environment well. But they don't exist in every instance of their optimal environment.

This geographic restriction is apparent, not only from the perspective of the environment, but also from the perspective of species. Although zebras are restricted to Africa, they are not the only species whose only residence is south of the Sahara. Wildebeests, Cape buffalos, black rhinos, white rhinos, African elephants, and spotted hyenas all make their home in the same area. And nowhere else.

Why?

---

* A lion population also exists in a comparatively small geographic area in India.

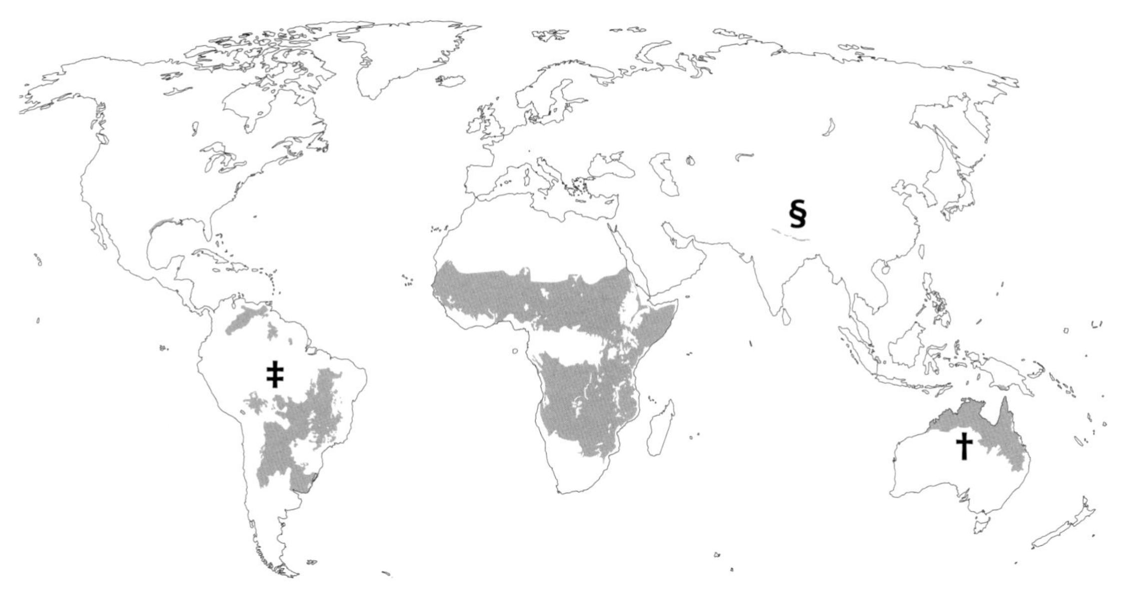

Figures 4.5. Regions of the world containing savannas. Gray shading depicts savannas, shrublands, tropical grasslands, and subtropical grasslands. The Terai-Duar is indicated by the "§" symbol, the Brazilian Cerrado by the "‡" symbol, and the savanna of northern Australia by the "†" symbol. Adapted from http://www.grida.no/resources/7556.

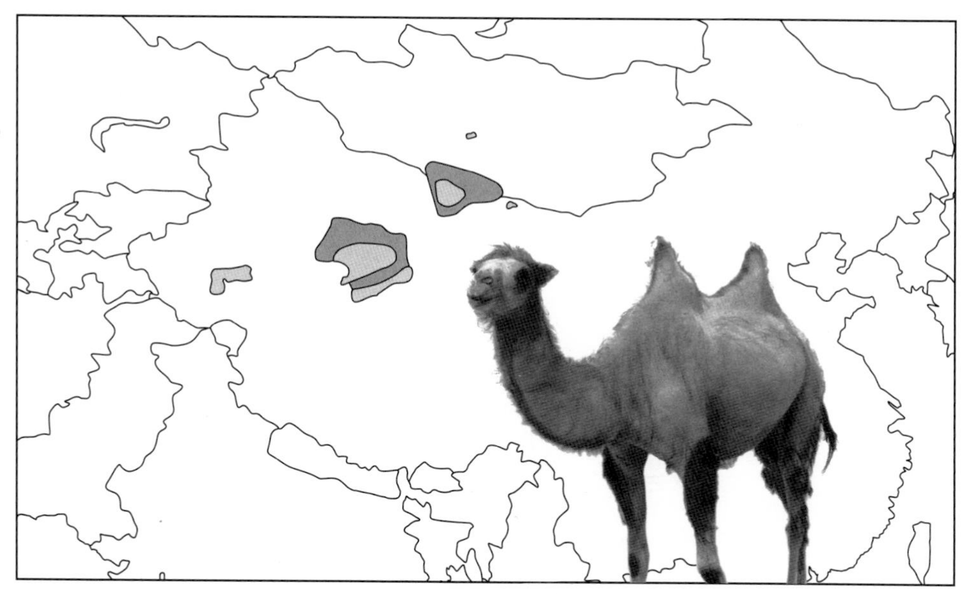

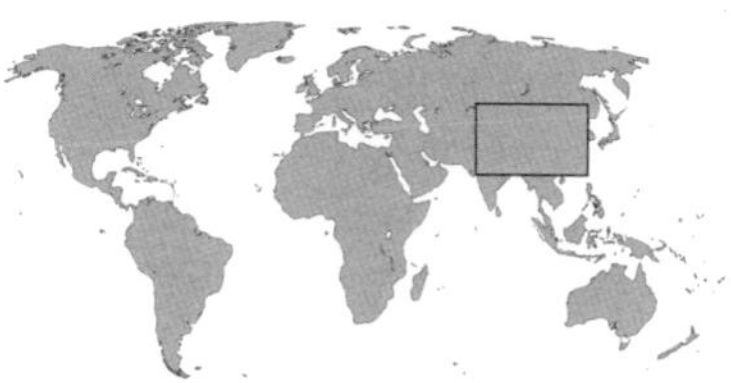

Figure 4.6. Geographic range of the wild camel. Known native range shown in light gray; probable native range in dark gray. Adapted from http://maps.iucnredlist.org/map.html?id=63543.

Figure 4.7. Map of dry shrublands and deserts around the world. Adapted from http://www.grida.no/resources/7561.

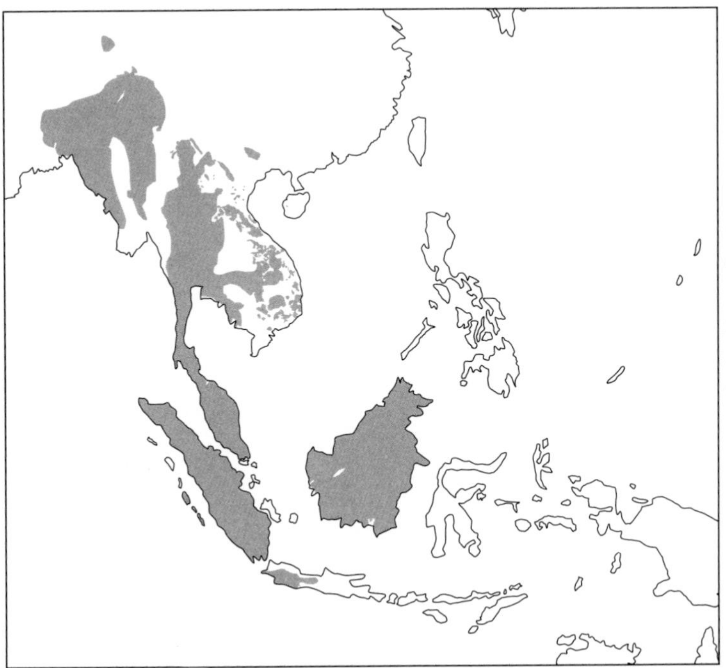

Figure 4.8. Geographic range of gibbons and orangutans. Combined map of native ranges for all species indicated by gray shading. Adapted from http://maps.iucnredlist.org/map.html?id=39780; http://maps.iucnredlist.org/map.html?id=17975; http://maps.iucnredlist.org/map.html?id=39876; http://maps.iucnredlist.org/map.html?id=39877; http://maps.iucnredlist.org/map.html?id=10543; http://maps.iucnredlist.org/map.html?id=39879; http://maps.iucnredlist.org/map.html?id=10547; http://maps.iucnredlist.org/map.html?id=10548; http://maps.iucnredlist.org/map.html?id=10550; http://maps.iucnredlist.org/map.html?id=10551; http://maps.iucnredlist.org/map.html?id=10552; http://maps.iucnredlist.org/map.html?id=39775; http://maps.iucnredlist.org/map.html?id=39776; http://maps.iucnredlist.org/map.html?id=41643; http://maps.iucnredlist.org/map.html?id=39895; http://maps.iucnredlist.org/map.html?id=41642; http://maps.iucnredlist.org/map.html?id=39896; http://maps.iucnredlist.org/map.html?id=39779; http://maps.iucnredlist.org/map.html?id=39879.

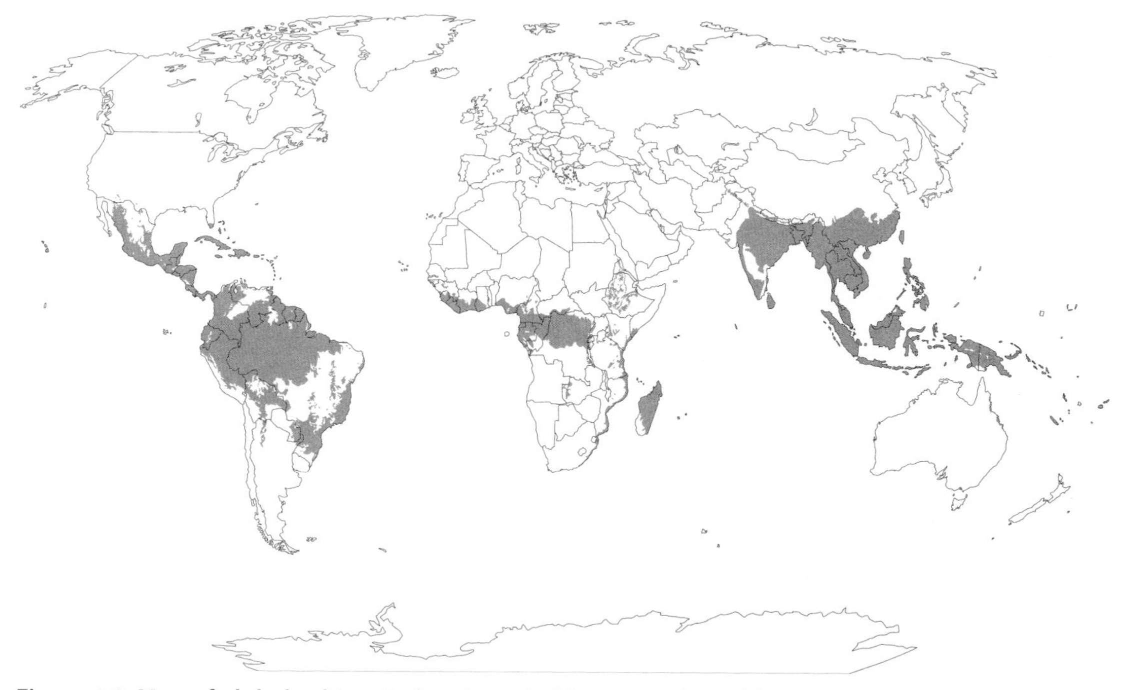

Figure 4.9. Map of global subtropical and tropical forests. Adapted from http://www.grida.no/resources/7568.

Australia is equally puzzling in this respect. Kangaroos, wallabies, numbats, and koalas (Figure 4.10) are all residents of the land Down Under. In fact, they're so well associated with the Australian region that they're iconic. Like so many other species, they're *indigenous* (i.e., native) to one continent — and nowhere else. In scientific terms, we'd refer to them as *endemic* — native to one place and only one place.

Why don't they exist elsewhere?

The species patterns on islands reveal an intriguing clue to this enigma. For example, today in Hawaii, nine species of mammals are indigenous. Six of the species are whales or dolphins. The Hawaiian Monk Seal, the North American River Otter, and a bat constitute the final three. No terrestrial mammals are indigenous to Hawaii.

Even though the Hawaiian Islands represent a small land area,[9] total landmass is not the explanation for the absence of land mammals. Rhode Island is six times smaller[10] than the Hawaiian Islands, yet this tiny northeastern state has 20 native species of mammals. Besides bats and otters, Rhode Island possesses over 10 species of rodents, several species of rabbits, a species of deer, and bobcats.

Since Hawaii is surrounded by the Pacific Ocean for thousands of miles, only creatures that can cross the ocean exist there. Whales, dolphins, seals, and otters can all swim there; bears, foxes, and deer cannot. Bats can fly to Hawaii; bobcats, squirrels, and rabbits cannot. In both Hawaii and Rhode Island, barriers (or lack thereof) to migration — not landmass — explain the geographic distribution of indigenous species.

Figure 4.10. Select marsupials of Australia.

The Hawaiian Islands are not the only example of oceans keeping terrestrial mammals out. American Samoa has 18 native mammals — 15 species of whales and dolphins, and 3 species of bats. Fiji has 25 native mammals — 19 species of whales and dolphins, and 6 species of bats. In New Zealand, 38 species of whales and dolphins are indigenous, as are 2 species of seals, 1 species of sea lion, and 3 species of bats. No terrestrial mammals are native to any of these islands.

Darwin and Alfred Russel Wallace synthesized these facts into a compelling explanation — one with striking implications for our geographic mystery. On a global scale, the two men noticed that barriers to migration defined regions of endemism. Wallace's map of the biogeographic zones of the world[11] (Color Plate 17) made this point obvious. His regions identified clusters of native fauna, and these clusters happen to correlate with geographic features that had strong implications for species' migration. For example, Wallace's "Ethiopian Region" is, essentially, sub-Saharan Africa. The hot and dry climate of the Sahara makes terrestrial migration out of sub-Saharan Africa a difficult — if not deadly — endeavor. The oceans surrounding sub-Saharan Africa make terrestrial escape to the west, east, and south prohibitive.

As compared to the size of the Ethiopian Region, Wallace's "Palæarctic Region" (e.g., Europe, the Middle East, and North and Central Asia) is larger (Color Plate 17). Not surprisingly, few barriers to migration exist in Eurasia. Unlike Africa, Eurasia contains no major sub-regions continuously bound by deserts and water. Instead, species can move almost unhindered on this large body of land.

However, the Himalayas in south-central Asia contain the tallest peaks in the world. Getting to Southeast Asia from East Asia or Central Asia requires crossing them. It should be no surprise that the boundary of Wallace's "Oriental Region" is defined by these mountains, as well as by surrounding water — two obstacles to free migration.

Traversing from the Oriental Region to Australia requires another water crossing. Hence, Australia and its surrounding islands constitute the "Australian Region."

From the Old World, travel to the Americas is currently impossible by land. Furthermore, were it not for the Isthmus of Panama, North America and South America would be cut off from one another. Thus, Wallace's "Nearctic" and "Neotropical" regions largely map to North America and South America, respectively (Color Plate 17). Thus, around the globe, barriers to migration seemed to naturally set the boundaries of regions with endemic fauna.

Applying these principles to the species we discussed — zebras, camels, wildebeests, orangutans, gibbons, etc. — their presence in their native locations appears to be due, in large part, to the natural barriers that keep them from migrating out.

But how did these species reach their homes in the first place? For example, if the Sahara is such an obstacle to migration, how did any animals arrive in sub-Saharan Africa at all?

Figure 4.11. Geographic location of the island of Madagascar.

Madagascar (Figure 4.11) is an island off the east coast of Africa that stands in stark contrast to the Hawaiian Islands. Over 35 times larger,[12] Madagascar is home to over 25 times more mammal species. An exception to the rule in the previous section, 100 of Madagascar's species are primates — terrestrial mammals without an obvious means to cross the 250 miles of water separating the island from the mainland.

Similarly, several hundred miles off the coast of South America, a chain of islands exists that seems to violate the general pattern of indigenous island fauna. Smaller in land area[13] than the Hawaiian Islands, the Galápagos Islands (Figure 4.12) nonetheless possess four species of rodents. More importantly, they are home to a giant tortoise — a reptile with a weight on the order of several hundred pounds (Figure 4.13). These large beasts would seem to challenge the rules of migration inferred from other islands.

European islands also display exceptions to the rule. Surrounded by water, the British Isles reside just over 20 miles from European shoreline

Figure 4.12. Geographic location of the Galápagos Islands.

(Figure 4.14). A little over half the size[14] of Madagascar, the British Isles possess three times fewer mammal species. Nonetheless, while home to over 20 species of aquatic mammals, the UK and Ireland harbor deer, foxes, weasels, rabbits, and rodents.

In a parallel vein, Australia itself is something of an enigma. Unlike Africa, it is not connected to Asia via land (i.e., the Middle East). Instead, Australia is cut off from the largest continental landmass on earth by a significant amount of water. Australia is an island.[15] Yet, as we observed, it contains scores of terrestrial marsupials.

In a related sense, the Americas are an island, too. The Bering Strait, the closest distance between the Old and New Worlds, represents a body of water 53 miles wide.[16] Everywhere else along their coasts, the Americas and the Old World are separated by an even greater distance traversable only by sea or air. Again, the Americas possess many native terrestrial mammal species, not just bats and whales.

Figure 4.13. Giant tortoise of the Galápagos Islands.

How, then, does migration explain the distribution of species around the globe?

Statistics offers some clues. The Hawaiian Islands are thousands of miles[17] from the nearest mainland. In contrast,

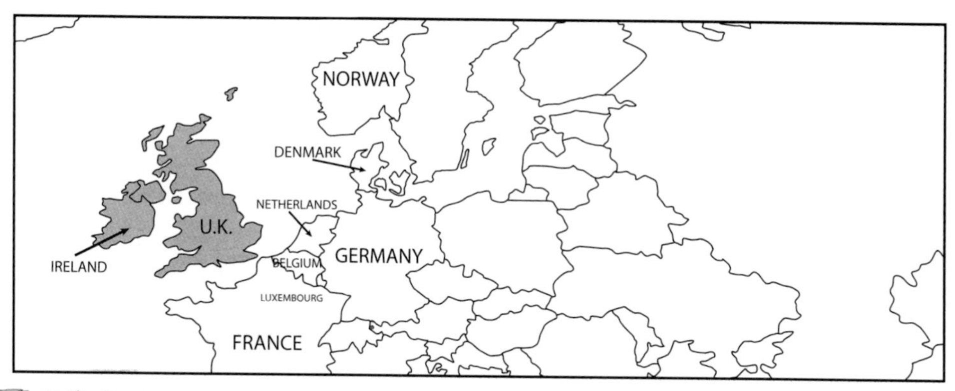

Figure 4.14. Geographic location of the British Isles.

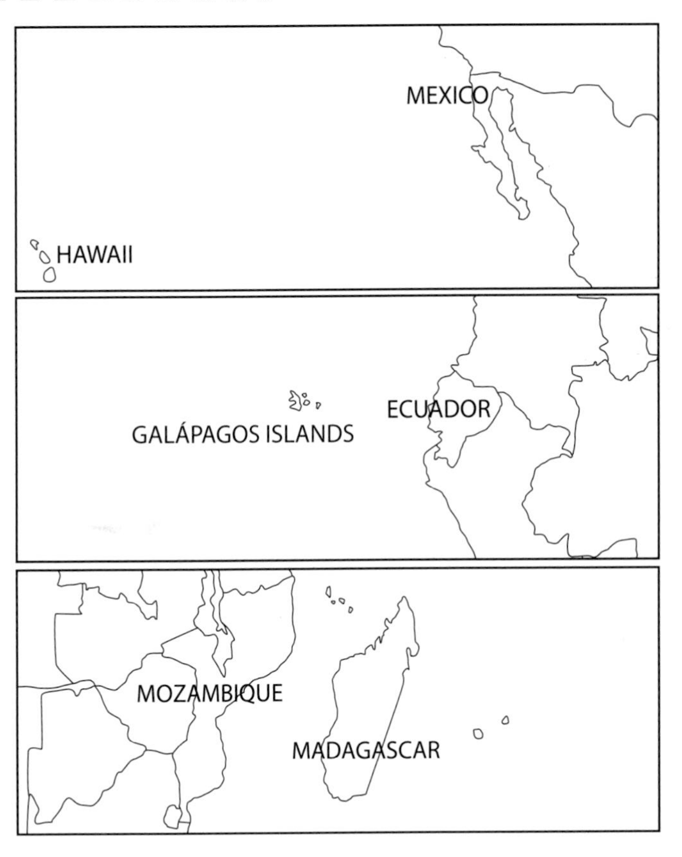

Figure 4.15. Relative distances from islands to mainland. The Hawaiian Islands are the farthest from the (North American) mainland. The Galápagos Islands and Madagascar sit closer to their respective mainlands (South America and Africa).

the Galápagos are only a few hundred miles[18] from the South American coast (Figure 4.15). If, by chance, a terrestrial mammal decided to set out to sea and swim to a new home, neither group of islands would be simple to reach. However, this mammal would have a much easier time reaching the Galápagos than it would reaching Hawaii. Despite the smaller size of the Galápagos as compared to Hawaii, the South American island chain would present fewer obstacles — because the journey would be shorter. Hence, the Galápagos Islands have a higher probability of landing an adventuresome terrestrial mammal than the Hawaiian Islands do.

Though simple distance seems to explain the Galápagos-Hawaiian Island discrepancy (Figure 4.15), the relationship between the distance from shore and the number of species might not be linear. Madagascar is two to three times closer to the African mainland than the Galápagos are to South America. Yet Madagascar has, not 3 times, but about 18 times as many mammal species as the South American islands do.

With a larger landmass, Madagascar might also be capable of supporting more species than the Galápagos.

In addition to distance and landmass, historical climate changes surely played a role in the success or failure of migrations as well. Part of the explanation for the European mammal patterns resides, literally, just below the surface. When examined closely, ocean depths (Figure 4.16) depict rather intriguing patterns — the British Isles sit in the middle of a shallow ocean shelf. Were ocean levels to drop significantly between Britain and Europe, the two continents would be connected, allowing land mammals to walk between the two. Once the mammals reached the British Isles, a restoration of today's ocean levels would presumably lock them in.

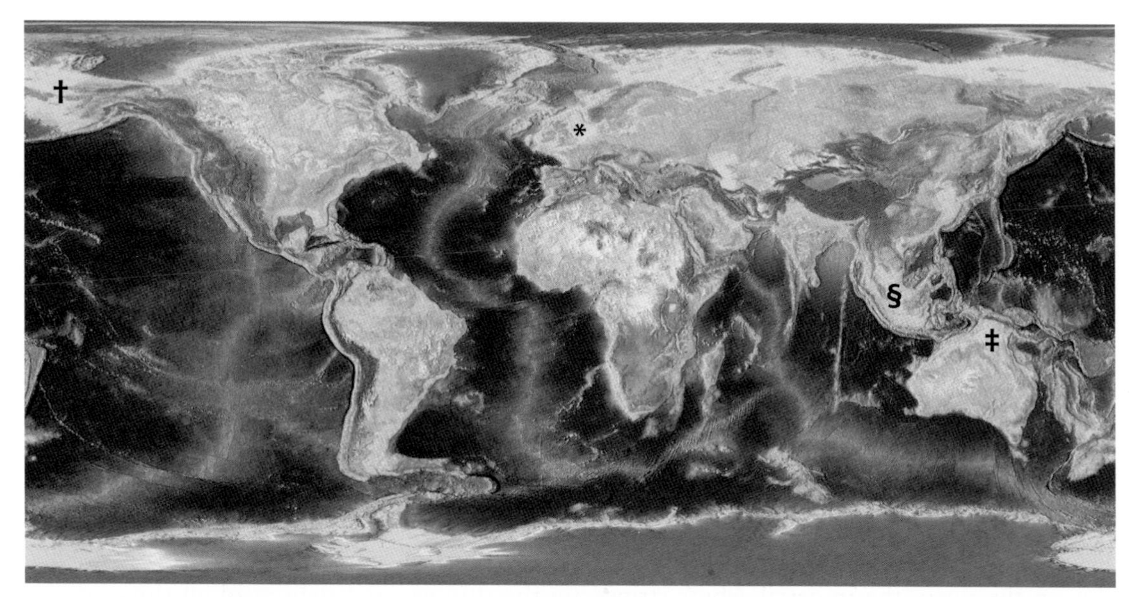

Figure 4.16. Map of the world showing mountain heights, ocean depths, and continental shelves. Light coloration (marked with "*") between European mainland and the United Kingdom, light coloration (marked with "§") among the islands of southeast Asia, light coloration (marked with "‡") between Australia and Papua New Guinea, and light coloration (marked with "†") between North America and northeastern Asia indicates shallow ocean depths separating each of these landmasses.

Historically, an ice age would be sufficient to lower and restore ocean levels in the pattern we just hypothesized. For ice to accumulate on land, large amounts of snowy precipitation would have to fall. For the clouds to accumulate sufficient snow for an ice age to occur, large amounts of ocean water would have to evaporate. Large amounts of oceanic evaporation would effectively lower sea levels and expose land bridges.

Today, no one denies the existence of an ice age. Evolutionists and modern creationists disagree on the timing and the number of ice ages, but both freely invoke the phenomenon. Thus, historical climatological events plausibly explain the existence of mammals on the British Isles.

An ice age could also explain several other global migration patterns (Figure 4.16). Though currently below water, the surface area under the Bering Strait off the coast of Alaska and eastern Russia is comparatively shallow. For terrestrial Asian animals, migration to the Americas could have happened on a solid surface rather than by swimming. Furthermore, since a shallow ocean shelf connects Japan to the Asian mainland, migration to Japan would have been equally straightforward during an ice age. In Southeast Asia, even though trekking from Asia to Australia might have still required passing through water, the swimming distance would have been much shorter than it is now.

Hence, among islands that appear, at first pass, to contradict our conclusions from the previous section, closer inspection of the details reveals broad agreement. In short, climatologically and ecologically informed statistics offer plausible explanations for the distribution of species around the globe today.[19]

For example, applying these principles to zebras, we can begin to piece together the puzzle of why zebras are African. Since an ice age would have little effect on Africa as a continent, historical climate changes are not needed as part of the explanation. However, a chance migration across the Sahara could explain how the zebra got to Africa. Furthermore, since their stripes fit the grassland environments of Africa, zebras likely migrated across the Sahara and thrived there.

To be precise, three different chance migrations would be required. Today, not one but three species of zebras roam sub-Saharan Africa — the plains zebra (Color Plate 18), the mountain zebra (Color Plate 19), and the imperial zebra (Color Plate 20). All three species possess characteristic features. For example, on the plains zebra, the stripes on the body are thick and extend all the way around its belly (Color Plate 18). On the mountain species, the stripes are thinner than on the plains species, and they do not reach all the way around the belly (Color Plate 19). Like the mountain zebra, the stripes of the imperial zebra also fail to cross the belly. But on the torso, the imperial zebra's stripes are the thinnest and densest of all three species (Color Plate 20).

Ecologically-informed statistics offer a plausible explanation for the African endemism of the three living zebra species. Though the Sahara presents a formidable barrier to migration, it appears that it is not an impossible barrier. On three separate occasions, by chance a zebra ancestor could have trekked across the desert and reached sub-Saharan Africa.

However, these same statistics fail to explain why none of these species were able to get back out.

Presumably, the same statistics that make three zebra entrances to sub-Saharan Africa an improbable — but plausible — event, would also make exit an equally improbable but plausible event. We could conclude that migration to Asia "just hasn't happened yet." But this explanation is hardly satisfying.

A closer look at several species' global distributions reveals a hint. For example, flightless birds exist in both Africa (Figure 4.17) (e.g., two species of ostriches, Color Plates 21–22) and South America (Figure 4.18) (e.g., the three species of rheas, see Color Plates 23–24 for two of the species). All five species occupy similar habitats. Nevertheless, the species in closest geographic proximity (i.e., the three species of rheas or the two species of ostrich) look much more alike than species at a great geographic distance.

It's possible that the five flightless bird species could have arisen separately. Then three of the species might have independently migrated to South America, while two migrated to Africa. By chance, the three rhea species could have joined their journeys, while the ostriches also just

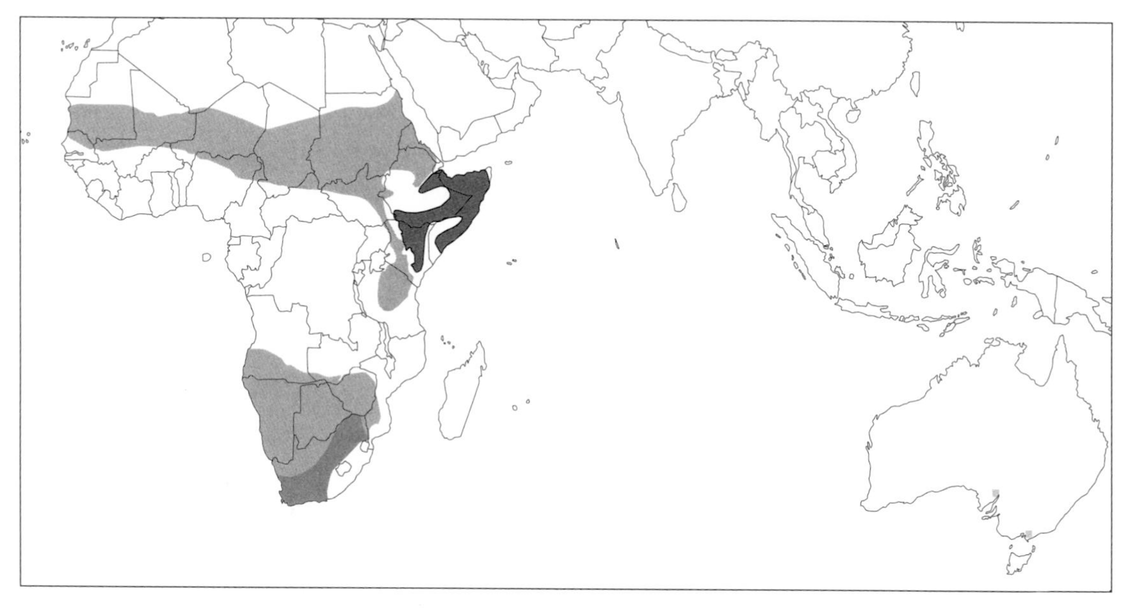

Figure 4.17. Geographic ranges of ostrich species. The range of the Somali ostrich is indicated by dark gray shading; the range of the common ostrich, by light gray shading. The intermediate gray shading represents regions where the common ostrich has been introduced. Adapted from http://maps.iucnredlist.org/map.html?id=45020636; http://maps.iucnredlist.org/map.html?id=22732795.

happened to join on a path to Africa. But this explanation would leave us wondering why the species with greatest outward similarity just happened to tag along.

Darwin proposed another explanation. Though he knew nothing about the relationship between DNA and heredity, he was able to draw rudimentary principles about the process of inheritance. For example, when you look at your own physical features, it's obvious that you inherited some of your parents' features. To put it in the negative, you don't look like your neighbors. The reason for the dissimilarity is the fact that you didn't inherit your neighbor's features. To a degree, biological similarity echoes common ancestry.

This fact is true of species around the globe. If you've ever kept pets, you're probably familiar with this as well. When cats

Figure 4.18. Geographic ranges of rhea species. Dark gray shading represents range of the lesser rhea; intermediate gray shading, the range of the greater rhea. The lightest gray shading represents the range of the Puna rhea (*Rhea tarapacensis*). Adapted from http://maps.iucnredlist.org/map.html?id=22728206; http://maps.iucnredlist.org/map.html?id=22678073; http://maps.iucnredlist.org/map.html?id=22728199.

produce a litter, they birth more cats. They don't birth dogs. When snakes reproduce, they don't make elephants. They produce more snakes.

However, common ancestry does not preserve biological *identity*. You don't look like carbon copies of your parents. If you've observed a litter of cats, you've probably witnessed a whole host of coat colors and patterns. Though common ancestry preserves biological similarity, some level of change happens each generation.

Darwin called this principle *descent with modification*. Though the identification of DNA as the substance of heredity was still 100 years in the future, this simple principle was all that Darwin needed to make his point — and eliminate competing explanations.

Since inheritance strongly maintains features while still permitting some change, Darwin argued[20] that a single rhea population migrated to South America, and then subsequently formed the three rhea species. Similarly, a single ostrich population migrated to Africa, and then subsequently formed the two ostrich species. This would reduce the number of chance events from 5 to 2. It would also plausibly explain why the three rheas look so similar.

Darwin's reasoning extends to more species than birds. For example, in the tropics of Southeast Asia, two species of great apes exist, the Sumatran orangutan and the Bornean orangutan (Figure 4.19; see also Color Plates 25–26). The tropics of Africa also possess two great ape species, the western gorilla and the eastern gorilla (Figure 4.20; see also Color Plate 27–28). In terms of visible features among these four species, their natural groupings track with their geography. The two Asian species look more like each other and less like the African species. The same pattern holds true for the African species — they share more in common with each other than with the two Asian species.

Like the flightless birds, a single population of each great ape likely migrated to each continent — a single population of orangutans into Asia, and a single population of gorillas into Africa. Subsequently, these separate populations diversified into the four species observable today. Inheritance from a common ancestor — rather than four independent migrations — is the more likely explanation for the resemblance among species on the same continent.

To explain the endemism of the three zebra species to Africa, we could invoke three separate migration events across the Sahara. Alternatively, since common ancestry maintains similarity (in this case, the existence of stripes) while still permitting some change (in this case, the specific stripe patterns), we could invoke a single migration event followed by a couple of speciation events. The latter scenario requires fewer statistically improbable migration events.

It also implies that zebras got their stripes from somewhere else.

Since the principles we've illustrated with zebras apply to so many other species, migration is the major explanation for the distribution of species around the globe. Historical geologic events add

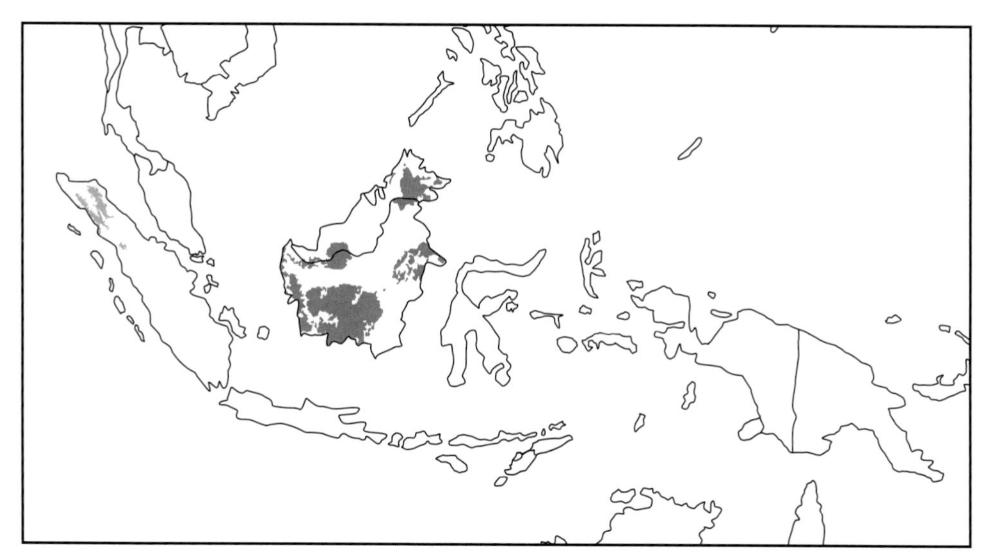

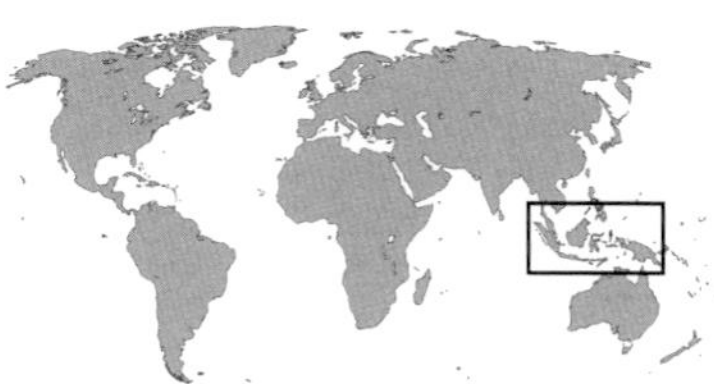

Figure 4.19. Geographic ranges of the orangutan species. Dark gray shading represents the native range of the Bornean orangutan; light gray, the native range of the Sumatran orangutan. Adapted from http://maps.iucnredlist.org/map.html?id=39780; http://maps.iucnredlist.org/map.html?id=17975.

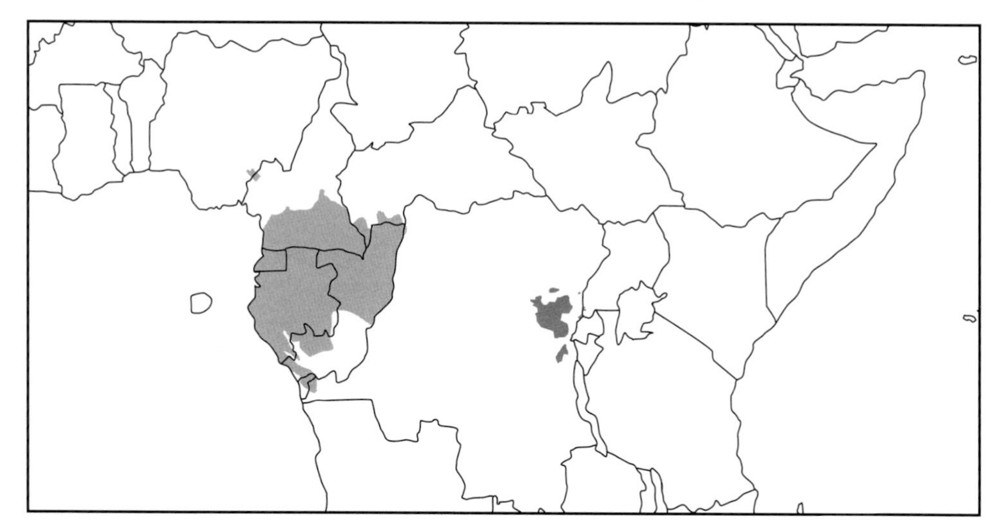

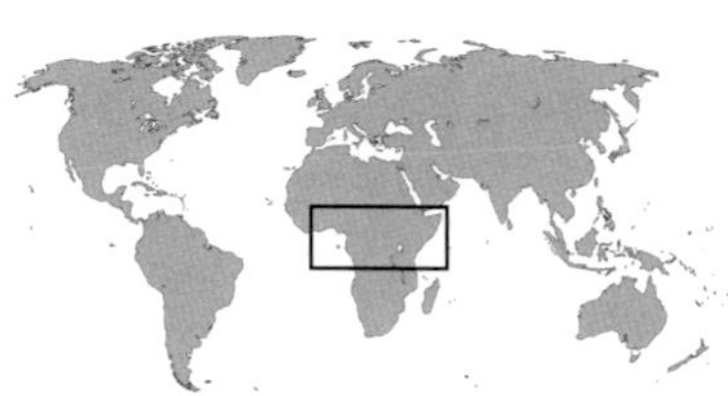

Figure 4.20. Geographic ranges of gorilla species. The native range of the Western gorilla is show by the large gray area on the left side of the callout. The native range of the Eastern gorilla is shown by the gray shading in the center-right of the callout. Adapted from http://maps.iucnredlist.org/map.html?id=39994; http://maps.iucnredlist.org/map.html?id=9404.

nuance to the overall explanation. But the dominant reason for why species are localized to their current habitats is movement from another location.

Thus, Darwin successfully eliminated the creationist hypothesis of the fixity of species' geography. Furthermore, as we observed in the preceding sections, the fact of migration also leads to the conclusion of common ancestry among some species. In other words, Darwin's geographic observations successfully eliminated the view of species fixity.

By analogy to a jigsaw puzzle, recall that species and the clues to their origins represent puzzle pieces. The hypothesis of species fixity predicts numerous independent, unconnected, isolated puzzles — one for each species. Geography represents an indirect clue to the origin of species, and it strongly suggests common ancestry of some. It's as if several geographic puzzle pieces could be connected. Since species pieces are connected to each of these geographic pieces, the geographical connection naturally tied species together. Thus, by virtue of the nature of the scientific method and despite lacking genetic data, Darwin could successfully argue for the scientific merits of his thesis. At least, he could do so in 1859.

Chapter 5
The Riddle of Ancestry

In an actual jigsaw puzzle, center pieces naturally lend themselves to multiple connections. Unlike edge pieces, center pieces don't typically have sharp boundaries that forbid certain arrangements. Instead, the irregular shapes of center pieces permit them to be linked in a variety of ways. In the process of puzzle assembly, the plasticity of these connections allows tentative arrangements to be explored and tested.

This characteristic can be both helpful and frustrating. Since every puzzle assembler must start the task somewhere, temporary connections can be useful stepping stones toward the final product. However, temporary connections can also represent a significant departure from the actual image. (I have made many such erroneous connections myself.) When they do, the apparent plausibility of these erroneous connections can hinder real progress toward the correct solution.

On the question of the origin of species, a similar principle applied. For example, Darwin used non-genetic data — the biological version of center pieces — to argue for the validity of his hypotheses. He made connections among species where others saw no link.

In more rigorous scientific terms, the method of inductive reasoning is what gave Darwin this sort of ability. For Darwin to argue for connections among species, all he needed to do was eliminate competing explanations. By successfully eliminating the hypothesis of species' fixity, Darwin made a plausible case for his model.

However, the strength of the method of inductive reasoning is also its Achilles' heel. The process of elimination is a very effective way of getting closer to the real answer. Yet, this same method lends itself to making premature (i.e., erroneous) conclusions.

For example, the body of knowledge that we call *competing explanations* is not fixed. New explanations can arise with time. Usually, these explanations are not discoveries of fundamentally new ways in which the world operates. Rather, these explanations arise from discoveries of ways in which the world has always worked, hitherto unrecognized.

Practically, this means that every scientific investigation is limited by the competing explanations known *at that point in time*. At later points in time, new competing explanations might arise that old tests failed to eliminate. Thus, conclusions reached at earlier points in time might look plausible in the moment, only to yield to different explanations at later time points.

This principle raised its head in the realm of other arguments[1] that Darwin used to refute the fixity of species.

Darwin derived his additional arguments from a deeper consideration of the rudimentary genetic observations that we discussed in the previous chapter. Again, to anyone who has glanced at their own family tree, similarities are obvious. Few, if any, of us have gone through life without having someone comment on how much we look like one or both of our biological parents. That inheritance produces similar offspring is an inescapable fact.

Equally inescapable is the fact that inheritance produces slightly different offspring. No children are carbon copies of their parents.[2] Even without any knowledge of how inheritance happens at the cellular and molecular levels, we know that something changes in offspring.

Furthermore, changes seem to accumulate with time. Offspring look most like their parents, but when compared to grandparents, the similarities start to decline. When you compare yourself to relatives more distant — either by the number of generations that have passed or by the number of intervening relatives — the differences are stark.

We've already observed an echo of this pattern in our comparisons of zebra species. In fact, the patterns are so strong that they've been formalized into a classification scheme — one that resonates with genealogical hints.

Based on the similarities and differences among creatures, the Linnaean* classification system groups creatures progressively in various categories. For example, the characteristic differences among the three zebra species that we observed in the previous chapter (Color Plates 18–20) result in their classification as separate *species*. However, since these species are all still obviously zebras, above the level of species they share the same category of classification — *genus*. Under the Linnaean classification scheme of giving species two names, all three species share the same genus name (*Equus*) but differ in the species name. The imperial zebra goes by *Equus grevyi*; the mountain zebra by *Equus zebra*; and the plains zebra by *Equus quagga*.

Since all three species of zebras bear significant resemblance to the one living horse species and the three living ass species, all seven species are grouped together in the next higher level of

* This is the same Linnaeus that we discussed in chapter 1.

classification — *family*. All seven species are members of the family *Equidae*. In fact, the similarities are so strong that the Przewalski's horse (Color Plate 29), the African wild ass (Color Plate 30), and the two wild Asian asses (the onager and the kiang; Color Plates 31–32) are all classified in the same genus as the three zebra species. The formal Linnaean terms for the one horse species and three wild ass species are *Equus przewalskii*, *Equus africanus*, *Equus hemionus*, and *Equus kiang*, respectively.

Equids, rhinos (Color Plates 33–37), and tapirs (Color Plate 38–41) are all different enough from each other that, in the Linnaean system, they are separated into different families. However, all three groups of creatures have an odd number of toes. This fact sets them apart from cattle, sheep, goats, deer, antelopes, giraffes, and other even-toed, hoofed mammals. Consequently, the equid, rhino, and tapir families (Equidae, Rhinocerotidae, and Tapiridae, respectively) together constitute the next higher level of classification, *order*. These three families form the order *Perissodactyla*. The even-toed, hoofed mammals form the order *Artiodactyla*. Nevertheless, both the even-toed and odd-toed species we have discussed suckle their young. So do monkeys and apes, aardvarks and armadillos, anteaters and pangolins, sloths and seals, and a whole host of other creatures. Despite their recognizable differences, together these creatures form the next higher level of classification, *class*. These species belong to the class of mammals — *Mammalia*.

Mammals and other vertebrates such as reptiles, amphibians, birds, and fish (along with a few other obscure creatures) possess a developmental precursor to a backbone. Thus, all belong to the same level of higher classification, *phylum*. These species constitute the phylum *Chordata*.

In the Linnaean system, the highest level of classification[3] is *kingdom*. The kingdom *Animalia* includes all species that are animals.

In descending order, the Linnaean hierarchy is kingdom-phylum-class-order-family-genus-species. Going up the hierarchy joins more and more diverse creatures into large groups which happen to share a small set of characteristics. Moving the opposite direction reveals smaller and smaller groups that each share an increasing number of features.

This classification pattern is reminiscent of something very familiar. If you were to classify your relatives based on their degree of relatedness to you, you would discover a similar grouping scheme. Though you might not use the formal Linnaean terms, a hierarchy of categories would be the natural result. For example, your parents would form one level of classification, your grandparents another, your great-grandparents another, etc. The closer the genealogical relationship, the fewer the members in the classification category. The more distant the genealogical relationship, the more members in the classification category.

Is this correlation simply a coincidence?

One major difference between the Linnaean system and a family tree is the time element. The family tree contains an intrinsic temporal stamp. For example, you know how long ago your relatives lived, and you use this information to create your groupings. In contrast, among living species, the Linnaean system appears to contain no such time stamp.

However, the Linnaean system isn't limited to living species. The fossil record contains numerous representatives of both extant (i.e., living) and extinct species. Furthermore, these fossils sit in geologic layers, which have their own implicit time stamp. For example, it's easy to cover one layer with another, much like the waves of the sea crash onto the beach, adding silt and sand on top of existing sand. In contrast, it's much harder to insert new layers under existing ones. Ocean waves don't normally crawl under the existing layers of sand on the beach, insert a new layer of sand, and then retreat, leaving the prior sand layers intact. Therefore, in general, the deepest layers in the fossil record represent the oldest deposits of fossil-rich sediments, whereas the shallowest layers are the most recent. Consequently, fossils contain their own relative time stamp.

Equid fossils are found in specific layers. The shallowest layers contain representatives of the living species. Deeper layers contain fossils of extinct equid species.[4]

This patterns holds true for other families.[5] For example, species from the other two families within the order Perissodactyla are also present in the fossil record. Living rhino species are found in the shallowest layers; deeper layers contain extinct species. Living tapir species exist in the uppermost layers; lower layers contain extinct species.

This arrangement implies a time sequence.

Let's review the concepts we just covered. Individuals on family trees can be grouped in hierarchical categories based on degrees of relatedness. These categories naturally contain temporal information as well as ancestry information. Conversely, animal species can be grouped in hierarchical categories. Based on the relative ordering of fossils in the fossil record, these categories can be placed on a relative time sequence. Together, these facts manifest a parallel between family trees and the structure of life.

Family trees contain yet another detail on genealogical relationships. If you were to compare your brother to your shared great-great-great-great-great-grandfather, the resemblance would likely be small. If the two stood side by side, the average onlooker might not place them on the same family tree. However, if you stood your great-great-great-great-great-grandfather next to his son, the resemblance would be obvious. And if you stood his son next to his son's son, a resemblance would again be obvious. In short, the genealogical connection between your brother and his distant relatives would become clear through a series of intervening relatives.

Today, when you stand a zebra next to a rhino, the resemblance between these species is much less striking than the differences. However, zebras can be connected back to a less equid-like ancestor via a series of extinct species (Figure 5.1). Rhinos can be connected back to a less rhino-like species in the same way. The same holds true for tapirs. In the deeper fossil layers, you can even find creatures like *Cambaytherium* that seem to blend the features of primitive rhinos, tapirs, and equids. To be sure, the visible differences between species might be bigger than the visible differences between generations on a family tree. But the resemblance is still present.

Darwin argued that this type of resemblance betrayed a genealogical link among these species.

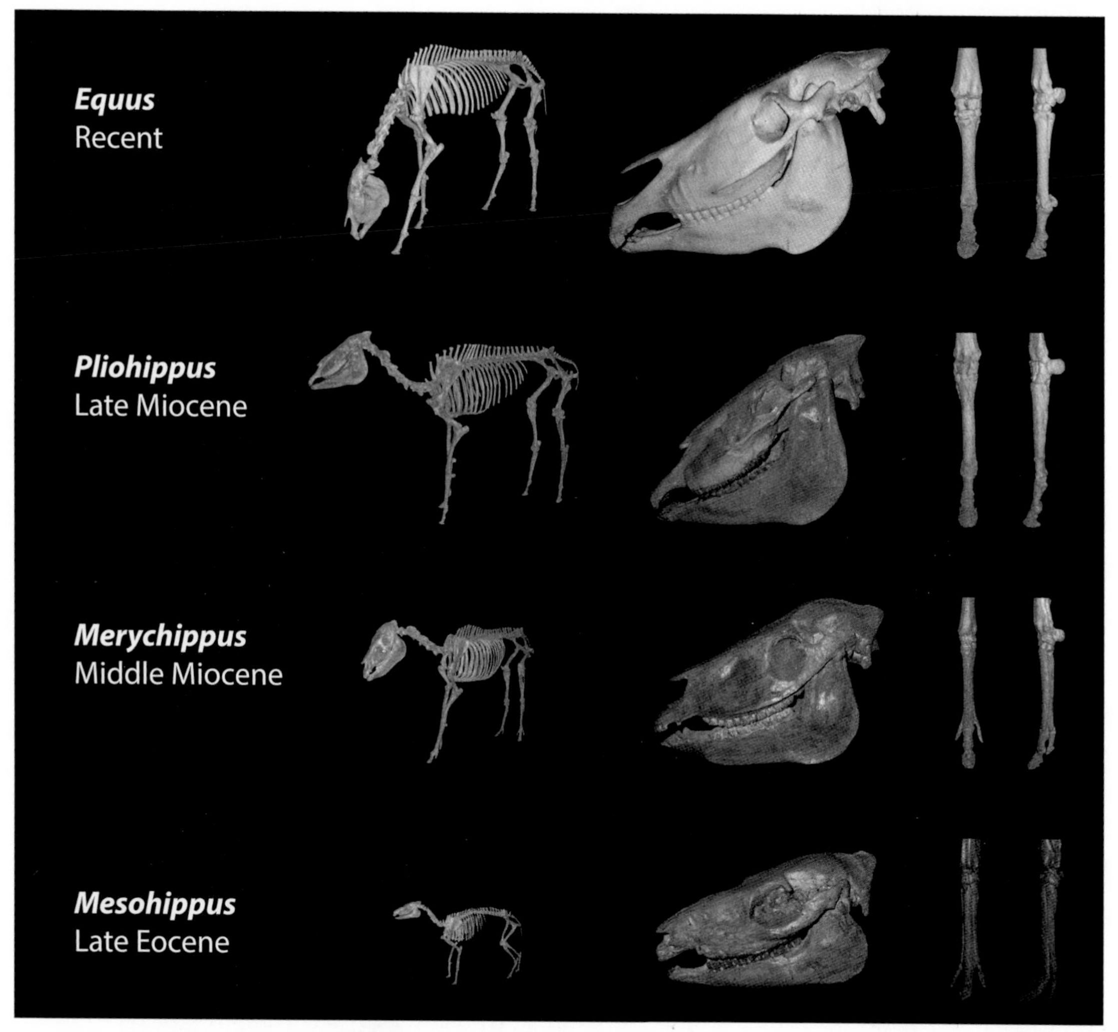

Figure 5.1. Fossil connections between modern equids and extinct equids. Miocene and Eocene represent geologic layers.

❖ ❖ ❖ ❖

Does zebra ancestry stop with tapirs and rhinos? We've already observed that connecting distant relatives on a family tree challenges intuition. The further back in time you go on the tree, the fewer the similarities to the present. It's no easy task to look for your relatives among the people who lived 3,000 years ago.

Similar challenges confront our investigation of the zebra family tree. The deeper you go in the fossil record, the starker the contrast between modern species and extinct ones. What could zebras possibly have in common with creatures like *Archaeopteryx* (Figure 5.2)?

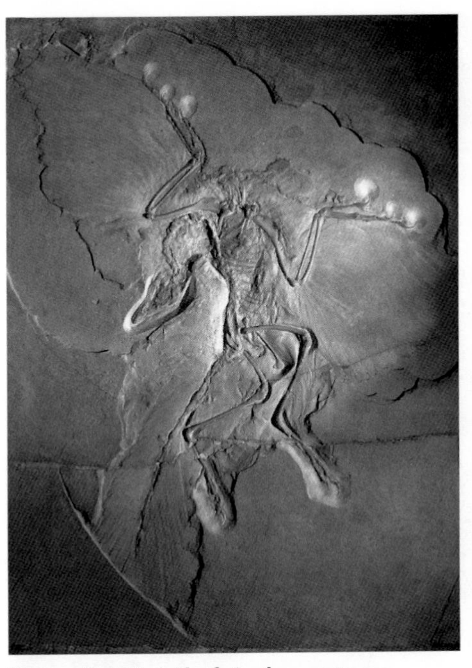

Figure 5.2. Fossil of *Archaeopteryx*.

Rather than answer this question with more fossils, let's examine living species. If two modern species share a family tree, then they share a common ancestor at some point on the tree. Two genealogical lineages would trace from this ancestor — one lineage to one of the modern species, and another lineage to the other species. Each lineage is a line of descent. The process of descent involves inheritance of certain features. Descent also involves modification, but descent preserves some similarity. Even if the genealogical lineages connecting two species ran deep on the family tree, an echo of similarity might still exist between the two.

For example, consider the anatomy of several diverse species. The structure of the human forelimb is well known. Descending the shoulder to the finger tips, you find one bone (the humerus), then two bones (radius and ulna), then a group of bones (the carpals),

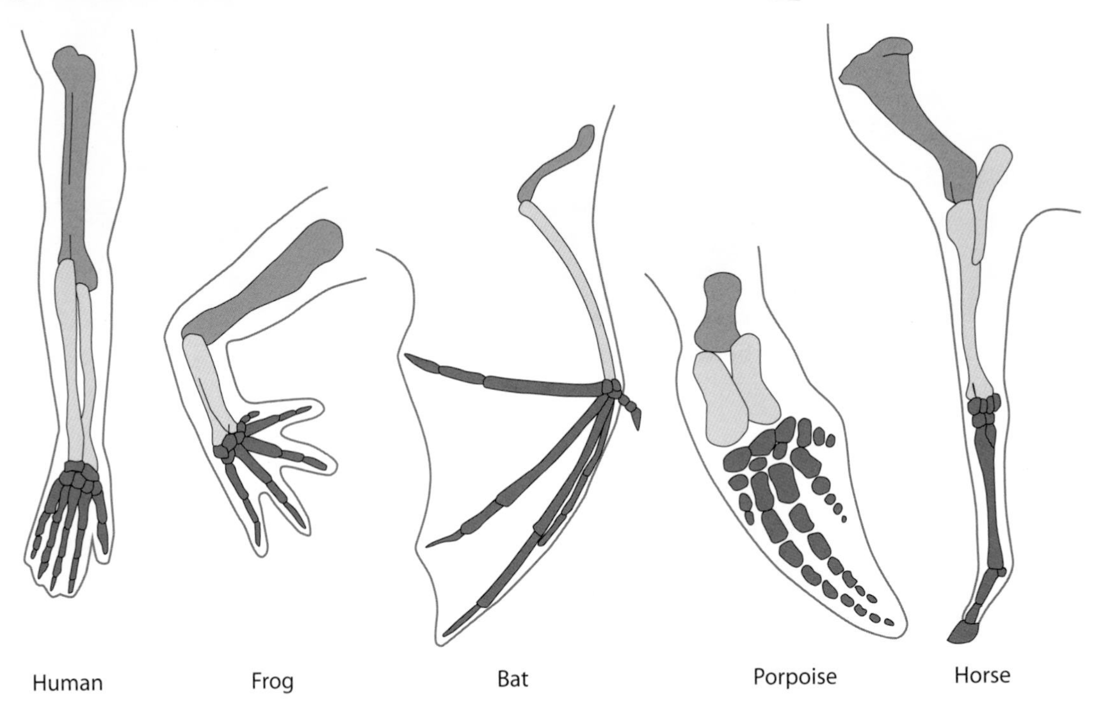

Figure 5.3. Shared forelimb structure across diverse species. Corresponding bones indicated by levels of gray shading. Adapted from https://online.science.psu.edu/sites/default/files/biol011/Fig-8-5-Vertebrate-Limbs.gif.

and then the metacarpals and digits (phalanges) (Figure 5.3). A similar order or pattern is found across very different vertebrate species — one bone, two bones, a group of bones, metacarpals and digits[6] (Figure 5.3). Even horses (i.e., a good representative of zebras) possess this pattern (Figure 5.3).

This shared pattern fits the process of descent and inheritance. Obviously, the specific details of the pattern vary from species to species (i.e., the *modification* part of *descent with modification*). But the pattern is the same (i.e., the *descent* part of *descent with modification*).

Consequently, since these species belong to different biological orders (e.g., humans = Primates; bats = Chiroptera; porpoises = Cetacea; horses = Artiodactyla) and, in some cases, to different classes (frog = Amphibia; humans, bats, porpoises, and horses = Mammalia), Darwin argued that this type of evidence extends the zebra family tree beyond rhinos and tapirs.

Shared patterns exist at the embryological level as well. Remarkable similarity exists in the early developmental forms of various species. For example, in creatures as diverse as turtles, chickens, and fish, a very similar embryonic stage appears — the *pharyngula* stage (Figure 5.4). Because each of these creatures has a characteristic mode of reproduction, the stages preceding the pharyngula stage are diverse. Conversely, since the process of development eventually produces something

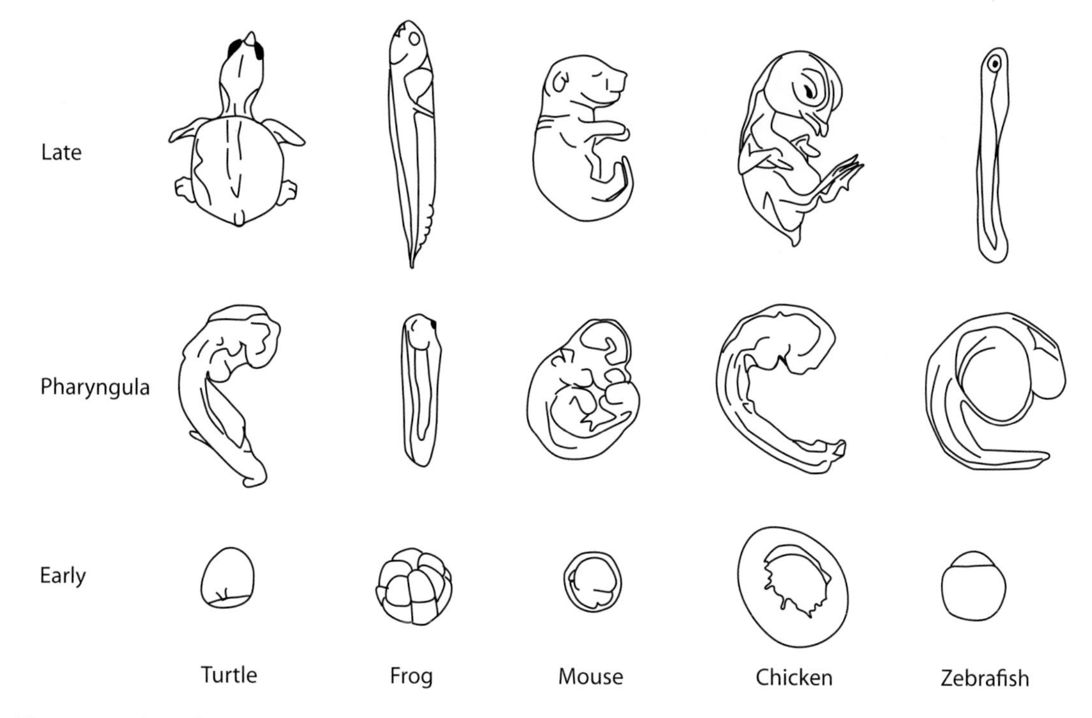

Figure 5.4. Development stages of vertebrate species. Early, pharyngula, and later stage embryos shown for five different species. Adapted from http://dev.biologists.org/content/develop/141/24/4649/F2.large.jpg.

resembling the characteristic form of each creature, the stages after the pharyngula stage are obviously different.[7] But, from Darwin's perspective, the fact that development can be cross-correlated at the pharyngula stage across very dissimilar species harkens to a shared genealogy.

Using the anatomical comparisons that we just discussed to extrapolate these developmental observations to zebras, Darwin argued this type of evidence ties zebra ancestry not only to frogs, but also to reptiles, birds, and fish.

All the principles that we've discussed apply across the Linnaean classification system. Diverse species fall into hierarchical groups, and these groups fall along a time sequence in the

Figure 5.5. Fossil (top) and artist's reconstruction (bottom) of *Tiktaalik.*

fossil record. Furthermore, the extinct species can be roughly connected to modern species via species that blend the features of distinct groups.

This fact is true even deep in the fossil record. For example, in 2006, the discovery of *Tiktaalik* (Figure 5.5) was published.[8] With scales like a fish, a head like a land-dweller, and limbs that looked like intermediates between fins and legs, *Tiktaalik* blended the features of fish and land-dwelling species. Blended species exist even deeper in the fossil record.

Darwin took these types of observations to an apparent conclusion: The family tree of zebras includes all species that have ever lived.

❖ ❖ ❖ ❖

The biological phenomena that we've explored parallel the properties of family trees. But they also have strong parallels to another concept. In other words, a competing explanation exists for all the evidence we've just discussed. This competing explanation becomes clear with an example. For instance, humans have extended considerable effort for thousands of years in attempting to move themselves from one place to another — and at faster rates. In doing so, they have produced characteristic patterns in their designs.

Figure 5.6. Similarity of sedans despite diverse geographic origins. Though each of the three car manufacturers hails ultimately from a different continent, the sedans from all three automakers share a number of features — four wheels, four doors, sloping windshields, headlights in the front, rearview mirrors, etc.

Take the sedan as an example. Whether sedans are manufactured by automakers based in Asia, North America, or Europe, all share a common set of features (Figure 5.6). You won't find cars without wheels, engines, doors, or windshields. Despite the geographic distance separating their locations, sedans made by Honda, Chevrolet, and Mercedes have four wheels, doors on the side, windshields in the front, brake lights in the back, and a host of other shared characteristics.

The reason for this similarity is not common ancestry. Honda, Chevys, and Mercedes did not evolve from a common sedan ancestor in the Arctic. Rather, because these cars were designed for similar functions, these purposes put constraints on the final product.

For example, when the purpose is transportation of several individuals and their luggage, four wheels are superior to either three or five.* Conversely, without windshields in the front, the passengers would be accosted with an unending attack of bugs and debris when traveling at high speeds. In addition, climbing into these cars from the top is challenging. Having doors on the side makes entry and exit far easier. Thus, human designers naturally produce similar patterns — due to shared functional purposes, not common ancestry.

This logic extends even further into the manufacturing process. For example, if you visited the assembly plants of these auto makers, you'd probably find a temporal stage in the manufacturing process that is shared. Despite the obvious differences at the final product level among Hondas, Chevrolets, and Mercedes, basic rules of efficiency and function result in similar stages of manufacturing. In fact, you'd probably even find shared elements in the blueprints for each of these cars. Why? Since the end products have similarities, the blueprints will, of necessity, have them as well.

Conversely, these means of transportation naturally fall into a hierarchical classification scheme. For example, we immediately recognize one category that we call *cars*. Even though each car is unique, all cars share a set of features — lack of a cab bed, a certain range of height from the ground, etc. No one has any trouble identifying a car.

If we shrink the list of shared features, pickup trucks and SUVs can be brought into the same group with cars. Since most trucks possess a cab bed, and since trucks tend to be higher off the ground than cars, the number of features shared among cars and trucks is less than those shared among cars alone. If we try to incorporate motorcycles into our classification scheme, the features that all these vehicles have in common is even smaller. A minimum of two wheels — rather than a total of four wheels — defines the group. In short, the more vehicles that we try to lump into our transportation classification system, the smaller the list of shared features becomes — just like what we observed with the Linnaean classification system.

---

* Balance and stability are easier to achieve with an even number of wheels, and four wheels are more stable than two.

Figure 5.7. (Opposite page) Biological life and the designed means of transportation share a similar pattern of classification. On the left, biological populations can be classified in a nested hierarchy, as per Linnaeus' classification system. For example, dragonflies and birds are part of the same *Kingdom* (Animalia). But dragonflies are part of one *Phylum* (Arthropoda), and birds, another (Chordata). Birds and elephants are part of the same *Phylum*, but they exist in different *Classes* (birds in Aves; elephants in Mammalia).

On the right, the means of transportation also fall into a natural nested hierarchical system of classification. For example, power and unpowered means of transportation would be classified in the same *Kingdom* (i.e., they are both means of transportation). But hang-gliders would fall in one *Phylum* (i.e., unpowered means of transportation), and jets, another (i.e., powered means of transportation). Jets and tanks are part of the same *Phylum*, but they exist in different *Classes* (i.e., jets in the class of airplanes; tanks in the class of land-based means of transportation).

SPECIES
GENUS
FAMILY
ORDER
CLASS
PHYLUM
KINGDOM

Eventually, the list of shared features among modes of transportation becomes very small. To include boats and planes with land vehicles, wheels aren't even on the list. Naturally, all boats form a category, as do all planes. But if we join these two groups with the land-based means of transportation, the similarities narrow considerably.

In essence, means of transportation naturally fall into Linnaean-style categories. Cars occupy the genus and species levels. Trucks, SUVs, and cars form a family. Orders represent the different purposes for land-based vehicles — e.g., tractors would naturally form one order; semi-tractor trailers would form another; and all trucks, SUVs, and cars (whether used for racing or leisure) would form yet another. The environments in which means of transportation operate would constitute the class level — boats, airplanes, and land-based vehicles occupying separate classes. Powered (e.g., 747s) and unpowered (hang-gliders) means of transportation fall into separate phyla. Together, these phyla would constitute a kingdom. Conversely, the power sources (nuclear power plants, wind energy, solar energy, etc.) for these means of transportation would constitute a separate kingdom.

The resemblance between the classification of life and the classification of the means of transportation is strong. The major vertebrate biological classes (mammals, reptiles, amphibians, birds, fish) fall along largely environmental divisions, just like vehicles (Figure 5.7). Furthermore, both types of classification can be made rigorously mathematical to generate unequivocal classification categories. We loosely observed this fact in both realms when we noted that the number of shared features decreases as the hierarchy ascends to more inclusive categories of classification. Had we performed more in-depth quantification, we would have reached the same conclusion.[9]

In light of these parallels, we would be justified in claiming that the hierarchical pattern of life strongly suggests that it was the result of deliberate design.[10]

These parallels between design and biology extend even further. In our discussion of the deeper fossil layers, we observed the presence of fossils that seem to blend the features of two different categories of creatures. Some products of human design also seem to span two categories. As we've observed, boats and tanks would fall in entirely different classes. Yet vehicles exist that blend the features of these two very different means of transportation. For example, amphibious assault vehicles resist classification. They are the perfect example of a designed "transitional form" — designed for the transitional environment between water and land (Figure 5.8).

In fact, if you were following my classification scheme for vehicles, you might have been mentally protesting that I left certain vehicles out of my classification example. In particular, I made no mention of the increasing number of hard-to-classify vehicles. With the invention of crossover vehicles, the distinction between cars and SUVs is becoming blurred. What is currently labeled a "car" and what is labeled an "SUV" is now a matter of debate.

Human engineers have designed — and are still designing — "transitional forms."

As added support for this conclusion, it's instructive to keep in mind that the Linnaean classification system was originally based, not on ancestry, but on *function*. Species fall into a hierarchical pattern because their characteristic functional features fall into a hierarchical pattern. This fact alone should immediately evoke thoughts of design.

Furthermore, any time we observe a biological pattern that fits the Linnaean classification system, we would be justified in wondering what function-based explanation accounts for this. For example, where the order of the fossils in the fossil record matches the Linnaean hierarchy, we should immediately begin asking what functional explanation accounts for the order of the fossils.

In light of these parallels, what can we conclude about the size of the zebra family tree — or the family tree of any other species alive today?

The inductive method of reasoning shows the way. As we observed in the previous chapter, the most important step in this method is the elimination of competing explanations. To conclude that the parallel between the Linnaean classification system and family trees implies common ancestry, we have to first eliminate any competing explanations. As we've just discussed, one competing explanation for the fact that species can be organized into a Linnaean-style classification system is the explanation of design.

But design implies a *designer*. Obviously, no one believes that *humans* designed species. So who or what might this designer be?

Since 1859, the scientific models espoused by creationists have significantly matured. In particular, the design element in the creationist framework has been derived more rigorously. This development has arisen from a careful interpretation and application of the text of Genesis.

Figure 5.8. Humans design transitional forms. Military engineers have designed diverse products for moving firepower through diverse environments. They have created boats for aquatic environments, tanks for land, and amphibious assault vehicles for the transitional environment between water and land.

For example, in Genesis 1, humans were created as separate and distinct from the rest of life. They were the only creature specifically designated as created in the image of God:

So God created man in His own image; in the image of God He created him; male and female He created them (Genesis 1:27).

Theologians still debate what the image of God entails. However, at a minimum, they understand the phrase to denote the fact that humans, in some way, reflect something about God.

This fact has massive consequences for the origins debate. For example, Genesis 1 does not list the design principles that God used when He created the world and the life forms in it. However, since Genesis 1 describes humans as made in the image of God, examining human activity might reveal something about the divine. To understand how God might have created, creationists are increasingly turning to the products of human invention. If humans design things in a certain way, then perhaps God might have as well. This logic leads to testable scientific predictions about the pattern and relative hierarchy among extinct and extant species.

We've just examined the predictions that follow from a careful consideration of the design principles derived from the realm of transportation. For example, diverse vehicles share a basic structure. You could say that they have a similar anatomy. Consequently, in light of Genesis 1, we should not be surprised to find that God reused similar structures and designs when He created the various creatures — such as humans, frogs, bats, porpoises, and horses. In other words, the creation model predicts the shared anatomical pattern in the forelimb of these creatures.

We also examined the design analogy for the developmental process. Basic rules of efficiency and function result in similar stages during the manufacturing process. Furthermore, since the end products have similarities, the blueprints will, of necessity, have them as well. Thus, shared developmental stages among turtles, chicken, and fish are no surprise to the careful student of design and engineering.

Finally, we observed that the means of transportation naturally fall into a nested hierarchical system of classification. We also observed that humans design "transitional forms." The existence of a nested hierarchy of classification and of "transitional forms" in biology fits exactly what the design model expects.

In short, the creation and design model predicts with equal force each of the biological evidences we examined. Thus, by the standards of the scientific method, the evidences that Darwin used to argue for universal common ancestry fail — because Darwin's evidences fail to eliminate competing explanations.

To be fair, both Darwin and his scientific descendants have attempted to eliminate the design explanation. In theory, the way in which to do this would be to look for things in nature that contradict the expectations of the hypothesis that life is designed — and that fit the expectations

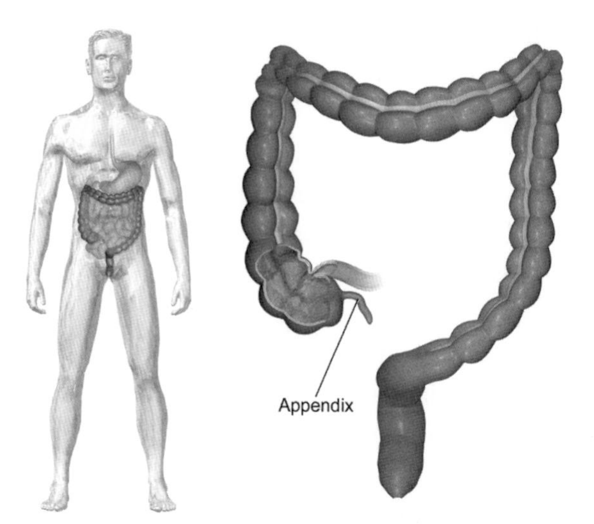

Figure 5.9. Anatomical location of the human appendix.

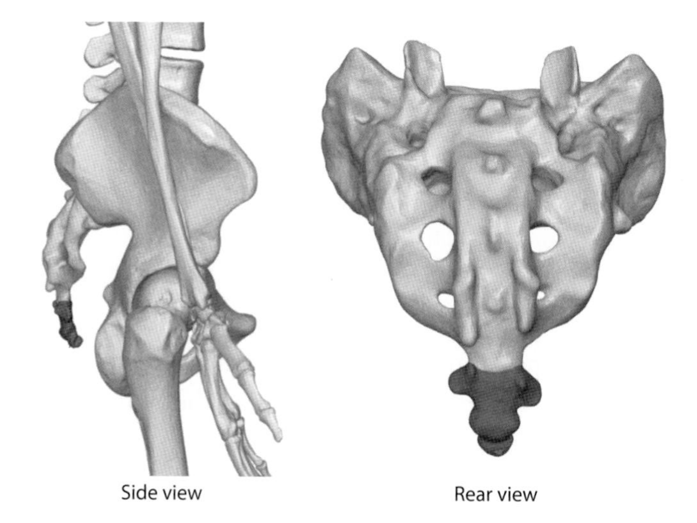

Figure 5.10. Anatomical location of the coccyx. The location of the coccyx with respect to the pelvic bones. Coccyx shaded in black.

of the hypothesis that life arose from the purposeless, undirected process of evolution. In practice, evolutionists have pursued this goal by pointing to biological structures or systems that seem poorly designed — that seem to lack the signature of intelligence. Evolutionists reason that, if a poor biological design exists, then surely an all-powerful and all-knowing Designer cannot be responsible for the origin of this structure or system.

For example, evolutionists have frequently cited organs in the human body as "vestigial" — purposeless leftovers of evolution. For many years, the existence of the human appendix (Figure 5.9) has been mocked as serving only a single purpose — lining the pockets of the surgeons whose services are required to remove it. The coccyx (Figure 5.10) has been the object of similar derision. Our "tailbone" has been viewed as a leftover "stump," a testimony to our supposedly primate, tailed past.

As another example, evolutionists frequently cite the existence of vestigial whale legs (Figure 5.11) as evidence of poor design. Whales are obviously well suited to movement through aquatic environments. In this ecological context, a

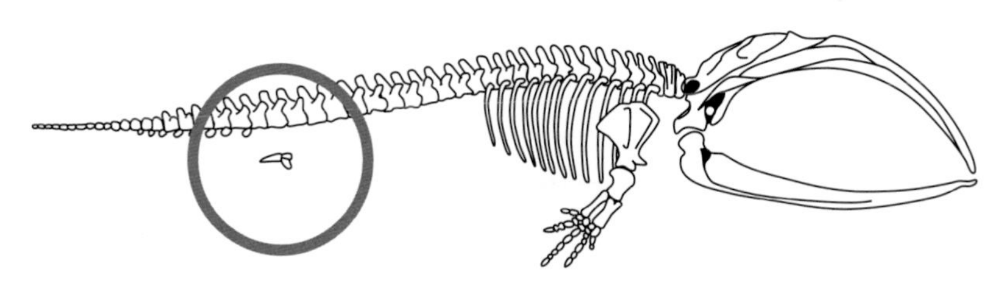

Figure 5.11. Location of purported vestigial whale "legs." Pelvic bones commonly cited as vestigial legs highlighted with circle.

rear appendage would seem to serve no purpose. Instead, if whales evolved from a four-legged terrestrial ancestor, then we might expect them to bear the scar of this process. Evolutionists claim that the leg stumps are this scar.

Many other examples could be cited, such as the supposedly backwards wiring of our retina, the seemingly poor design of the human pharynx, and the circuitous route both of the left recurrent laryngeal nerve and of the vas deferens. All of these examples follow the same logical path — where the design of the structure doesn't make sense, evolutionists see evidence against the design hypothesis.

However, all of the "bad design" arguments are, essentially, arguments from silence. In other words, evolutionists have looked for evidence of good design, and have found none. By their nature, arguments from silence are weak. Absence of evidence doesn't necessarily mean evidence of absence.

Furthermore, by the nature of the silence, these arguments are even weaker. Historically, the evolutionary arguments from silence have not arisen from years of intense investigation, which failed to identify positive evidence for function despite exhaustive searches. Rather, the absence of evidence arose *because of* the absence of investigation.

For example, recent studies have revealed a function for the human appendix in the immune system. Thus, its "uselessness" was a consequence of scientific ignorance, not of poor original design. Similarly, the coccyx is the site of attachment for many pelvic floor muscles. Remove it, and you might wish you hadn't the next time you need to empty your bowels.[11] Our "tailbone" is really a functional participant in our physiology, not a relic of history.

Conversely, with respect to the vestigial whale legs, research has revealed that the "legs" are not scars of once-terrestrial locomotion, but are functional appendages — except not for the purpose of movement. Instead, these pelvic bones perform an important function in copulation. In the buoyant environment of the water, mating is more challenging than on the land. Hence, some extra help near the posterior end of the whale make this process easier.[12]

A common thread in each of these examples is a shallow understanding of what design entails. For example, to rigorously assess the design (or lack thereof) of a structure or system, all possible purposes need to be investigated. In the realm of engineering, engineers know that good designs often balance multiple purposes. Even non-engineers intuitively know this to be true. Women wear high-heeled shoes, not because they're incredibly comfortable, but because they fulfill the aesthetic purpose for which they were made. Comfort is often sacrificed for aesthetics. But no one would claim that high-heeled shoes evolved via undirected processes from a common ancestor with a baseball cleat.

Furthermore, for a designed structure to be the product of intelligence, it need not be essential for life. In other words, though essential structures are the product of design, not all designed structures are essential. The human appendix and coccyx can be removed, and you'll survive. They are not essential to life. Nevertheless, they both still perform important functions in the body.

To make the point even more obvious, the human hand is entirely non-essential. Cut it off, sew up the wound with stitches, and you can move on with your life. However, no one could possibly deny that it is one of the most well-designed machines on the planet. From the tender affection expressed when a couple holds hands, to the violent competition of a boxing match, to the precision and skill of a 90 mile-an-hour strike thrown by a Major League Baseball pitcher, the abilities of the human hand are unparalleled. Just ask the engineers who are trying to create replacements for military amputees.

Not surprisingly, each of the remaining examples of bad design that we've discussed have not withstood rigorous scrutiny. Upon careful examination, and with a more full-orbed appreciation of the principles of engineering and design, the supposedly backward wiring of our retina, the seemingly poor design of the human pharynx, and the circuitous route both of the left recurrent laryngeal nerve and of the vas deferens all actually turn out to balance several competing design purposes.[13]

In some cases, when the argument for non-function can no longer be sustained in the face of new research, evolutionists have emphasized a different element of the anti-design argument. In other words, rather than point to non-function as evidence of bad design, they have emphasized certain elements of the biology that seem to harken more to evolution than to any other explanation. For example, evolutionist Jerry Coyne concedes that the human appendix is functional. But he claims that the size of the human appendix matches the expectations of evolution.[14] As our evolutionary ancestors evolved from an herbivorous diet in the trees to a more carnivorous diet on land, Coyne claims that our appendix size would have changed consistent with this dietary progress.

Recent studies have shown that there is little correlation among mammals between diet and appendix size.[15] Coyne's counter-explanation has been effectively rendered invalid.[16]

Given the trajectory of the evolutionary argument against design, we can predict the types of examples that evolutionists will likely use in the future to try to undermine the creationist model. Specifically, since the anti-design argument draws on ignorance rather than evidence, we can predict that future examples of "poor design" will arise from the least-studied areas of biology. For example, embryological development is still, to this day, a large black box. How a single cell transforms itself into a breathing, moving, crying newborn continues to be the subject of intense investigation (see chapter 3). In fact, this biological building project is unlike any building project with which we're familiar. Unlike skyscrapers, embryos must respire, take in energy, export waste, and be living at every moment of development. In contrast, skyscrapers don't need to be functional until the final structure is complete. Construction workers can put in an 8-hour day, and then turn off the lights and go home. They don't need to be present 24 hours per day to keep the skyscraper from collapsing.

Because of the uniqueness of the biological construction project that we call the process of development, some structures in the embryo are functionally necessary in the embryo but no longer in the adult. Many evolutionists are apparently unaware of this. Thus, it wouldn't be a shock if future examples of "vestiges" are drawn from ignorance of the developmental process.

In summary, despite the evolutionists' attempts to practice inductive reasoning, they still have not been able to eliminate creationist explanations. At least, they haven't been able to do so in the realm of the popular evidences for universal common ancestry.

Based on the evidence we've discussed, we cannot scientifically conclude anything about the full size of any species' family tree. So, which species share a genealogy?

Lost amidst the popular arguments for universal common ancestry are more subtle hints on the scope of species' family trees. The popular arguments can be found in the later chapters of *On the Origin of Species*. The subtle hints were suggested in Darwin's opening arguments. The very first sentence of chapter 1 reads:

> When we look to the individuals of the same variety or sub-variety of our older cultivated plants and animals, one of the first points which strikes us, is, that they generally differ much more from each other, than do the individuals of any one species or variety in a state of nature.

Applying Darwin's observation today, we can quickly appreciate Darwin's point. For example, among horses, the "cultivated animals" — the different *breeds* (i.e., domesticated varieties within a single species) — possess tremendous variation in size, coat pattern, coat color, hair length, and body proportions (Color Plates 42–57). Donkey breeds are also strikingly diverse (Color Plates 58–62).

When you compare this diversity to the amount of variety among the one horse and three ass species in the wild, the diversity among breeds far outstrips the variety in the wild (compare Color Plates 29–32 to Color Plates 42–62). In more rigorous quantitative terms, we can easily reach the same conclusion. If we use the existence of a breed or species as a marker of diversity, far more diversity exists in breeds than in species. Over 850 breeds of horses and donkeys exist today,[17] yet only four species of these creatures exist in the wild.

In Darwin's day, the origin of these breeds was unknown. Yet Darwin saw the potential significance of discovering their origin:

> When we attempt to estimate the amount of structural difference between the domestic races [i.e., breeds] of the same species, we are soon involved in doubt, from not knowing whether they have descended from one or several parent-species. This point, if it could be cleared up, would be interesting; if, for instance, it could be shown that the grey-hound, bloodhound, terrier, spaniel, and bull-dog, which we all know propagate their kind so truly, were the offspring of any single species, then such facts would have great weight

> in making us doubt about the immutability of the many very closely allied and natural species — for instance, of the many foxes — inhabiting different quarters of the world.[18]

Today, the origin of these breeds is uncontroversial. Both evolutionists and creationists accept that horse breeds trace to a common ancestor. Both positions accept that donkey breeds trace to a common ancestor.

If we accept the common ancestry of horse breeds (or donkey breeds), can we deny the common ancestry of wild horse and ass species? If breeds could vary so widely yet still have a common ancestor, why couldn't species — which have less variety than breeds — also have a common ancestor?

What we've just observed is analogous to a human family tree. Let's say that one branch has 100 members. Let's also say that, together, these members have diverse features — curly hair, straight hair, blond hair, brown hair, black hair, various eye shapes, a spectrum of skin colors, short and tall bodies, and numerous sizes and shapes in noses and lips are all present. If you ran into someone who had a subset of these features, you would be hard pressed to exclude them from the family tree. The diversity of traits in the existing members captures the features in the candidate member.

The same scenario holds true in the equid family. Zebras share many traits with horses and donkeys. In fact, far more visible diversity exists among donkey and horse breeds than among ass, horse, and zebra species combined. Could all of these species share a common ancestor?

The stripes on zebras present a slight barrier to reaching this conclusion. Among horses and donkeys, no modern breed exists that has stripes like a zebra. At first pass, it might be difficult to imagine how a striped zebra could share a common ancestor with unstriped horses and asses.

If we step back into Darwin's day, these concerns disappear. Until 1883,[19] a partially striped creature termed the *quagga* existed (Figure 5.12). Today, researchers are attempting to recreate the quagga, and their efforts have generated several partially striped individuals (Color Plates 63–64).

Furthermore, the African wild ass possesses partial striping. In fact, when donkeys and zebras hybridize, some of their offspring (Color Plate 8) have a strong resemblance to the striping pattern on the African wild ass (Color Plate 30). If going from a striped to an un-striped or partially striped creature is as easy as these breeding successes imply, then our analogy between breeds and species is all the more persuasive.

Breeds exist for more species than just equids.[20] In other living families, the same pattern holds true — a vast excess of breeds over species. For example, the family Bovidae possesses 138 living species.[21] From cattle-like species to goat-like species to antelopes and duikers (Figure 5.13), Bovidae is a very diverse family. Yet over 3,000 breeds of cattle, sheep, and goats exist in this family.

Similar patterns hold true in the camel and llama family (Camelidae), pig family (Suidae), dog and wolf family (Canidae), cat family (Felidae), and rabbit family (Leporidae) — far more breeds than living species[22] (Table 5.1).

Order	Family	Common names	Breed #	Wild species #	Breed excess
Artiodactyla	Bovidae	cattle, sheep, antelopes	3,218	138	3,080
Artiodactyla	Camelidae	camel, llama	106	3	103
Artiodactyla	Suidae	pigs, swine, boars	669	17	652
Carnivora	Canidae	dogs, wolves, foxes, coyotes	335	35	300
Carnivora	Felidae	big cats, small cats	71	38	33
Lagomorpha	Leporidae	rabbits, hares	231	62	169
Perissodactyla	Equidae	horses, asses, zebras	873	7	866

Table 5.1. Mammal families in which the number of breeds exceeds the number of living species. Modified with permission from Table 4 of N.T. Jeanson and J. Lisle, "On the Origin of Eukaryotic Species' Genotypic and Phenotypic Diversity: Genetic Clocks, Population Growth Curves, and Comparative Nuclear Genome Analyses Suggest Created Heterozygosity in Combination with Natural Processes as a Major Mechanism," *Answers Research Journal,* 2016, 9:81–122.

Order	Family	Common names	Breed #	Wild species #	Breed excess
Galliformes	Phasianidae	chickens, pheasants	1,555	188	1,367
Anseriformes	Anatidae	ducks, geese	452	172	280
Struthioniformes	Struthionidae	ostriches	16	2	14

Table 5.2. Bird families in which the number of breeds exceeds the number of living species. Modified with permission from Table 5 of N.T. Jeanson and J. Lisle, "On the Origin of Eukaryotic Species' Genotypic and Phenotypic Diversity: Genetic Clocks, Population Growth Curves, and Comparative Nuclear Genome Analyses Suggest Created Heterozygosity in Combination with Natural Processes as a Major Mechanism," *Answers Research Journal,* 2016, 9:81–122.

Order	Family	Common names	Breed #	Wild species #	Breed deficiency
Artiodactyla	Cervidae	deer	12	54	–42
Rodentia	Caviidae	guinea pigs	17	20	–3
Columbiformes	Columbidae	doves, pigeons	68	367	–299
Struthioniformes	Casuariidae	emus, cassowaries	2	6	–4

Table 5.3. Mammal and bird families with breeds, in which the number of breeds is less than the number of species. Modified with permission from Tables 4–5 of N.T. Jeanson and J. Lisle, "On the Origin of Eukaryotic Species' Genotypic and Phenotypic Diversity: Genetic Clocks, Population Growth Curves, and Comparative Nuclear Genome Analyses Suggest Created Heterozygosity in Combination with Natural Processes as a Major Mechanism," *Answers Research Journal,* 2016, 9:81–122.

Birds also follow suit.[23] The chicken family (Phasianidae) and the duck family (Anatidae) both possess hundreds (in one case, over a thousand) more breeds than species (Table 5.2). Breeds in the ostrich family exceed the number of species (Table 5.2).

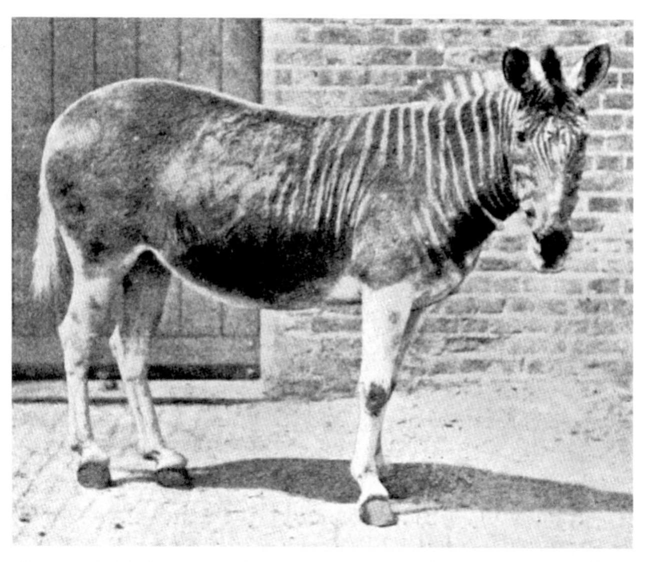

Figure 5.12. Image of a quagga, a partially striped zebra relative. Now extinct.

Could this eventually hold true in general across vertebrate classes, as we continue to domesticate and breed species? In both the mammalian and bird datasets, three to four *orders* were represented. In other words, these patterns were true across broad classification ranks.

Figure 5.13. Example of a duiker species.

On the other hand, some mammal and bird families have fewer breeds than species (Table 5.3). Some have no breeds at all. But some species are also notoriously difficult to tame — so the potential for breed formation might still exist. If mankind could tame species in every mammal, reptile, bird, amphibian, and fish family on the planet, would we always end up with more breeds than species? If we did, then the family trees of vertebrate species would extend to the *family* level of classification (i.e., in the formal Linnaean sense of the term).

The structure of this breed-species comparison that we've just considered conforms perfectly to the rules of inductive reasoning. It doesn't make sense to claim that hundreds of horse and donkey breeds came from a common ancestor, but that seven wild species of equids did not. The latter must have originated via descent with modification. Effectively, this argument eliminates the hypothesis of the fixity of species.

However, these same rules of inductive reasoning prevent over-application of this logic. In other words, the elimination of the hypothesis of species fixity does not automatically require the acceptance of the hypothesis of universal common ancestry. Since 1859, creationists have approached the question of the origin of species with a keener eye on the biblical text. Though the Scriptures leave many questions unanswered, modern creationists take the explicit statements from Genesis as a starting point from which to ask research questions. These practices have produced a creationist view that rejects species fixity while not embracing universal common ancestry. In other words, another competing explanation exists for Darwin's breed-species argument.

The modern creationist view derives primarily from several key phrases in Genesis 1–11. For example, in Genesis 1, the phrase "after their kind" or "after its kind" is used repeatedly to describe the units of God's creative work in biology. The Hebrew word translated as "kind" is *min* (transliterated). Though the exact definition of *min* is debated among Hebrew scholars, its use in other biblical passages intimates an answer. In Genesis 6–7, Noah is commanded to build an ark and to bring two of every land-dependent, air-breathing *min* on board. Specifically, he is commanded to bring male and female of each *min*. The purpose is the propagation of these *min* after the Flood. Together, these details suggest that *min* can be identified by testing which individuals can hybridize. In other words, if two individuals can successfully produce live offspring, this would be good evidence that they're probably part of the same *min*.[24]

Over time, hybridization studies have revealed a general pattern for *min*. Surprisingly, some species can still produce live adult offspring. We already observed an example of this — the products of horse-zebra and donkey-zebra crosses (Color Plate 8–9). Consequently, modern creationists do not equate *min* with *species*. Instead, whether mammals, reptiles, or birds, *min* appear to be best approximated by the classification level of family or order.[25] Since this rule of thumb seems to apply across vertebrate classes, the fish and amphibian *min* would also appear to be best approximated by the classification level of family or order. Thus, applying this principle back to the text of Genesis, modern creationists conclude that Noah brought on the Ark representatives of each family or order, not of each species.[26]

This inference has profound ramifications for the modern creationist understanding of the origin of species.* Since vertebrate families and orders today are typically composed of more than one species, modern creationists endorse the formation of new species within vertebrate families and orders (at least within those families and orders where hybridization tests have tied species together).[27] In other words, they have no problem with the breed-species argument that Darwin articulated — recognizing that it extends only up to the level of family or order.[28]

In addition, modern creationist views differ from the creationist views of 1859 on the question of geography. Again, modern creationists have approached science with a keener eye on the biblical text. In Genesis 8, the narrative describes the fate of Noah's Ark — it lands on the mountains of Ararat. From this location, the *min* disembarked. Scholars still debate the exact location of the Ararat of Genesis 8, but they agree that it is likely somewhere in the Middle East.

* It also makes testable predictions in genetics, a subject explored in later chapters.

From the Middle East, the *min* spread out over the earth. In other words, they migrated — just as Darwin argued. Modern creationists would have no qualms with the conclusions of chapter 4.

In summary, Darwin's breed-species comparison and his inferences from geography together argue for common ancestry among species within vertebrate families, whose current locations were determined by migration. But evolution isn't the only explanation that predicts this result. Modern creationist views do as well.

With respect to Darwin's evidences for universal common ancestry, all fail to eliminate modern creationist views. His evidences do not distinguish between the hypothesis of common ancestry and the hypothesis of common design. Furthermore, though his breed-species comparison successfully eliminated the hypothesis that each and every species has been independently designed, his comparisons failed to eliminate the hypothesis that original ancestors of each family were designed. In other words, all of Darwin's evidences fail to eliminate modern creationist views.

This fact finds a parallel in the assembly of a massive jigsaw puzzle. Consider one which consists of a million pieces, but which lacks a box cover. Furthermore, consider one in which the pieces are standard jigsaw puzzle shapes. With a puzzle this large, it's almost inevitable that you will find a seemingly plausible connection between two pieces — but then, in the end, discern it to be incorrect. As a puzzle grows even larger, some pieces will almost certainly seem to naturally connect together — despite their proper (eventual) placement elsewhere in the puzzle. Similarly, Darwin's clues initially seemed to plausibly connect all species together. However, these same pieces worked equally well in positions that connected species within a family, while simultaneously rejecting connections between different families. Thus, species within a family were connected on the same puzzle. But beyond the level of family, uncertainty reigned.

In light of the history of genetics, this uncertainty was natural. The only direct scientific record of a species' ancestry is found in DNA. Since the DNA sequences of various species (i.e., the edge pieces to the puzzle) weren't available until recently, it's no surprise that the debate over ancestry would persist to this day.

Chapter 6
A Stitch in Time

One of the most important pieces in the puzzle of the origin of species wasn't discovered by Darwin. It was assumed by him.

In the 150 years since Darwin's seminal work, his assumption has been undermined from an unexpected source.

The critical piece was the timescale over which species originated. At first pass, the timescale might not seem to carry much significance. What difference does a timescale make on questions of species' ancestry?

However, on the question of how species arose, the timescale carries tremendous importance. It quickly reveals which clues fit and which ones don't. By extension, the fit — or lack thereof — among species pieces has strong implications for how connected or disconnected they are.

Darwin knew the importance of the timescale and its relationship to the question of *how* species originate. Darwin's chief answer to the question of *how* was natural selection — the survival of the fittest. In the context of evolution and of universal common ancestry, the mechanism of natural selection was inextricably bound to eons of time:

> That natural selection will always act with extreme slowness, I fully admit. . . . I do believe that natural selection will always act very slowly, often only at long intervals of time, and generally on only a very few of the inhabitants of the same region at the same time. I further believe, that this very slow, intermittent action of natural selection accords perfectly well with what geology tells us of the rate and manner at which the inhabitants of this world have changed.[1]

Darwin's reasoning is easy to follow. Just look at your own family tree. Changes are visible each generation — but the differences are usually not drastic. Descent with modification occurs at a nearly imperceptible pace. The same principle applies to species in the wild.

To explain the same point with a negative example, creationists often challenge evolutionists about the pace of biological change. Creationists wonder why, today, we don't see fish-like creatures evolving into mammals. Because, the evolutionists reply, large-scale evolution takes much longer than a human lifespan.

Under the evolutionary model, the *how* of speciation and the *when* are tightly linked.

When Darwin wrote *On the Origin of Species*, his central goal was not the derivation of a timescale. Prior to Darwin, the scientific community had already settled on a span much longer than several thousand years. Charles Lyell's *Principles of Geology* had already laid down arguments for millions of years, and Darwin simply followed suit — as the quote above illustrates.

But is one field — geology — sufficient evidence to conclude an unimaginably long span of time? Or are multiple independent lines of evidence required to verify millions of years of earth — and species' — history? Have Lyell's arguments stood the test of time?

Had Darwin taken a more critical view of his own arguments, he might have found ample reason to question the accepted geologic timescale.

As we discussed in chapter 5, Darwin observed that more visible variety exists in breeds than in species. For example, much more variety is visible among horse and donkey breeds than among living equid species in the wild (compare Color Plates 18–20 and 29–32 to Color Plates 42–62). We also observed that this fact naturally rejects the hypothesis of the fixity of species. How could massive breed variety arise from a common ancestor, while minimal species variety could not? If breeds have a common ancestor, then species must, too.

In 1859, Darwin was not able to make as firm a conclusion about the common ancestry of breeds as I just articulated. The common ancestry of breeds was not as well established a fact as it is now. Nevertheless, we observe that Darwin knew where this sort of breed-species logic might lead, and he endorsed this sort of thinking.

The same sort of thinking and reasoning applies with equal force to the question of the timescale over which species arise. For example, proponents of the millions-of-years (i.e., evolutionary) timescale freely concede that breeds have a recent origin. This concession follows from two lines of evidence. First, since humans are directly responsible for producing breeds, then these breeds must have arisen contemporary with the origin of modern humans. Since evolutionists put the origin of modern humans within the last 200,000 years, then breeds must be no older than 200,000 years. Second, some evolutionists have published specific timescales for the origin of breeds. These timelines put the origin of breeds within the last 12,000 years.[2]

Consider again the visible variety in breeds and species: The former has far more diversity than the latter. According to the evolutionists' own timescale, in just 12,000 years, humans have produced hundreds of horse and donkey breeds. Long hair, short hair, all sorts of coat colors, ponies, Clydesdales — the amount of variety is remarkable. Yet the proponents of this timescale turn around and stretch the origin of just seven equid species over several million years. This position is as logically deficient as the species fixity position of 1859. If the greater variety in breeds took 12,000 years, then surely the lesser variety in species took the same amount of time — or less. By the evolutionists' own logic, species must have arisen in 12,000 years or less.

Because this argument is derived from Darwin's own reasoning, attempts to rebut it strike with equal force at Darwin's original thoughts about ancestry. For example, someone might criticize the fact that I'm using events on a farm as an analogy for events in the wild. They might find fault with equating the two. How could processes used on domestic animals be compared to processes in the wild?

The problem with this objection is that it applies equally well to Darwin's original argument. If events on the farm cannot be compared to the wild, then Darwin's most persuasive argument for common ancestry falls apart.

Furthermore, to this day, evolutionists rely very heavily on the analogy between artificial selection on the farm (breeding) and natural selection in the wild, in order to argue for the latter:

> The main point I want to draw out of domestication is its astonishing power to change the shape and behavior of wild animals, and the speed with which it does so. Breeders are almost like modellers* with endlessly malleable clay, or like sculptors wielding chisels, carving dogs or horses, or cows or cabbages, to their whim. . . . The relevance to natural evolution is that, although the selecting agent is man and not nature, the process is otherwise exactly the same. This is why Darwin gave so much prominence to domestication at the beginning of *On the Origin of Species.* Anybody can understand the principle of evolution by artificial selection. Natural selection is the same, with one minor detail changed.[3]

If artificial selection is not a good analogy for natural selection — if events on the farm are not good analogies for the wild — then *On the Origin of Species* has to be gutted.

As another example, perhaps the fact that humans were involved invalidates the argument for a recent timescale. After all, in the wild, no humans exist to direct the process of speciation. In the wild, intelligent entities do not pick and choose which traits to save; foreknowledge and purpose are not part of the equation. Wouldn't this difference destroy the analogy?

Once again, if we allow this objection to stand, then we simultaneously invalidate Darwin's claims about common ancestry. Darwin tried to argue from the greater to the lesser — if diverse

---

* Dawkins' British spelling.

breeds had a common ancestry, then not-so-diverse species must also be related. But what if the diversity in breeds stems from the fact that intelligent agents are purposefully manipulating domesticated creatures? Absent this intelligent and purposeful process in the wild, perhaps no new species could form. If no new species could form, then species must not be related.

Thus, if we object to the involvement of humans in our analogy, then we quickly lose confidence in the common ancestry of species.

As a last objection, some might claim that humans accelerate the process of selection; therefore, no conclusions can be drawn about the speed of speciation in the wild. This assertion assumes the very point in question. In a debate, arguments cannot be won by simply asserting the truthfulness of one side. Furthermore, the time argument that I just presented has a built-in acknowledgement of the differences between breeds and species. I'm not arguing that hundreds of equid species formed in just 12,000 years or less. I'm arguing that this timescale produced just seven equid species. If over 850 equid breeds can form in a few thousand years, then surely seven species require no longer. If the number of species were equal to or in excess of the number of breeds, then perhaps breeds could not be used as an analogy for species. Then again, if species outnumbered breeds, Darwin's claims about ancestry within families would never have gotten started. Thus, the recent origin of species is a difficult conclusion to escape.

What's remarkable about the argument I just articulated is its scope. Recall from the last chapter that breeds exist for more species than just equids. In the cattle, sheep, goat, and antelope family (Bovidae), the pig family (Suidae), the rabbit family (Leporidae), the camel and llama family (Camelidae), the dog and wolf family (Canidae), the cat family (Felidae), the chicken family (Phasianidae), and the duck family (Anatidae), breed numbers far exceed species numbers. In light of this trend, could this pattern eventually hold true across vertebrate classes, as we continue to domesticate and breed animals? If it does, then the origin of species within vertebrate families must have taken place recently. Why? Evolutionists invoke the 12,000-year timescale for more than just equids. They apply it to virtually every domestic animal.[4]

The implications of this thought experiment are enormous. Consider the properties of the vertebrate classification system. As we discussed in a previous chapter, the narrowest classification category is species. When we speak of a zebra species, we mean a very narrowly defined type of creature. As we ascend upward in this classification system, the category of genus includes more types of creatures (e.g., several species of equids). For other species (i.e., cat species), the level of family includes even more species than the level of genus. In other words, families tend to include multiple genera. The categories of order, class, and phylum are even more inclusive.

Consequently, at each higher level of classification, the sheer number of different representatives at each level decreases. For example, over 5,400 mammal species exist today.[5] They belong to less than 1,300 genera. These genera belong to less than 200 families.[6] As each classification level becomes more expansive in terms of the species it contains (i.e., within the class Mammalia, 5,400 mammal species exist), it also shrinks in terms of the total number of different representatives or instances at that level (i.e., within the class Mammalia, less than 200 families exist).

Thus, far more vertebrate species exist than vertebrate families. If species within vertebrate families have a recent origin, then only the origin of families would be left to explain. Since far fewer families, orders, classes, and phyla exist as compared to species, the recent origin of species within a family has tremendous consequences for the overall timescale. If species within vertebrate families arose within the last 12,000 years, then recent history has seen the origin of over 98% of the living vertebrate diversity[7] on earth (Table 6.1).[8]

Number of vertebrate species:	68,926
Number of vertebrate families:	1,123
Species + families:	70,049
Species, as % of total:	98.4

Table 6.1. The percentage of vertebrate diversity whose origins are explicable over a recent timescale.

What I've just derived might seem to contradict something I stated earlier in this chapter. Near the beginning, I cited a common creationist objection to evolution — the fact that, today, we don't see major transformations in species. I also cited the popular rebuttal — that evolution occurs very slowly, much too slowly for us to observe. In light of what I just claimed about the recent origin of 98% of the living vertebrate diversity on earth, one of these two claims would appear to be wrong. If 98% or more of vertebrate species arose within 12,000 years or less, then massive amounts of new species must have formed contemporary with recorded human history. Yet everyone would seem to agree that little observable change is happening. Which is it?

Upon deeper examination, a different perspective emerges. Since we've been dealing with vertebrates, and since we're dealing with the question of observable speciation, let's focus on the vertebrate classes that are easiest to observe with the naked eye — mammals, birds, reptiles, and amphibians. Today, over 30,000 living species exist in these four classes.[9] If all of them arose within 12,000 years, then at least two and half new species have formed, on average, every year (30,000 species / 12,000 years = 2.5 new species per year).* In other words, about every five months, one new species would form.

Could such a rapid pace of speciation have gone undetected? A new species every five months sounds far-fetched. How could the ancients not have noticed the diversity of life unfold before their eyes? In our own era, why don't we hear reports of new species forming every few months? Where is the evidence for this modern speciation sprint?

Historically, this evidence would have been difficult to obtain. The formation of 30,000 species wouldn't have happened only in, say, Rome. Rather, it would have been distributed around the globe. Conversely, the geographic limits and isolation of the ancients would have made detection of speciation prohibitive. For example, until Columbus' voyage in 1492, the residents of Europe would have had no idea that the Americas existed. Until the 1600s, Australia was a

* I used round numbers to keep the math simple.

non-entity for European scientists.[10] Before the Renaissance, detecting speciation on these continents would have been impossible for European scientists.[11]

The origin of species on other continents would have been equally mysterious. Though Europeans might have enjoyed the exports from the Orient, few would have personally traveled the Great Silk Road.[12] Speciation in East and Central Asia would not have been accessible to Europe. Conversely, before the 1800s, virtually no Europeans ventured into the jungles of central Africa.[13] Formation of new species in Africa would have gone unnoticed by European naturalists.

In other words, if pre-Renaissance Europeans rarely ventured to any of the five other habitable continents, the opportunities to observe the formation of new species would have dropped dramatically. For sake of simplicity, let's assume that all species are evenly distributed among the six major habitable continents. If the species on all continents formed at the same rate, then the rate of 2.5 new species per year would have dropped by 80% to an observable rate of 0.4 new species per year *per continent* (2.5 / 6 = 0.4). In other words, one new species per continent every two and a half years. Even this number is generous since Europe is rather small compared to the other continents.

Actually, the potential for Europeans to observe speciation would have been even less. Travel between different parts of Europe would have been much slower in pre-Renaissance times as compared to the present. We take for granted the ease with which a person can move from Portugal to Poland (Figure 6.1). In pre-Renaissance times, transportation was far less efficient. Effectively, this would have limited the observations of any individual to an even narrower geographic area. The same factors would have limited communication of new discoveries between far-flung sites. Scientific consortiums dedicated to monitoring species' changes would have been rare — if they occurred at all.

Using 30 countries as a rough approximation of geographic subdivisions in pre-Renaissance Europe, we can recalculate the observable speciation rate. For an average European, the rate of local speciation would have been 0.014 species per year per European country (2.5 / 6 continents / 30 countries = 0.014 new species per year) — if not lower. In other words, one new species every 72 years. Given the life expectancy of pre-Renaissance Europeans, effectively this rate of speciation would have been invisible. Even by today's standards, scientific careers are shorter than whole lifespans. Historically, detecting the formation of new species would have been hard.

Figure 6.1. Geography of Europe. West-to-East distances in Europe illustrated by Portugal and Poland (both highlighted with dark shading).

But what about the very recent past? And the present? If we took the numbers from the preceding section at face value, these questions would become fairly pressing. Surely the collective scientific community of the past few centuries would have discovered new species forming. If the recent speciation timescale were true, then speciation events must have registered with someone.

When we dig deeper into the numbers, these conclusions are harder to support. For speciation to be observed, "species" must exist. Of course, as biological entities, species have existed for as long as life has been on this planet. But as recognized units of classification that can be systematically studied and observed by scientists, "species" did not enter the vocabulary of the scientific community until a very recent date.

As we discussed in chapter 1, Linnaeus formalized the term with his publication of *Systema Naturae* in the 1700s. In other words, when we speak of "species" today, we're basically speaking in a Linnaean term. Our conversation about the origin of species implicitly expresses a very specific definition that pre-Renaissance people would not have used — at least not as globally consistent as we do today. Before the 1700s, the specific discussion that we're having would have been nonexistent. Until Linnaeus' concepts were published and accepted, formal investigations into the origin of species were not possible.

This fact has immediate ramifications for the 12,000-year timescale of speciation that we've been considering. Let's consider the known species in Linnaeus' day, and let's take mammal species as a representative illustration. In the 1758 version of *Systema Naturae*, only 154 of the roughly 5,400 currently recognized mammal species were listed.[14] Effectively, over 97% of modern mammal species were unknown in the 1700s. Since existing species are the ancestors of new species, how could speciation have been detected in Linnaeus' day? If 97% of the ancestors were invisible to science, how could the formation of new species from these ancestors have been observed?

The details of the actual mammals on this 1758 list reveal how simultaneously broad and narrow Linnaeus' observations were. Today, mammal species are classified into genera, genera into families, and families into orders. The roughly 5,400 living mammal species are grouped into 29 orders.[15] Surprisingly, the 154 species in Linnaeus' list captured 62% of these orders (18 of the 29).[16] Despite the severe numerical limitations of Linnaeus' species dataset, his list sampled a broad swath of mammal life.

Yet the mammal orders *not* represented in his list reveal the practical limits of observation in Linnaeus' era. The orders from which Linnaeus failed to include representative species are all at a significant geographic distance from Europe. Linnaeus missed entire orders of marsupials — primarily the ones containing species exclusive to Australia and Southeast Asia. He also omitted mammalian orders from Africa and South America. Why? Captain James Cook did not reach Australia until 1770.[17] Alexander von Humboldt did not explore the South American interior until the late 1700s and early 1800s. David Livingstone, the famous voyager through the African interior, wasn't born until the 1800s.[18] Linnaeus' list was short because his contemporaries' globe was small.

By the time that Darwin wrote *On the Origin of Species*, the world known to Europe had grown. The list of modern mammal species had as well. Around 90% of mammal orders and families were recognized with at least one species representative in the accepted list.[19] The only orders without species representatives were from three obscure marsupial orders. These three orders contain a total of only 9 species. Since only 1,780 of the more than 5,400 modern mammal species were recognized in 1859, the absence of 9 species from the list is easy to explain via statistics. In other words, with over 3,500 mammal species left to discover in 1859, it's no surprise that these particular 9 species were absent from the list. Thus, nearly all mammal orders and families were represented in the 19th-century datasets.

Despite this broad taxonomic representation, specific tallies were still deficient. If only one-third of all mammal species (1,780 of more than 5,400) were recognized in Darwin's day, what was missing? Over the next 150 years, species tallies in only three mammalian orders would contribute roughly 80% of the new species. Rodents (order Rodentia), bats (order Chiroptera), and moles and shrews (order Soricomorpha) alone would add nearly 3,000 mammal species to the global total.[20] Since these species tend to be small and harder to detect, it's no surprise that these categories would increase the most by our day.

Despite the matter-of-fact way in which I've communicated these mathematical realities, the numbers don't tell the whole story. I've told you a narrative based on a strict adherence to a specific timescale — the dates on which scientists first formally published the fact of a particular species' existence. But just because a scientist publishes doesn't mean that the rest of the scientific community is aware of it. Especially in the Internet era, ease of communication, content searching, and literature review are far easier than they were in previous decades. As an example, consider a standard mammalian reference text. In 1993, the second edition listed 4,629 total mammal species.[21] In fact, according to the strict timescale I've been using, a total of 5,186 mammal species had been published by this date.[22] Why the difference? Without the Internet, collaborative efforts and communication were slow. Only today are these processes separated by a very short time delay.

In summary, from 1758 to 2005, over 5,000 mammal species were formally recognized for the first time by the scientific community. In other words, on average, over 20 new species per year were added to the scientific consciousness. In contrast, if the origin of all mammal species occurred within the last 12,000 years, an average of 0.45 new species formed per year — or 1 new species every 2.2 years. The rate of species recognition far exceeds the theoretical rate of species formation.[23]

How can species recognition be distinguished from species formation? Were any of the recognition events of the last two and a half centuries actual formation events? How would we know?

Part of the challenge in distinguishing these processes is an implicit assumption I made about species. One textbook gives at least six different definitions for "species."[24] I ignored this ambiguity and acted as if *species* were a uniform term.

The challenge that this ambiguity presents is most abundantly obvious at the level of visible traits. For example, compare the African Cape buffalo (Color Plate 65) to the African forest buffalo (Color Plate 66). Now compare the two Asian wild asses — the Onager and the Kiang (Color Plates 31–32). Only one of these pairs contains two different species. The other pair contains two subspecies (a level of classification below species) within a single species.

One of the key factors in defining species is the concept of isolation. If two groups of creatures are isolated from one another, they might be considered separate species. However, isolation can be assessed various ways. Classically, reproductive isolation was the determinant of species boundaries. With new genetic tools available to the scientific community, isolation can now be defined without a physical test for hybridization between two populations.

In our example, the African Cape buffalo and the African forest buffalo have not yet demonstrated sufficient levels of isolation to be considered separate species. By contrast, the Onager and the Kiang have starkly different genetic content — they have different chromosome numbers. Obviously, this genetic separation seems to fly in the face of what their visible traits depict. Yet this is the current state of affairs in the modern scientific era.

Conversely, the current classification state of these four groups of creatures is not likely to be permanent. New, more comprehensive genetic data from the two buffalo subspecies might reveal genetic isolation, which could, in turn, lead to the reclassification of these subspecies as separate species. At present, of the two buffalo subspecies, only the African Cape buffalo possesses a published genome sequence.[25]

A similar phenomenon recently occurred with giraffes. Before 2016, the scientific community recognized a single species. Nevertheless, obvious visible differences separated different subpopulations of giraffes (see Color Plates 67–70 for representatives of some of the many giraffe subpopulations), and these distinct groups also happened to be fairly well-separated geographically (Color Plate 71).[26] The scientific community recognized these facts with a designation of separate subspecies — not of separate species.

Then, in 2016, a group of investigators published new genetic data on these various subspecies. At the DNA level, isolation among four of these groups was obvious. In fact, the genetic isolation among four of the giraffe subpopulations was so strong that it suggested reproductive isolation. Hence, these investigators reclassified the single giraffe species into four.[27]

The term "species" is a moving target.

Even in zebras, the tidy categories that I've described don't represent the whole story. To be sure, standard mammalian taxonomy sources depict three zebra species. Yet, even within these specific categories, obvious visible variety exists. The mountain zebra species contains two subspecies;[28] the plains zebra, six subspecies.[29] Some of the latter bear strikingly distinct features. The Burchell's zebra subspecies possesses brown "shadow" stripes (Color Plate 72). Another subspecies is maneless (Color Plate 73). The Crawshay's zebra subspecies has fairly thick black striping on its neck (Color Plate 74). All three of these subspecies are distinct from the Grant's zebra subspecies (Color Plate 18).

These nuances of classification make the distinction between species detection and species formation very difficult to discern. In theory, to detect species formation, all we need is the full accounting of the morphological, genetic, and reproductive separation in two candidate species populations at two different time points. A speciation event would manifest itself as an overlap (i.e., non-isolation) between the populations at one time point followed by isolation at another.

For example, once the genome sequence of the African forest buffalo is determined, and once the reproductive isolation (or lack thereof) is quantified between these two groups, we can begin to track the change in this population over time. Currently, it appears that these groups overlap. If genetics underscored this conclusion, we could start tracking changes with time. After a decade or two of observation, we could reassess the genetic and reproductive isolation in these two groups. If separation and isolation have occurred, then we can confidently say that a new species has formed. Instead, if we detect persistent overlap and not isolation, we would have good evidence for the continuing subspecies status between these two groups.

However, if we failed to detect a speciation event in these two groups after 20 years, what would this imply about the timing of speciation? Would this result disprove the rapid speciation hypothesis?

To be clear, the speciation rates that I've predicted are broad estimates. With a theoretical timescale of 12,000 years and a modern mammalian dataset of roughly 5,400 species, the average rate of mammalian speciation is still 0.45 new species per year — or 1 new species every 2.2 years. However, as averages, these numbers represent very general, mammal-wide patterns, not rates specific to particular groups.

Specifically, recall that, from Darwin's own logic, we derived an argument for the recent origin of species *within a family*. When multiple families are forming species simultaneously, we can calculate an overall rate for the entire group. Together, all 153 mammal families combined would form an average of 1 new species every 2.2 years.

But each family is not necessarily speciating at this rate. Instead, each family will likely have its own individual rate. We can estimate the rate from the total number of living species within each family. For example, the equid family has only 7 living species.[30] Over 12,000 years, this amounts to a rate of 1 new species every 1,714 years. In contrast, the bovid family (e.g., cattle, sheep, goats, antelopes, etc.) has 138 living species.[31] In 12,000 years' time, a new species will have formed about once every 87 years — much faster than the equids. The most species-rich mammal family, Muridae (i.e., one of the mouse families) has 712 total species.[32] In 12,000 years, one new species will have formed, on average, every 17 years. All of these individual rates acting simultaneously and in concert are what give rise to the average mammal-wide speciation rate estimate.

Consequently, the rapid speciation hypothesis must be tested and evaluated for each family individually. If we observed African buffaloes for two decades and discovered no new species, this would say nothing about the rapid speciation hypothesis in equids. To test the equid speciation hypothesis, they would have to be observed for nearly 2,000 years before the rapid speciation hypothesis could be scientifically rejected. Conversely, equid tracking would say nothing about speciation in mice. Mice would also have to be observed directly for this conclusion to be reached.

Even if we limit our focus to a single family, our observations would have to be very comprehensive within the family. Careful tracking of just one species in a family is insufficient to test the rapid speciation hypothesis. In particular, nothing in our discussion precludes the formation of modern species in a stepwise manner (Figure 6.2). For example, consider a hypothetical series of speciation events in mice. What would the original mouse ancestor spawn initially? Would it give rise to more ancestors? Or to one of the many modern mouse species? Is it possible that the ancestor of mice existed and spawned a modern species — but without the disappearance of the ancestral population? If so, the ancestor could continue producing more species. The repetition of this process might continue until 712 species were formed.

If modern species could form stepwise over 12,000 years, this would make detection of speciation very difficult in any one species. Based on the numbers above, a modern mouse species

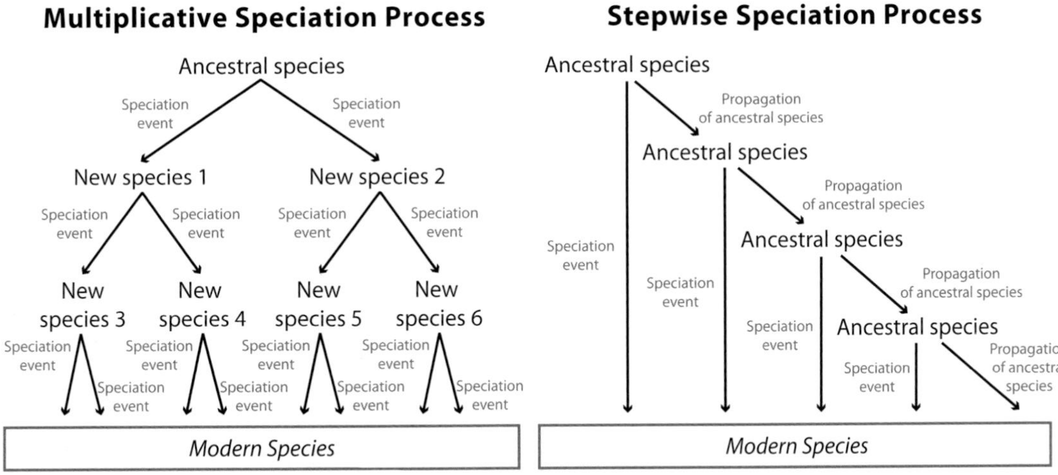

Figure 6.2. Comparison of multiplicative versus stepwise speciation processes. Under the multiplicative model, each new species becomes the ancestor of another set of new species. In contrast, under the stepwise model, most new species are dead ends. They do not give rise to additional species. Instead, the ancestor keeps spawning new species, without losing its identity.

might form every 17 years. But the new species that formed might not *itself* give rise to any more species (Figure 6.2). Instead, the still-existing ancestor might be the one giving rise to the next species. In other words, the ancestor might be the active participant in the speciation process, and the descendants might participate very little, if at all. If so, then our hypothetical experiment with African buffaloes is almost meaningless. Perhaps the African buffalo species formed 1,000 years ago. It might not form any new species again. In 87 years, a new bovid species might form from one of the other 137 bovid representatives alive today. In other words, the only way to test the rapid speciation hypothesis is to *simultaneously watch every species within a family.*

For families with a single species, this task is harder than it first appears. With only one species, global tracking is easy. But families with single species have the slowest predicted rates of speciation. These families would have to be tracked for millennia before even one new species would form.

For species-rich families, the task doesn't get easier. On one hand, their predicted rates of speciation are very fast. In one human lifetime, the most species-rich families would be predicted to form new species. For example, if a mammal family possesses 700 species or more, the rapid speciation model predicts a speciation rate of 1 new species every 17 years, if not faster (12,000 years / 700 species [or more] = 17.1 years per species [or less]). Not until the number of species within a family drops to less than 171 does the predicted rate of speciation climb to 1 new species every 70 years or longer (12,000 years / 70 years per species (or higher) = 171 species (or lower)).

On the other hand, global tracking within species-rich families is much more complex. Four percent of mammal families (7 out of 153 families) have more than 171 species.[33] Comprehensively tracking each individual species within each of these seven families is a monumental task. The seven families include two mouse families, a shrew family, a squirrel family, and three bat families. These small creatures tend to reproduce quickly, making continuous tracking of their genetic and reproductive isolation very difficult. In addition, these parameters would need to be followed on a global scale. At various time points, the isolation *in each species* would need to be quantified and documented. Obviously, these sorts of experiments have not been done.

In short, tests of the rapid speciation hypothesis against historical and modern data are just beginning. Despite claims on both sides of the origins debate about the rates of biological change in the present, no one yet has a good handle on what the rate actually is.

❖ ❖ ❖ ❖

Today, I suspect that most people are unfamiliar with Darwin's (unintentional) argument for a recent species origin. Historically, geology has been the central discipline in which questions on speciation timescales have been debated. In fact, as I've walked through the data in this chapter, you might have been mentally reviewing the common geologic arguments for the millions-of-years timescale. You might have been rejecting my arguments by mentally citing multiple independent geologic lines of evidence. Effectively, this form of reasoning rejects Darwin's argument *a priori*

— that it can't possibly be true because it contradicts so much geologic evidence. The problem with this rejoinder is that multiple independent mammal and bird families all show the same breed-species pattern.

Nevertheless, you might claim that geology represents a direct scientific record of the past. In contrast, the central argument from this chapter uses modern comparisons to extrapolate into the past. Stood next to scientific disciplines dedicated to the study of earth history, the breed-species analogy might seem weaker.

If these breed-species arguments were the strongest evidences for a recent timescale for the origin of living species, then we might simply chalk them up as a weird anomaly in a sea of evidences for ancient history. However, though these new connections were weak, they were not the end of the controversy on the timescale of species' origins. They were the beginning.

Part Three

Dawn of a New Era

Chapter 7
Turning the (time) Tables

By themselves, the edge pieces of a jigsaw puzzle do not reveal the puzzle's shape and size. Disconnected from one another, it is impossible to know if the edges will eventually form a square, rectangle, or some other shape. In fact, it's impossible to know if the puzzle will even have four sides — it might have three, or even five. Or more. Only corner pieces determine these parameters.

Without the shape, the size of the puzzle remains a mystery.

If I'm assembling a standard jigsaw puzzle, I rarely find the corner pieces first. Numerically, they are swamped by center pieces and edge pieces. For example, in a rectangular 1000-piece puzzle, only 4 corner pieces (i.e., 0.4% of the total pieces) exist. Despite their importance, corner pieces are notoriously difficult to track down.

The discovery of the biological corner pieces was the third major development after Darwin. Functionally, this development mirrored my typical experience with actual jigsaw puzzles. As compared to the total number of species and clues, these biological corner pieces were numerically rare. In other words, if the DNA sequences in each species represented the edge pieces, the number of edge pieces were as abundant as the number of species. In contrast, the clues to the connections among these edges were harder to find.

Yet these clues were as critical to the whole as actual corner pieces are to real jigsaw puzzles. In fact, given the lack of a box cover, the uncertain number of total pieces, and the unknown shape of the puzzle of the origin of species, the biological corner pieces were even more critical to the assembly.

Naturally, the corner pieces were genetic discoveries. Since the genetic differences among species represented the edge pieces, the corner pieces were critical insights into how these genetic differences related to one another.

Furthermore, given the youth of the field of genetics (see chapters 2–3), it should come as no surprise that the discoveries of these corner pieces were very recent. In fact, some of the content of this chapter involves discussion of initial findings and of ongoing research.[1]

Despite the preliminary nature of some of these discoveries, their implications were immense.

Over the last few years, one of the most critical genetic observations came from analysis of DNA sequences, not between species, but *within* species. Recall from chapter 3 that the first complete DNA sequences were from separate species. As technology has progressed, investigators have also obtained the DNA sequences from multiple individuals within a particular species. In fact, several studies have sequenced the DNA from individuals of known genealogical relationships.

For example, among humans, researchers have sequenced the DNA of parent-offspring pairs.[2] In animal,[3] plant,[4] fungal,[5] and bacterial[6] species that can be grown easily in the lab, similar experiments have been performed. At the start of these experiments, the DNA from the original population was obtained. Then, after letting these creatures undergo multiple rounds of reproduction, the DNA sequence of the later generations was obtained.

The results were striking. In the parent-offspring pairs, the DNA sequence in the parents differed from the DNA sequence in the offspring. In other words, the offspring contained DNA sequences that could not be traced to either parent. Among the animal, plant, fungal, and bacterial experiments, the DNA sequence of the original population differed from the DNA sequence of the later generations. In all cases, the amount of difference was very slight — a small percentage of the total sequence. But a difference was detectable nonetheless. In short, researchers discovered that DNA mutates.*

The types of mutations varied. Some were switches in the identity of single base pairs — *single nucleotide polymorphisms (SNPs)* (Figure 7.1). Others were insertions or deletions — *indels* — of several base pairs (Figure 7.1). On occasion, whole chunks of DNA sequence were inserted ("large insertions") or deleted ("large deletions") (Figure 7.1). Some DNA sequences switched locations in the genome (a process termed *translocation*) (Figure 7.1).

The discovery of the fact of DNA mutation had another ramification for our discussion. Upon reflection, it should be fairly easy to see that these experiments didn't just document the fact of mutation. They also measured the rate. Since the reported units of measurement from

* The primary reason for these mutations is unknown. It appears that the protein machine that copies DNA occasionally makes mistakes. In fact, cells have DNA repair machines that fix known errors and mutations. Nevertheless, it appears that some mistakes still slip through.

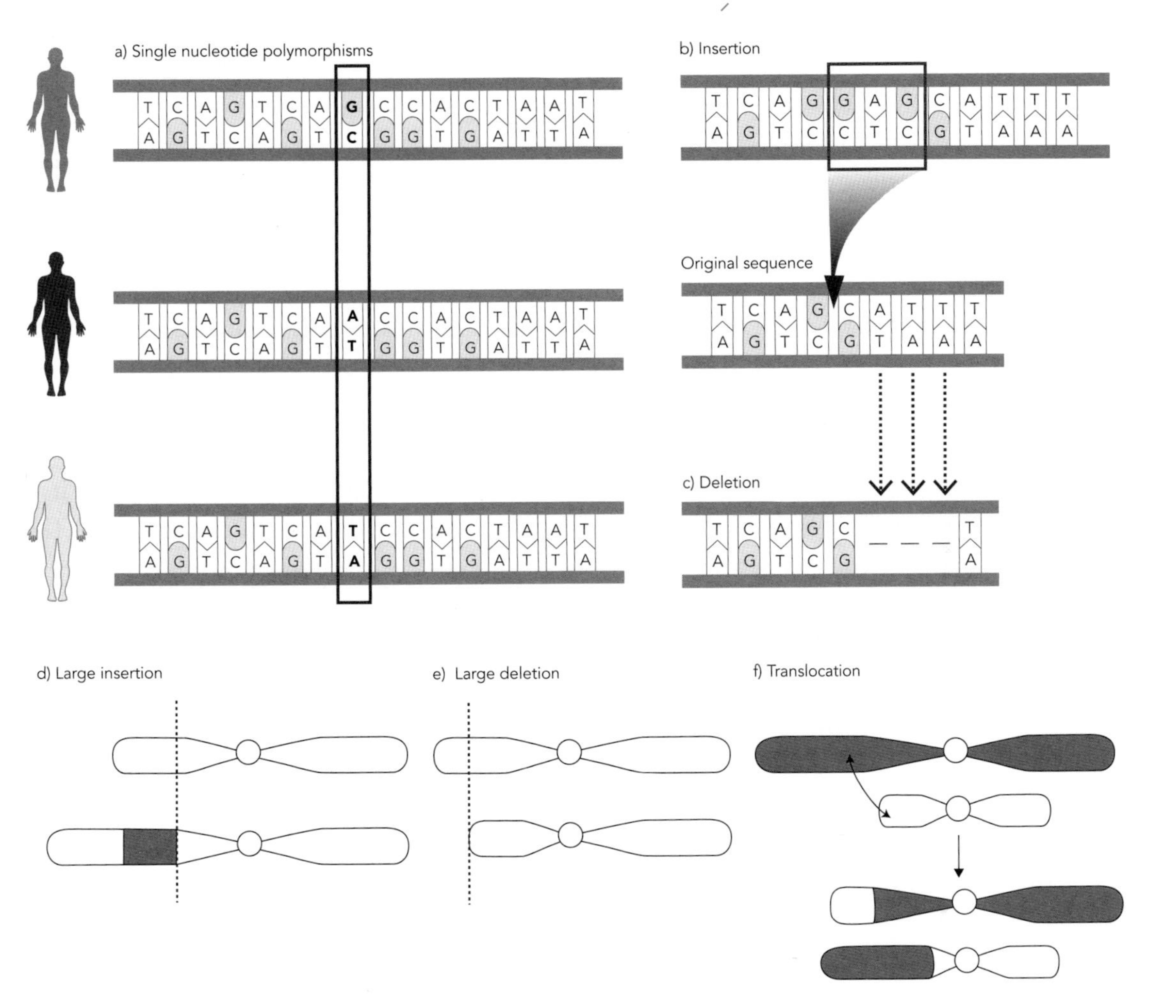

Figure 7.1. Types of mutations. (a) Mutations which switch the identity of single base pairs are "single nucleotide polymorphisms" (SNPs). (b) Mutations which involve the addition of one or a few base pairs are "insertions." (c) Mutations which remove one or a few base pairs are "deletions." (d) Mutations which add whole chunks of DNA sequence are "large insertions". (e) Mutations which remove whole chunks of DNA sequence are "large deletions." (f) Mutations in which DNA sequences switched locations in the genome are known as "translocations."

these studies were *mutations per generation*, these experiments naturally had implications for the timescale over which DNA differences arise.

On the surface, these discoveries suggested a very simple explanation for the connections among edge pieces: DNA differences within or between species were the product of the mutation rate and the time of the origin of each individual or species. If we know the number of DNA differences and the mutation rate, then we can calculate when two species originated. Conversely, if we know the mutation rate and the time when a population split into two species, then we can calculate the number of DNA differences between two species.

Actually, the latter calculation represents a prediction. In light of the information discussed in chapter 6, we now recognize that the timescale over which species originate is a matter of scientific investigation, not settled fact. Mutation rate predictions are one way to test the various views of the timescale.

Perhaps not surprisingly, the results of these tests turned several fields of science upside-down.

As we discovered in chapter 3, DNA exists in a variety of compartments. Among animal, plant, and fungal species, DNA exists in more than just the nucleus. It also exists in the tiny energy factories of the cell, the mitochondria.*

Though only about 0.001% of the size of the nuclear genome in humans, the sequence of the mitochondrial DNA (*mtDNA*) genome holds several advantages over nuclear DNA. First, its small size means that it is much cheaper and easier to sequence than the nuclear genome. The timeline of mtDNA sequencing bears out this fact. For example, the human mtDNA sequence[7] was obtained two decades before the human nuclear DNA sequence.[8]

Furthermore, the human mtDNA sequence was just the initial drop in a torrent of mtDNA discoveries. The publication of the mouse[9] and cow[10] mtDNA genomes followed later in 1981 and 1982, respectively. By 1999, mitochondrial genomes from 56 vertebrate species had been published.[11] Today, over 6,800 curated animal mitochondrial genome sequences exist in the public databases.[12] Over 880 are from mammals.[13] When all sequences are counted, including those that have yet to be curated,[14] over 50,000 vertebrate mitochondrial genomes exist.

The mtDNA dataset represents one of the most biologically diverse genetic datasets in existence.

Second, the mtDNA content from these thousands of species was shared. In other words, though the mtDNA sequence was obtained from thousands of very diverse species, the actual gene set among these species was highly similar.[15] In fact, it was so similar that mtDNA sequences from very different species could be directly compared to one another at the base pair level.

In contrast, the nuclear DNA content of various species was comparatively dissimilar. Though comparative nuclear DNA analyses have been performed (see chapter 8), these analyses have been technologically more challenging to pursue than mtDNA comparisons.

Historically, many mtDNA comparisons have been done — with intriguing results. For example, when mtDNA genome sequences are compared across diverse mammal species, a hierarchical grouping naturally forms.[16] Using familiar species as illustrations, the seven living equid species (family Equidae) are genetically closest to one another. All seven differ from one another by less than

* Other genetic compartments exist — for example, chloroplasts in plants. However, the fact that mitochondrial DNA is found across three kingdoms — animals, plants, fungi — makes it more useful for our purposes here. (The fact that chloroplast DNA is not found in humans makes comparisons of human and plant chloroplast DNA impossible.)

1,100 base pairs (Table 7.1).[17] Species* from the other families (Rhinocerotidae, Tapiridae) within the order Perissodactyla follow next. Rhinos and tapirs differ from equids by about 2,400 to 2,500 base pairs. Yet rhino species differ from one another by less than 1,900 base pairs. Conversely, all three families are separated from other mammalian species by even more mtDNA differences. For example, bighorn sheep belong to a different mammalian order (Artiodactyla), and this species is over 3,000 base pairs away from equid, rhino, and tapir mtDNA sequences.[18]

As we observed in chapter 5, evolutionists see this pattern as evidence of common ancestry. Darwin expected life to form a branching, tree-like structure. This groups-within-groups (nested hierarchical) mtDNA pattern fits this expectation.

Furthermore, in light of the discovery of the fact of DNA mutation, it would seem that evolution had a ready-made mechanism by which to explain this pattern — the accumulation of DNA differences with time. As species diverged from common ancestors, they would continue to mutate their DNA. The longer two species had been separated, the more DNA differences would accumulate between them. The mtDNA patterns that we just observed fit the evolutionary model of the mammal ancestor evolving first, then the ancestors to Artiodactyl and Perissodactyl species diverging from one another, and then the various ancestors to modern Perissodactyl families diverging from one another, with the tapir-rhino split occurring first, followed by the rhino-equid split. Lastly, the species within each of these families would have diverged from one another.

If evolution were the only explanation for nested hierarchies, then we might conclude that all mammals — from echidnas to equids (Figure 7.2) and every mammal in between — have a common ancestor.

However, at least one other competing explanation exists. Again, as in chapter 5, the mtDNA hierarchy shows strong parallels with the hierarchy present within the Linnaean classification system.[19] Since this system is based on biological function, the parallel between the mtDNA hierarchy and the Linnaean categories suggests that the mtDNA hierarchy has something to do with function. In other words, the creation/design model predicts the fact of mtDNA nested hierarchies as much as evolution does.

More specifically, I have taken these design expectations one step further and derived a very detailed, testable model on mtDNA genome function.[20] For mtDNA differences among species within families, my model treats these as functionally neutral changes — the result of mutation over time. However, with respect to the differences between families — those mtDNA positions that are identical among species within a family but different to species outside of the family — my model views these as having been created. Thus, my model predicts that these mtDNA differences play a functional role specific to each family.

---

* In Table 7.1, only some of the species within the rhino and tapir families are shown because mtDNA sequences have not yet been obtained from all species within these families. Comparisons were done among all species with mtDNA sequences available at the time of analysis.

Table 7.1. Pairwise mitochondrial DNA differences among Perissodactyl species. The differences between any two species can be found at the intersection of each row and column. Shading added to highlight levels of genetic differences.

	Mountain zebra (*Equus zebra*)	Imperial zebra (*Equus grevyi*)	Plains zebra (*Equus quagga*)	Kiang (*Equus kiang*)	Onager (*Equus hemionus*)	African wild ass (*Equus africanus*)	Przewalski's horse (*Equus przewalskii*)	Black rhinoceros (*Diceros bicornis*)	White rhinoceros (*Ceratotherium simum*)	Sumatran rhinoceros (*Dicerorhinus sumatrensis*)	Javan rhinoceros (*Rhinoceros sondaicus*)	Indian rhinoceros (*Rhinoceros unicornis*)	Malayan tapir (*Tapirus indicus*)	Bighorn sheep (*Ovis canadensis*)
Mountain zebra (*Equus zebra*)		753	755	804	813	912	1089	2464	2486	2484	2536	2467	2526	3170
Imperial zebra (*Equus grevyi*)	753		524	733	740	838	1002	2486	2467	2510	2518	2470	2515	3131
Plains zebra (*Equus quagga*)	755	524		744	750	829	1000	2452	2449	2502	2498	2443	2511	3134
Kiang (*Equus kiang*)	804	733	744		29	795	956	2436	2421	2442	2492	2435	2475	3104
Onager (*Equus hemionus*)	813	740	750	29		791	967	2439	2424	2454	2503	2449	2485	3104
African wild ass (*Equus africanus*)	912	838	829	795	791		1046	2473	2466	2523	2537	2484	2516	3159
Przewalski's horse (*Equus przewalskii*)	1089	1002	1000	956	967	1046		2456	2466	2496	2504	2439	2492	3137
Black rhinoceros (*Diceros bicornis*)	2464	2486	2452	2436	2439	2473	2456		1106	1795	1808	1760	2357	3182
White rhinoceros (*Ceratotherium simum*)	2486	2467	2449	2421	2424	2466	2466	1106		1766	1789	1717	2331	3153
Sumatran rhinoceros (*Dicerorhinus sumatrensis*)	2484	2510	2502	2442	2454	2523	2496	1795	1766		1806	1805	2435	3110
Javan rhinoceros (*Rhinoceros sondaicus*)	2536	2518	2498	2492	2503	2537	2504	1808	1789	1806		988	2418	3118
Indian rhinoceros (*Rhinoceros unicornis*)	2467	2470	2443	2435	2449	2484	2439	1760	1717	1805	988		2375	3104
Malayan tapir (*Tapirus indicus*)	2526	2515	2511	2475	2485	2516	2492	2357	2331	2435	2418	2375		3125
Bighorn sheep (*Ovis canadensis*)	3170	3131	3134	3104	3104	3159	3137	3182	3153	3110	3118	3104	3125	

In contrast, since the evolutionary model treats all mtDNA differences as the result of mutation, it generally predicts that most mtDNA differences are functionally neutral.

In the early days of the central dogma, my creationist explanation might have been quickly dismissed. After all, nearly the entire mtDNA genome consists of protein-coding genes, and nearly the same set of mtDNA genes exist in very diverse species. Furthermore, we know the functions of the proteins that are encoded by these genes — these proteins perform the final steps of energy transformation and metabolism (i.e., of sugars, fats, etc.). From a design perspective, there would seem to be no functional reason to design these proteins differently in different species. Why should the breakdown of sugars happen one way in a zebra and another way in a sheep?*

Conversely, the singular functions (i.e., catalysis at very specific metabolic steps) of mitochondrial proteins appear to present the perfect argument for common ancestry. We know from laboratory experiments that some proteins can tolerate slight changes to their sequence. In some parts of the protein, substitutions of one amino acid for another do nothing to affect the chemical reaction catalyzed by the enzyme. As long as the critical amino acids remain intact, the protein functions normally.[21]

Consequently, the mtDNA differences among mammal species might seem to be better explained by the accumulation of mutations via evolution, as we've just discussed.

Since the 1970s, the advances in molecular biology have upended the overly simplistic and early views of protein function. For example, evidence is accumulating for the phenomenon of protein "moonlighting."[22] Rather than perform a single function in a cell, proteins perform an unexpected number of additional functions. As an illustration, some proteins that were classically thought to catalyze a single metabolic reaction now appear to function in information flow as well. Conversely, the proteins encoded by mtDNA genes might also moonlight and function in additional subcellular processes.

To date, moonlighting has not been documented for proteins encoded by mtDNA. However, this phenomenon has also not been rigorously tested. Once these experiments have been performed, we can directly compare the predictions of the evolutionary model to the creationist one.

Thus, the nested hierarchy of mtDNA differences among mammals doesn't reveal anything new about species' ancestry. Rather, it represents an experiment waiting to be performed.

---

* To clarify, the early steps of sugar breakdown are not catalyzed by proteins encoded in mtDNA; instead, proteins encoded by mtDNA participate in the later steps.

FOLLOWING TWO PAGES: Figure 7.2. Major divisions of mammals. The three major divisions of mammals, with representative species shown. The divisions are defined primarily by the mode of reproduction (i.e., placentals nourish their young in the womb via a placenta; marsupials nourish their young in a pouch; monotremes lay eggs).

PLACENTALS

Whale
Tiger
Elephant
Aardvark
Mole
Lemur
Zebra
Moose
Manatee

MONOTREMES

MARSUPIALS

If the *pattern* of mtDNA differences does not distinguish between the hypothesis of universal common ancestry and the hypothesis of design, what aspect of mtDNA biology does? A detailed examination of the mtDNA of our own species revealed a critical clue.

Of all the species studied at the mtDNA level, none compares to our own. Among the over 50,000 mtDNA genome sequences in the uncurated public databases, over 34,000 belong to *Homo sapiens*.[23] German, Hungarian, Mbuti, San, Moroccan, Cherokee, Guarani, Han Chinese, Tuvan — an enormous number of ethnolinguistic groups possess a published mtDNA sequence.

Furthermore, pedigree-based measurements of the rate at which human mtDNA mutates have been the subject of over 15 studies — which span two decades.[24] Though the results of these studies have been hotly contested,[25] part of the reason for this controversy is the fact that mutations are statistically rare events. Consequently, mutations require a certain minimal sample size to detect. Not surprisingly, the 15+ studies differed dramatically in their statistical resolving power. Nevertheless, once each of these results were weighed appropriate to their statistical resolving power, a clear result emerged: One mtDNA base pair mutates every 5 to 8 generations.[26] Could this rate inform the question of human ancestry?

At a minimum, this measured rate implies that DNA acts like a clock. Each time a new mutation occurs, another tick of the clock marks the passage of time. Since ticks don't happen every year or every generation, the DNA clock is less precise than our watches. Nonetheless, DNA can still measure time.

Could the mtDNA clock elucidate the timescale over which our own species appeared?

To answer this question, we have to answer a more specific, technical question: Has the clock always ticked at the same rate?

Long before the discovery of the mtDNA clock, the timescale over which species originated had been discussed extensively. Naturally, the scientific methods used to establish a timescale involved measurement of some rate, and then an assumption about how fast or slow the rate was in the distant past.

These assumptions have massive implications for the analysis of the human mtDNA clock.

Under the evolutionary model, the timescale of human origins has been placed within the larger framework of the evolutionary origin of the universe and earth. According to evolution, the universe evolved 13 to 14 billion years ago, and the earth formed around 4 to 5 billion years ago.[27] The first major animal lineages (phyla) appear several hundred million years ago,[28] and great ape lineage about 18 to 20 million years ago.[29] Modern *Homo sapiens* arrived on the scene very late in evolution — around 200,000 years ago.[30]

Traditionally, the evolutionary arguments for this timescale came from geology and other physical sciences. For example, today the oceans are salty. But they haven't always been this salty. Currently, salt enters the ocean via a variety of processes, including the erosional processes that bring minerals from the land to the sea. The rate at which all of these salt-depositing processes occur can be estimated. Collectively, the overall rate is slow. Assuming the ocean had no salt to begin with, we can calculate when the oceans first began accumulating salt. It's definitely not thousands of years ago. Rather, the timescale is in the tens of millions of years.[31]

Similar calculations could be done for the rate at which mud enters the ocean and settles on the ocean floor.[32]

Modern methods for establishing a timescale involve radioactivity. One of the more common radioactive "clocks" involves the element carbon. For our purposes, carbon exists in two forms — a radioactive form (i.e., *carbon-14*) and a non-radioactive form (i.e., *carbon-12*). The radioactive form is produced when cosmic rays hit nitrogen in the atmosphere. Over time, this radioactive carbon decays back to nitrogen. Specifically, after around 5,700 years, half of all the carbon-14 atoms in a sample will have decayed back to nitrogen.

Carbon-14 doesn't remain exclusively in the atmosphere. Eventually, it enters the biosphere — primarily via plants. Since plants harvest carbon dioxide from the atmosphere, plants naturally assimilate carbon into their structures while they are living. Since they do not discriminate between radioactive carbon and non-radioactive carbon, both types enter their system. Currently, the low levels of carbon-14 in the atmosphere imply that the carbon in plants is primarily non-radioactive carbon, with a low level of carbon-14.

Via the food chain, carbon-14 eventually makes its way into animals. Herbivores consume plants, digest them, and use the nutrients to build and maintain their bodies, thereby incorporating carbon-14 into their bodies (Color Plate 75). Carnivores eat carbon-14-containing herbivores. Thus, some radioactive carbon reaches the top of the food chain.

Although the carbon-14 in an organism is constantly decaying back into nitrogen, the organism continually replenishes its supply. Plants take in new carbon-14 from the atmosphere, and animals obtain it by eating the plants or other animals. Consequently, the carbon-14 to carbon-12 ratio in a living organism is relatively constant.

Once a plant or animal ceases to harvest energy from the environment, entry of carbon-14 into an individual stops, and the loss of carbon-14 is no longer balanced by an influx of new carbon-14. As a result, death starts the ticking of the carbon-14 clock. As time passes, radioactive carbon-14 decays into nitrogen, and the amount of carbon-14 relative to carbon-12 decreases with time (Color Plate 75). When a scientist measures the ratio of carbon-14 to carbon-12, the date of a carcass, artifact, or archaeological site can be estimated.

At some point, the carbon-14 will decay to such an extent that it is no longer detectable. Exponential decay processes are alike in this manner. Consequently, each radioactive element has an upper limit on how much time can be recorded via its decay.

Compared to other radioactive elements on earth, carbon-14 decays fairly quickly. Regardless of the initial size of the carbon-14 sample, all of the carbon-14 will decay to below detectable limits within 100,000 years.[33] Consequently, to determine the ages of substances thought to be older than 100,000 years, geologists and paleontologists turn to chemical elements that decay more slowly, like uranium and rubidium.

Other methods are also used to establish the evolutionary timescale. For example, the plates in the earth's crust are currently moving away from one another. South America is moving farther and farther away from Africa. If these two continents were once a single landmass, the current rates of plate movement suggest that a supercontinent existed millions of years ago.[34]

Some methods aren't conducive to measuring millions of years of time. Nevertheless, they still indicate that more than 12,000 years were required to produce the result in question. For example, measurements of sedimentation in lakes, of ice layer formation at the poles, and of stalactite and stalagmite formation in caves all lead to this conclusion.[35]

In the field of astronomy, the size of the universe and the speed of light argue for an ancient universe. Though light travels at 186,000 miles per second, the size of the universe is far bigger than our wildest imaginations. For light to travel from one end of the visible universe to the other, it would take billions of years at this speed.

Together, all of these methods rest on a foundational assumption. Whether the method involves the rate of salt flow into the ocean, the rate of radioactive decay, the rate of tectonic plate movement, the speed of light, or any of these other processes that we discussed, present rates have been assumed to have been largely constant for all of earth's history. To be sure, evolutionists acknowledge past environmental changes, such as changes in cosmic ray intensities. These slight perturbations have been used to make slight modifications to the assumption of constant rates of change. However, as a rule, evolutionary geologists and astronomers assume that massive alterations to present rates have not occurred in the distant past.

This assumption is critical for a number of reasons. First, without it, the geologic and astronomical arguments for millions and billions of years collapse. Second, as already mentioned, these discussions largely predate the discovery of DNA clocks.

Third, creation scientists have long questioned this assumption. Specifically, *young-earth creationist (YEC)* geologists* propound the idea that the universe and earth are just 6,000 to 10,000 years old, and that species have formed within this timeframe. They also hold to a world-wide flood (the Flood of Noah) about 4,500 years ago. In theory, this world-wide flood would dramatically alter global rates of geologic change. At a minimum, these creationists would argue that the rates at the Flood were very different from rates we can measure today.

These YEC geologists have developed testable scientific models and have documented phenomena in the field that are consistent with their ideas. For example, they've made and experimentally tested hypotheses on accelerated rates of radioactive decay,[36] accelerated rates of plate tectonic movement,[37] and accelerated rates of geologic deposition and erosion.[38]

The astronomers of the YEC community have followed suit in their own fields of study. They have successfully predicted the magnetic fields for planets in our own solar system,[39] and

* To distinguish them from old-earth creationists (OEC) who hold that the universe and earth are billions of years old.

they have revisited astronomical data in light of Einstein's general relativity. Specifically, with respect to the latter, YEC astronomers have uncovered the directionality limits of the speed of light I cited above. Because of Einstein's discoveries, the measured speed of light represents a round trip speed. Measurements of light in a single direction are physically impossible without circular assumptions.[40] Since our view of the universe is illuminated by the travel of light in one direction — from distant stars and galaxies to our telescopes — the great size of the universe is not a sound argument for an ancient universe.

In response, the evolutionary community has largely dismissed these data. They've insisted that invoking a catastrophic phenomenon like a global flood is unscientific. Regarding the assumption that rates have been constant over millions (or billions) of years, they've asserted that this assumption must be accepted. Despite the fact that these YEC and flood models make testable scientific predictions, evolutionists have ignored or rejected these creationist claims.

To be consistent with evolutionary practice, we should assume constant rates of change in genetics. In other words, when using the mtDNA clock to trace a species' history, we should assume that the clock has ticked at a largely constant rate.

Examination of current evolutionary literature reveals that the assumption of constant rates of change is largely followed. When discussing molecular clocks, evolutionists typically measure the DNA difference between two species, assign the time of origin from the evolutionary geologic timescale, and then calculate a rate of mtDNA mutation from these parameters. Implicitly, this methodology assumes constant rates of mtDNA mutation.

However, very few evolutionary clock analyses invoke the *measured* rates of mtDNA change. By analogy, the typical evolutionary molecular clock methods parallel the following (theoretical) geologic practice: Let's say a geologist wants to know the rate of erosion in the Grand Canyon. Rather than measure it directly, the geologist first determines the ages of the layers in the Grand Canyon. Then the geologist determines the depth of the Grand Canyon. By dividing the depth by the ages, the geologist calculates how fast (or slow) the Colorado River has been eroding the gorge. Obviously, this "rate" is simply a prediction, not an actual measurement. In practice, geologists determine the rate of erosion by directly measuring it in real time. This measurement directly tests the prediction we just made.

Similarly, the rates of mutation in typical evolutionary molecular clock discussions represent a prediction, not an actual measurement. This prediction can be tested with the human pedigree-derived rate that we just discussed.

Using these experimentally derived rates, we can make predictions on the origin of humans. For example, by taking the evolutionary time of origin for humans or for other species from the fossil record and by multiplying the time by the mutation rate, we can predict how many mtDNA differences should be present today. For comparisons between individuals in the same species, this math and methodology is the same as that which the evolutionists have been using for years. In technical terms, the equation is a *coalescence* calculation.[41] When we're comparing mtDNA differences between two separate species, we multiply our calculation by 2 — to account for the fact that mtDNA differences have been accumulating independently in both species. In technical terms, this second equation is a *divergence* calculation.[42]

With respect to humans, evolutionists have proposed that chimpanzees are our closest living relatives. They have put the time of divergence between the human and chimpanzee lineages around 4.5 to 17 million years ago.[43] Using this timescale, along with the measured human mtDNA mutation rate, we can predict how many mtDNA differences should exist between humans and chimpanzees today.

Before we can perform this calculation, the mutation rate that I reported earlier must be converted to an absolute timescale. To convert units of *mutations per generation* to units of *mutations per year*, we need to know the ages at which humans and chimpanzees give birth.

In technical terms, the length of time from conception to reproductive maturity is referred to as the *generation time*. Specifically, since mtDNA is inherited primarily — if not exclusively — through the maternal lineage, we need to know the generation times for female humans and female chimpanzees. For chimpanzee females, the average generation time is around 25 years.[44] In humans, the generation time varies. Some women give birth early in life; others, late in life. Since we're calculating mutations over many generations, the safest approach is to predict mutations over a whole range of generation times — from 15 years to 50 years. In practical terms, this means that humans mutate one mtDNA base pair every 76 to 419 years.[45]

Using this rate, we can predict how many mtDNA differences should exist between humans and chimpanzees after 4.5 to 17 million years of mutation. Though the chimpanzee mtDNA mutation rate has not yet been empirically measured, we will assume that it is the same as the human mutation rate.* Since we're comparing the DNA of two species to one another, a divergence calculation is most appropriate. At a mutation rate of one base pair per 76 to 419 years, a minimum of 21,480 mtDNA differences (1 mutation per 419 years * 4.5 million years * 2 = 21,480) and a maximum of 447,368 mtDNA differences (1 mutation per 76 years * 17 million years * 2 = 447,368) would arise. Today, only 1,483 mtDNA differences separate these two species.[46] (See also Figure 7.3, which uses more precise calculations, based on previously published work.[47]) The evolutionary timescale predicts mtDNA differences far in excess of what is observed.

These results also raise an important question. In humans, the total length of mtDNA sequence is less than 17,000 base pairs. How could over 447,000 mtDNA differences arise between humans and chimpanzees?

In practical terms, the 447,000 result is the number of predicted *mutations*. Since the total mtDNA genome size is far less than 447,000 base pairs, each mtDNA position would have been

* In chapter 8, we will find that the nuclear DNA mutation rates in these two species differ by only around 12%. Thus, our assumption in this chapter appears to be reasonable.

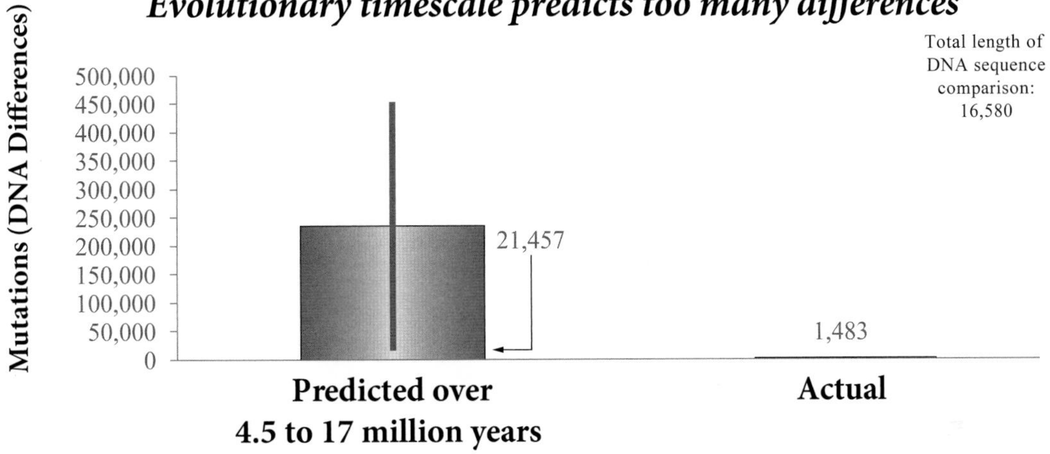

Figure 7.3. The evolutionary timescale predicts too many mitochondrial DNA differences between humans and chimpanzees. Mutations were predicted from the empirically determined human mitochondrial DNA mutation rate. The height of the column represents the average prediction. The black line approximates the 95% confidence interval. In short, if all chimpanzee and human females were consistently age 50 when they gave birth, then only around 21,000 mitochondrial DNA differences (the lower end of the black line) would have accumulated over 4.5 million years. Instead, if all chimpanzee and human females were consistently age 15 when they gave birth, then around 450,000 mitochondrial DNA differences would have accumulated over 17 million years. Today, only 1,483 mitochondrial DNA differences separate the human and chimpanzee mitochondrial DNA sequences.

mutated multiple times over. In other words, the mtDNA genome would have been mutationally saturated. Today, a comparison of human and chimpanzee mtDNA reveals two genomes that are far from mutational saturation — the 1,483 differences represent just 9% of the total human mtDNA genome length.

These evolutionary predictions improve little if we narrow our focus to living and extinct members of the genus *Homo*. For example, Neanderthals are classified within the *Homo* genus, and a Neanderthal mtDNA sequence has been published. Evolutionists put the split between the Neanderthal and modern human lineages about 400,000 to 700,000 years ago.[48] Treating them as members of the same species,[49] we can use a coalescence calculation to predict how many mtDNA differences should exist today between Neanderthal sequences and sequences from living humans. At a mutation rate of one base pair per 76 to 419 years, a minimum of 955 mtDNA differences (1 mutation per 419 years * 400,000 years = 955) and a maximum of 9,211 mtDNA differences (1 mutation per 76 years * 700,000 = 9,211) would arise. Today, only 213 mtDNA differences separate Neanderthals and modern humans.[50] (See also Figure 7.4, which uses more precise calculations based on previously published work.[51]) Again, the evolutionary timescale predicts mtDNA differences far in excess of what is observed. The discrepancy between

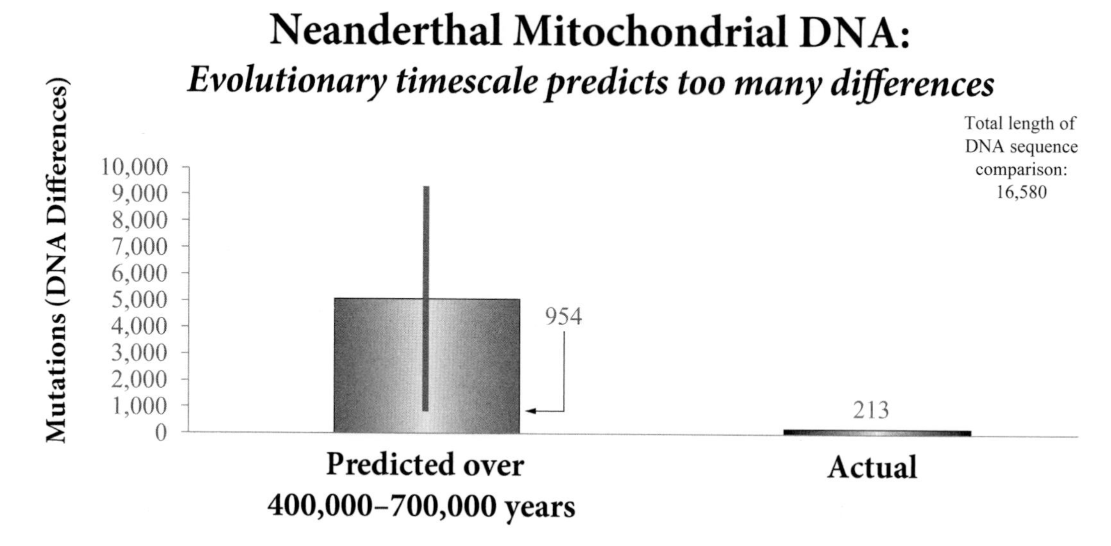

Figure 7.4. The evolutionary timescale predicts too many mitochondrial DNA differences between modern humans and Neanderthals. Mutations were predicted from the empirically determined human mitochondrial DNA mutation rate. The height of the column represents the average prediction. The black line approximates the 95% confidence interval. In short, if all Neanderthal and human females were consistently age 50 when they gave birth, then only around 950 mitochondrial DNA differences (the lower end of the black line) would have accumulated over 400,000 years. Instead, if all Neanderthal and human females were consistently age 15 when they gave birth, then around 9,200 mitochondrial DNA differences would have accumulated over 700,000 years. Today, only 213 mitochondrial DNA differences separate the human and Neanderthal mitochondrial DNA sequences.

predictions and reality is less than what we observed for the human-chimpanzee calculations. But it still fails to capture actual differences.

When we focus just on differences among modern humans, the discrepancy becomes even smaller — but still fails to result in a successful prediction. As mentioned above, evolutionists put the origin of *Homo sapiens* in Africa about 200,000 years ago. Since we're examining differences within a single species, a coalescence calculation applies. At a mutation rate of one base pair per 76 to 419 years, a minimum of 477 mtDNA differences (1 mutation per 419 years * 200,000 years = 477) and a maximum of 2,632 mtDNA differences (1 mutation per 76 years * 200,000 = 2,632) would arise. Today, an average of 77 mtDNA differences separate African mtDNA sequences from other mtDNA sequences. An average of 39 mtDNA differences separate non-African sequences from other mtDNA sequences[52] (see also Figure 7.5). The evolutionary timescale still fails to accurately predict reality.

If these predictions are unable to account for mtDNA differences that we see today, what model can accurately predict them? If we expand our analysis further back into evolutionary time and include more primate species, then the number of differences in the "Actual" column would

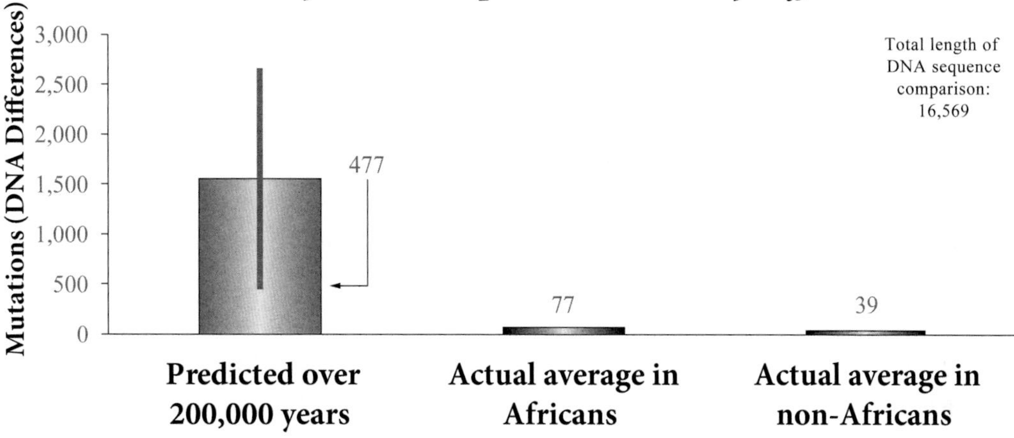

Figure 7.5. The evolutionary timescale predicts too many mitochondrial DNA differences among modern humans. Mutations were predicted from the empirically determined human mitochondrial DNA mutation rate. The height of the column represents the average prediction. The black line approximates the 95% confidence interval. In short, if all human females were consistently age 50 when they gave birth, then only around 480 mitochondrial DNA differences (the lower end of the black line) would have accumulated over 200,000 years. Instead, if all human females were consistently age 15 when they gave birth, then around 2,600 mitochondrial DNA differences would have accumulated over 200,000 years. Today, only around 80 mitochondrial DNA differences exist in African populations, and around 40 differences in non-African populations. Modified with permission from Figure 2 of N.T. Jeanson, "On the Origin of Human Mitochondrial DNA Differences, New Generation Time Data Both Suggest a Unified Young-Earth Creation Model and Challenge the Evolutionary Out-of-Africa Model," *Answers Research Journal*, 2016, 9:123–130.

increase. However, the longer timescale would necessarily lead to a higher number of predicted differences. Since the mutation predictions for the human-chimpanzee timescale already exceed the mtDNA genome size, this lengthening of the timescale would only make the predictions even more at odds with reality.

On the other side of the timescale spectrum, we might be able to make accurate predictions for a very narrow group of modern humans. Perhaps the recent origin of one of the European ethnolinguistic groups will be explicable by the mutation rates we've discussed. But if this is all that the evolutionary timescale can explain, what do we do with the rest of the timescale for human evolution?

Can the timescale itself be changed? In theory, perhaps this is possible. However, in practice, this would require significant reinterpretation of the conventional evolutionary geologic model — an action which could produce significant disarray in this discipline.

In a similar vein, perhaps the assumption of constant rates of change could be altered. However, as we observed above, evolutionists have insisted for years that changing rates must not be invoked to explain the majority of phenomena observed in geology and astronomy. Instead, they have claimed that present rates are the key to the past, and that the world we see today has arisen primarily by slow, constant rates over time. Invoking changing rates in genetics would be logically inconsistent with the practice of evolutionary geology and astronomy.

Perhaps the explanation involves natural selection. At first pass, this might seem plausible. After all, mtDNA encodes proteins with critical functions in the cell. If you interrupt basic metabolism, cellular death is sure to result. Surely most of the thousands of mtDNA mutations that have occurred over the last several million years of evolutionary time were lethal to the possessors of these mutations. Consequently, natural selection would surely have eliminated these mutations (and individuals) from the mtDNA pool.

How might we evaluate the natural selection hypothesis? The scientific community has a long-established practice of dealing with scientific controversies. We've already discussed in chapter 4 how to advance a scientific debate towards resolution. The scientific method operates like a process of elimination. When two hypotheses offer competing explanations for the same phenomenon, one must be eliminated before scientific inferences can be made.

Naturally, this logic assumes that two competing hypotheses actually make *testable* predictions. We assumed as much in our discussion of the history of genetics (chapter 2–3) and in our discussion of Darwin's arguments from biogeography. For example, Mendel was successful as a scientist because he inferred rules that made testable, accurate predictions about the mathematical ratios of traits among offspring in each pea plant generation. As another example, in our discussion of whether DNA or proteins were the substance of heredity, we observed that both of these hypotheses made testable predictions. If proteins were the substance of heredity, their chemical elimination in the experiments of Avery and colleagues should have eliminated the transforming ability of the heat-killed smooth cells. The same prediction follows from the hypothesis that DNA is substance of heredity. Conversely, if species were created in their present locations, then you might expect the fauna on islands to possess more terrestrial species. You wouldn't expect the native fauna to be so skewed towards aquatic and aerial species. In other words, the hypothesis of the fixity of species' geography makes testable predictions.

Hypotheses that fail to make predictions do not qualify as science. As evolutionists maintain to this day:

> Science is . . . a process of acquiring an understanding of natural phenomena. This process consists largely of posing hypotheses and testing them with observational or experimental evidence. . . . Scientific research requires that we have some way of testing hypotheses based on experimental observational data. *The most important feature of scientific hypotheses is that they are testable* [emphasis his].[53]

The importance of this fact to the evolutionary community is manifest in the way in which it has been applied to creationist ideas:

> Science differs in this way [see quote above] from creationism, which does not use evidence to test its claims, does not allow evidence to shake its a priori commitment to certain beliefs, and does not grow in its capacity to explain the natural world. Unshakeable belief despite reason or evidence (i.e., faith) may be considered a virtue in a religious framework, but is precisely antithetical to the practice of science.[54]

In other words, since the most important feature of a scientific hypothesis is that it is testable, the seeming un-testability of the existence of God, of the supernatural creation of various creatures, and of a global flood a few thousand years ago has typically removed creationist ideas from the realm of science.

Some evolutionists have even taken the criticism of the creation model one step further. They have summed up creationist views in a short phrase: "God did it." Besides rejecting this phrase as unscientific, they have denounced it as *anti*-scientific. For example, let's say that you were testing a potential anti-cancer drug in the lab. If you were laboring over a confounding experimental result, "God did it" wouldn't seem to reveal an answer. At least, it wouldn't lead to discoveries on how the natural world operated. Rather, testable hypotheses would be the only scientific way forward toward a solution.

In light of this historical practice, we can revisit the evolutionary explanation of natural selection. The elimination of thousands of mtDNA mutations by natural selection might seem plausible. But to be scientific, this explanation would have to make testable predictions. For example, the mtDNA mutation rate in the most divergent African people groups (San peoples, Biaka peoples, etc.) has not yet been measured. Can the evolutionary explanation of natural selection *predict* what this rate will be? In other words, before the rate is actually measured, will evolutionists publish a guess as to what it will be? If not, is the evolutionary explanation scientific?

⁂

Curiously, the human mtDNA data that we've just discussed fits a model that many have previously discounted. In a previous section, I discussed the YEC geologists and astronomers who hold to a 6,000-year timescale for the earth and universe. Predicting mtDNA differences for *Homo* individuals over 6,000 years exactly captures both the average mtDNA differences among non-Africans and among Africans (Figure 7.6). The non-African differences were best predicted by a moderate generation time (i.e., about 30 years), and the African differences by a fast generation time (i.e., about 15 years) (Figure 7.6).[55]

Historical data offered an explanation as to why. Since mtDNA is inherited primarily — if not exclusively — through the maternal lineage, data on female generation times are the most rele-

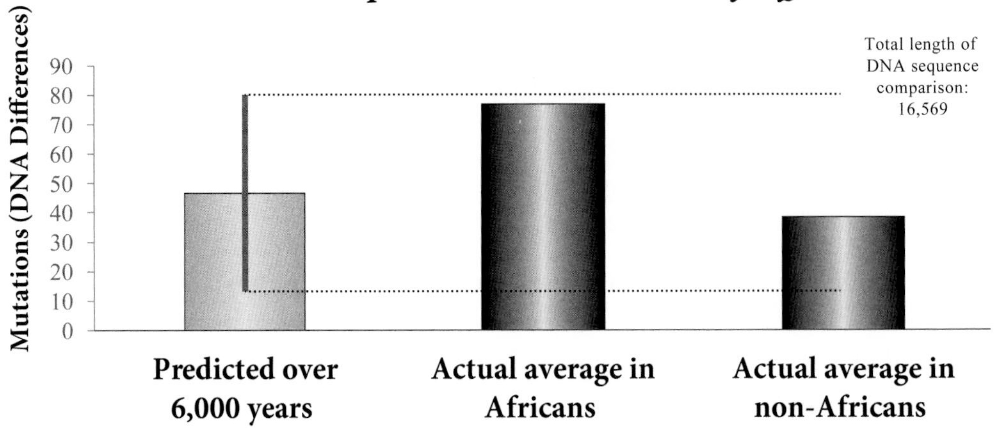

Figure 7.6. The 6,000-year timescale captures the mitochondrial DNA differences among modern humans. Mutations were predicted from the empirically determined human mitochondrial DNA mutation rate. The height of the column represents the average prediction. The black line approximates the 95% confidence interval. In short, if all human females were consistently age 50 when they gave birth, then only around 12 mitochondrial DNA differences (the lower end of the black line) would have accumulated over 6,000 years. Instead, if all human females were consistently age 15 when they gave birth, then around 80 mitochondrial DNA differences would have accumulated over 6,000 years. Today, around 80 mitochondrial DNA differences exist in African populations, and around 40 differences in non-African populations. Modified with permission from Figure 2 of N.T. Jeanson, "On the Origin of Human Mitochondrial DNA Differences, New Generation Time Data Both Suggest a Unified Young-Earth Creation Model and Challenge the Evolutionary Out-of-Africa Model," *Answers Research Journal*, 2016, 9:123–130.

vant to our analyses. United Nations marriage data from the 1970s revealed that women from African nations married younger than women from non-African nations (Table 7.2).[56] My mtDNA predictions suggest that this discrepancy was also true in the centuries preceding the 1900s.

Alternatively, these marriage data might simply be an artifact, and not a reflection of historical practices among African people groups. Conversely, some African lineages might mutate their mtDNA at a faster rate than non-African lineages. Measurement of a form of genetic change (recombination — see chapter 9) in a different DNA compartment (the nucleus — see chapter 8) suggests that Africans have faster rates of genetic change than non-Africans.[57] This might also be true in the mtDNA compartment.

As mentioned above, no direct measurement of the mtDNA mutation rate has been performed in the most divergent African people groups. I expect that the rate in these groups will be on the order of 1 mutation per 5 to 8 generations — or faster. In fact, I wouldn't be surprised if these divergent African lineages mutate twice as fast as the non-African lineages — 1 mutation per 2.5 to 4 generations.

Age bracket:	15–19	20–24	25–29	30–34	35–39	40–44	45–49	50–54	55–59	60–64	65+
% African women married	33	68	81	84	83	79	74	65	57	45	28
% non-African women married	11	47	70	77	79	77	74	68	62	52	32
Fold-difference	2.9	1.5	1.2	1.1	1.1	1.0	1.0	1.0	0.9	0.9	0.9

Table 7.2. Percentage of women who were married (data from 1970s), by age bracket and by geography. Adapted with permission from Table 1 of N.T. Jeanson, "On the Origin of Human Mitochondrial DNA Differences, New Generation Time Data Both Suggest a Unified Young-Earth Creation Model and Challenge the Evolutionary Out-of-Africa Model," *Answers Research Journal,* 2016, 9:123–130.

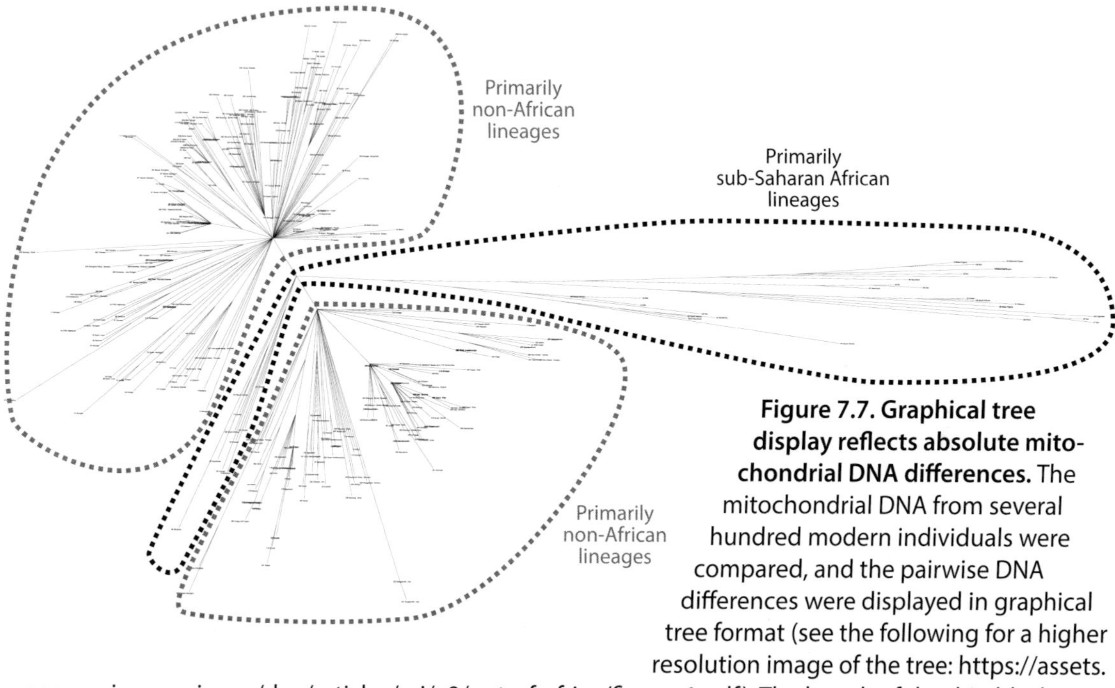

Figure 7.7. Graphical tree display reflects absolute mitochondrial DNA differences. The mitochondrial DNA from several hundred modern individuals were compared, and the pairwise DNA differences were displayed in graphical tree format (see the following for a higher resolution image of the tree: https://assets.answersingenesis.org/doc/articles/arj/v9/out-of-africa/figure-1.pdf). The length of the thin black lines connecting any two individuals is proportional to the number of mitochondrial DNA differences between them. In other words, long lines represent many DNA differences; short lines represent few. Consistent with what we observed in Figure 7.6, the longest lines were in the sub-Saharan African lineages. (Groups of lineages demarcated with thick dashed black and gray lines.) Adapted with permission from Figure 1 of N.T. Jeanson, "On the Origin of Human Mitochondrial DNA Differences, New Generation Time Data Both Suggest a Unified Young-Earth Creation Model and Challenge the Evolutionary Out-of-Africa Model," *Answers Research Journal*, 2016, 9:123–130.

In other words, the 6,000-year timescale makes testable predictions about the rate of mtDNA mutation.

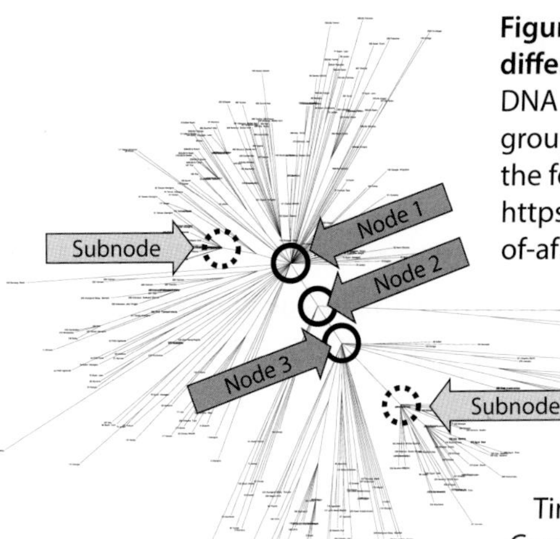

Figure 7.8. Three node structure of mitochondrial DNA differences among modern humans. The mitochondrial DNA differences displayed in Figure 7.7 fell into three major groups (nodes), with several subgroups (subnodes). (See the following for a higher resolution image of the tree: https://assets.answersingenesis.org/doc/articles/arj/v9/out-of-africa/figure-1.pdf.) Adapted with permission from Figure 1 of N.T. Jeanson, "On the Origin of Human Mitochondrial DNA Differences, New Generation Time Data Both Suggest a Unified Young-Earth Creation Model and Challenge the Evolutionary Out-of-Africa Model," *Answers Research Journal*, 2016, 9:123–130.

The 6,000-year timescale makes even more penetrating predictions. To see how, consider the structure of the mtDNA differences among various people groups. We can visualize differences between individuals in a tree-like diagram (Figure 7.7).[58] Just like a family tree, the length of the branches represent a particular distance. In family trees, the branch lengths correspond to genealogical distance — to how many generations separate two people. In the mtDNA tree (Figure 7.7), the branch lengths represent the number of mtDNA differences separating any two people. For example, we observed earlier that African lineages have about twice as many mtDNA differences as non-African lineages (Figure 7.6). Consistent with this fact, many African branches are about twice as long as the non-African branches (Figure 7.7).

Another fact is readily observable in the human mtDNA tree (Figure 7.8): Three major groupings are visually identifiable. In fact, these nodes are so iconic that the evolutionary community has assigned specific labels to them. In technical terms, they are referred to as *haplogroups*. The *L*, *M*, and *N* haplogroups[59] have been recognized for years.[60]

These haplogroups have characteristic features. For example, the L haplogroup consists almost exclusively of the sub-Saharan African lineages that we have already discussed. In contrast, the M and N haplogroups contain ethnic groups from the rest of the world. In addition, we just observed that the L haplogroup branch lengths are about twice as long as the M and N haplogroup branch lengths.

What do these features imply about our ancestors?

Again, since mtDNA genomes are inherited primarily — if not exclusively — through the maternal lineage, the mtDNA haplogroups reveal something about our ancient maternal ancestors.

Who were these women?

Our search for our ancestors thus far has focused exclusively on genetics. Since genetics is the only direct scientific record of our ancestry, this is appropriate. Any model of human origins must wrestle with what we've explored. Nevertheless, clues from other fields of science can inform the questions that remain.

In particular, several geologic clues intimate a major event in the last few thousand years, which would have had dramatic consequences for the human population. For example, when we look at the earth's surface on a global scale, we immediately recognize why our home is called the Blue Planet. Seventy percent of the earth's surface is covered with water. In fact, so much water exists on earth today that, if you were to take a gigantic bulldozer to the land and push the mountains into the seas, the entire earth would be covered with water to a depth of almost two miles. In other words, the oceans are deeper than the mountains are high.[61]

However, our ocean has not always existed in its current state. Some of this ocean appears to have once covered vast amounts of the continents. Today, on some of the highest peaks, fossil sea shells can be found. The explanation for the curious fact is straightforward. Before these shellfish were buried and fossilized, they existed in a shallow sea of some sort. Later, these fossil-rich sediments were pushed upward by tectonic forces, resulting in the fossils on high mountains, as we observe today.

Shellfish are not limited to mountaintops. The fossil record is filled with the remains of once aquatic creatures.[62] Even currently land-locked regions like Kansas were also once under water.[63] Moreover, whatever process caused the fossilization of creatures like these shellfish must have been extensive. For example, many of the earth's fossil-bearing layers stretch over entire continents. Some of the sandstone layers in North America reach from California to New England.[64] An effect as wide as the North American continent requires a cause at least equal in size.

The cause must also have been catastrophic in nature, rather than slow and gradual. Numerous peculiar fossils have been found that defy a tranquil fossilization event. For example, an *Ichthyosaur* mother in the process of giving birth was preserved in the rock layers (Figure 7.9). As another example, a fish was swallowing another fish before death ended its culinary endeavors

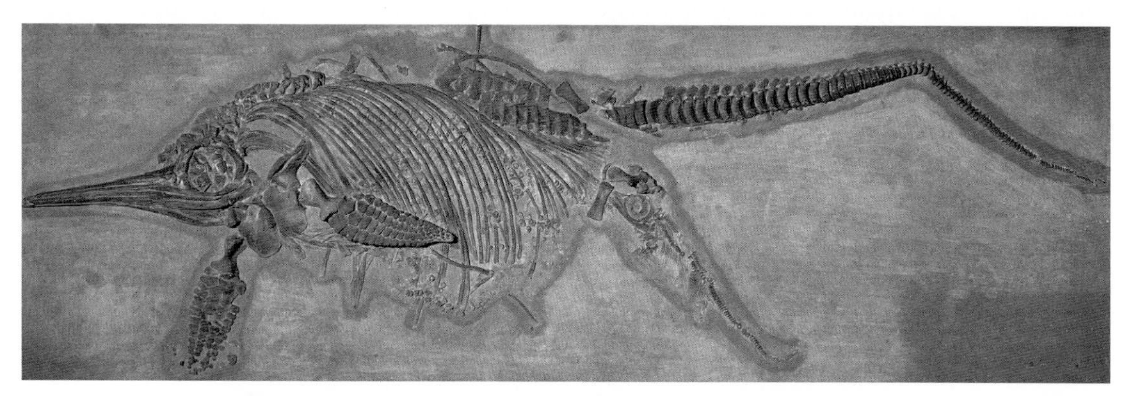

Figure 7.9. Fossil *Ichthyosaur* in the process of giving birth.

Figure 7.10. Fossil of a fish eating another fish.

(Figure 7.10). It's difficult to imagine that these creatures froze in position, slowly sank to the ocean bottom, and then were slowly covered with sediment in a manner that avoided scavenging or disturbance. Instead, a rapid catastrophic burial seems much more likely.

In addition to individual fossils, we now have data suggesting that entire fossil beds were the result of catastrophic burial.[65]

Together, these conclusions appear to conflict with traditional geologic interpretations of the earth's crust. Nevertheless, recent geologic research has augmented the above paleontological results with geologic evidence for catastrophic burial. For example, some of the layers in the Grand Canyon that were thought to be iconic examples of desert deposition now appear to be the result of catastrophic aquatic processes.[66]

A recent geologic catastrophe illustrated the types of mechanisms that may have been at work. When Mount St. Helens erupted in May of 1980, it began a series of geologic events that would span the next several years. The resultant volcanic flows and falling ash created a geologic landscape of remarkable character. Today, below the crater, a 100-foot canyon exists (Color Plate 76). Though a small stream runs through the canyon, the action of this slow water flow over innumerable ages did not carve the canyon. After 1980, a mud flow cut the gorge that uncovered these layers. Furthermore, the layers themselves are recent. They have all been laid down within the last several decades. In fact, despite the explosive nature of the eruption, very finely laminated layers resulted from events in 1980 and since then (Color Plate 77).[67]

The Mount St. Helens landscape is a 1/40 scale model of a much more familiar geologic landscape — the Grand Canyon (Color Plate 78). Though conventional geologic thinking puts the ages of the layers in the Grand Canyon at millions of years, the events at Mount St. Helens demonstrate the speed with which catastrophes can accomplish geologic work. The Grand Canyon doesn't require millions of years to form. It just requires a large enough catastrophe.

The catastrophic forces that formed the fossils layers around the world appear to have operated in the recent past. For example, soft tissue is recoverable from fossils.[68] Some of these fossils are dated by conventional means to hundreds of millions of years ago. Yet, we know from experience that soft tissue decays rapidly. It's difficult to imagine that dinosaur tissue could remain pliable after 65 million years of repose in the earth's crust.

The YEC geology community has offered an explanation for these phenomena. Deriving their model from a careful interpretation and application of Genesis 6–9, these YEC geologists invoke the global Flood of Noah for many of the observations we've just made. Obviously aquatic, globally extensive, catastrophic, likely accompanied by violent volcanic and tectonic activity,[69] the Flood matches the observations that we've made. Furthermore, the geologic model derived from the text of Genesis makes testable, accurate predictions.[70]

You might not agree with this geologic model. But consider the testable predictions that this model makes in the field of genetics. Chapters 6–11 of Genesis give detailed accounts of human population history around the time of the Flood. Specifically, the only human survivors were Noah, his wife, his three sons, each with his wife.[71] After the Flood, the latter three couples (i.e., the three sons and their wives) bore children, and their descendants repopulated the earth.

The implications of this model for human maternal ancestry are clear. Since mtDNA is inherited from mothers and *not* from fathers, the three sons of Noah would have inherited their mtDNA from Noah's wife. But then these sequences would have gone no further. These males would not have passed on their mtDNA sequence to their offspring. Instead, their descendants would have derived their mtDNA sequences from the wives of these three men.

The Book of Genesis offers no details on the identity of these three women. They might have been sisters. Or they might have been distant relatives. Regardless, there were three. Conversely, the fact that all mtDNA lineages trace to three major nodes (haplogroups) fits this fact (Figure 7.8).

Shortly after the Flood,* the post-Flood peoples were divided into major ethnolinguistic groups. Genesis 11 describes the post-Flood peoples as sharing one language, attempting to build a tower to heaven at a location called *Babel*, and then being forced to separate[72] as a result of sudden language confusion.**

Consistent with this post-Flood dispersal, many of the various major ethnic lineages genetically join one another in the mtDNA tree at the L, M, and N nodes, rather than forming a nested hierarchy back to these nodes (Figure 7.7).*** It's as if these lineages diverged suddenly from one another, rather than being related back to one another through a long series of mixing, intermarriages, and slow dispersal.

The relative branch lengths among these nodes also fits this narrative. In Genesis 5, the detailed genealogy from Adam to Noah's sons contains explicit time stamps for the number of years between each generation. Adding up these numbers yields a time from Adam to the Flood

* Likely just a few generations, based on the descendants listed in Genesis 10.

** Language diversification post-Babel would have resulted in the over 7,000 languages (https://www.ethnologue.com/) documented today.

*** This is more easily visible in the online pdf version of Figure 1 in the paper where it was originally published. Paper: N.T. Jeanson, "On the Origin of Human Mitochondrial DNA Differences, New Generation Time Data Both Suggest a Unified Young-Earth Creation Model and Challenge the Evolutionary Out-of-Africa Model," *Answers Research Journal,* 2016, 9:123–130, available online at https://answersingenesis.org/genetics/mitochondrial-dna/origin-human-mitochondrial-dna-differences-new-generation-time-data-both-suggest-unified-young-earth/; direct link to Figure: https://assets.answersingenesis.org/doc/articles/arj/v9/out-of-africa/Figure-1.pdf.

of around 1,650 years. Following the Flood, the genealogy of Genesis 11 traces the descendants of one of Noah's sons down to Abraham. With the explicit time stamps in Genesis 5 and Genesis 11, the time from Adam to Abraham is about 2,000 years. In the New Testament, Matthew 1 documents Abraham's genealogy to Jesus. From the Flood to Jesus, around 2,500 years passed.[73] From Jesus' birth until the present, about 2,000 years have passed. In other words, less than 1,700 years spanned the time period from Adam to the Flood, but around 4,500 years passed from the Flood until the present. This leads to an approximately 1.0:2.7 time ratio of pre-Flood time to post-Flood time.

However, the branch lengths of the mtDNA tree reflect DNA differences due to mutation, and mutations are transmitted *per generation*. In Genesis 5, the paternal generations between the act of creation and the Flood are listed, and very few generations pass. Because of the long lifespans and late child-bearing ages of the patriarchs, only around 10 generations passed in a period of 1,600 to 1,700 years. Similar long lifespans may have been true in the maternal lineages as well. If so, then the ratio of pre-Flood time to post-Flood time would be even more divergent.

This roughly 1:3 (or higher) time ratio fits the relative branch length proportions among the three nodes in the mtDNA tree. For example, if the three nodes represent the three wives of Noah's sons, then the branches connecting these women must represent mutations that occurred pre-Flood. Branches radiating out from these nodes would represent mutations that occurred post-Flood. Visually, it's immediately apparent that the proposed pre-Flood mutations are far fewer than the postulated post-Flood mutations (Figure 7.11).[74]

Thus, on three counts the mtDNA tree fits the young-earth creation model. First, the absolute DNA differences in this genetic compartment fits the 6,000-year timescale. Second, the existence of three nodes fits the fact of three surviving maternal ancestors at the time of the Flood. Third, the relative lengths of the branches connecting to and radiating out from these nodes, match the temporal expectations of Genesis 1–11, and the genealogies of the New Testament.

Now let's make the predictions of this model even more specific. If all modern human lineages trace their genetic differences to ancestors who lived only a few thousand years ago, then the history of civilization must be recorded in mtDNA. In other words, if what I've described is true, then the Roman and Mongol empires, the Greek conquests, the Persian Empire, and so many other events in recent history must have left their stamp in our mtDNA.

This hypothesis can be tested — and I'm just beginning to investigate this phenomenon. Some of these subnodes (Figure 7.8) might correspond to these recent historical events.

Up to this point, you may have noticed that I have said nothing about the YEC predictions for Neanderthal DNA. I did so deliberately. The explanation for these differences follows from what I just discussed. When Neanderthal and modern human sequences are visualized together in tree format, the Neanderthal sequences branch off of the sub-Saharan African lineages (Figure 7.12).[75] From the YEC perspective that I've just outlined, it would appear that this lineage derived from ancient Africans. Since some African people groups might mutate their

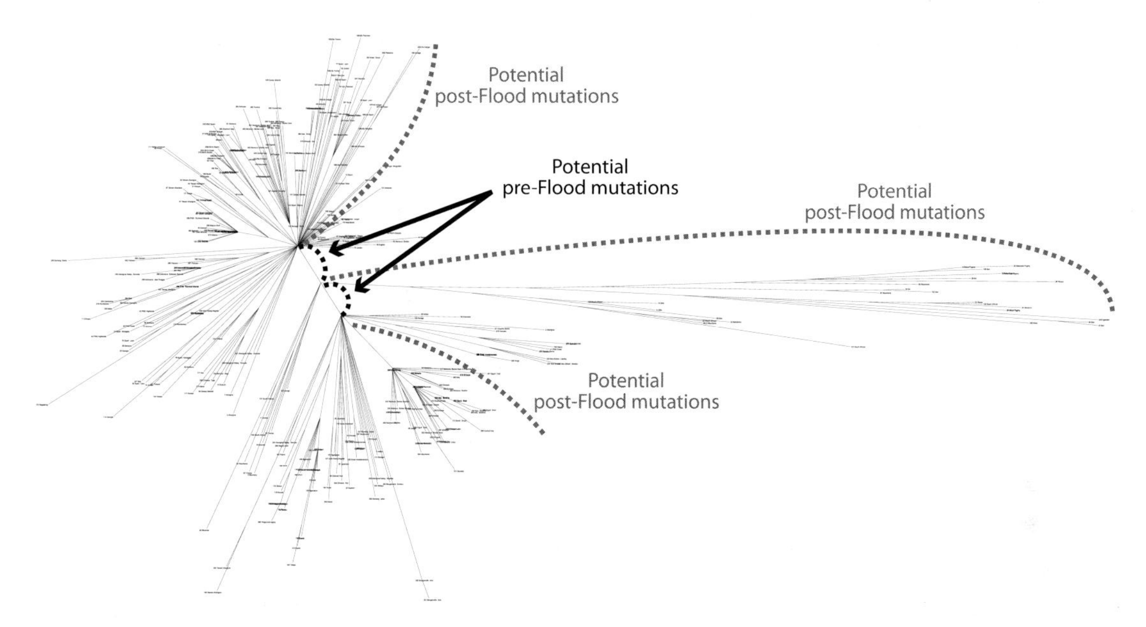

Figure 7.11. Relative branch lengths match expectations of young-earth creationist timescale. The lengths of the two branches that connected the three major nodes were relatively short, consistent with the shorter timescale (and likely fewer generations, due to long lifespans) pre-Flood. In contrast, the lengths of the branches radiating away from each of the three nodes were long, consistent with the longer timescale post-Flood. Adapted with permission from Figure 1 of N.T. Jeanson, "On the Origin of Human Mitochondrial DNA Differences, New Generation Time Data Both Suggest a Unified Young-Earth Creation Model and Challenge the Evolutionary Out-of-Africa Model," *Answers Research Journal*, 2016, 9:123–130. (See the following for a higher resolution image of the tree: https://assets.answersingenesis.org/doc/articles/arj/v9/out-of-africa/figure-1.pdf.)

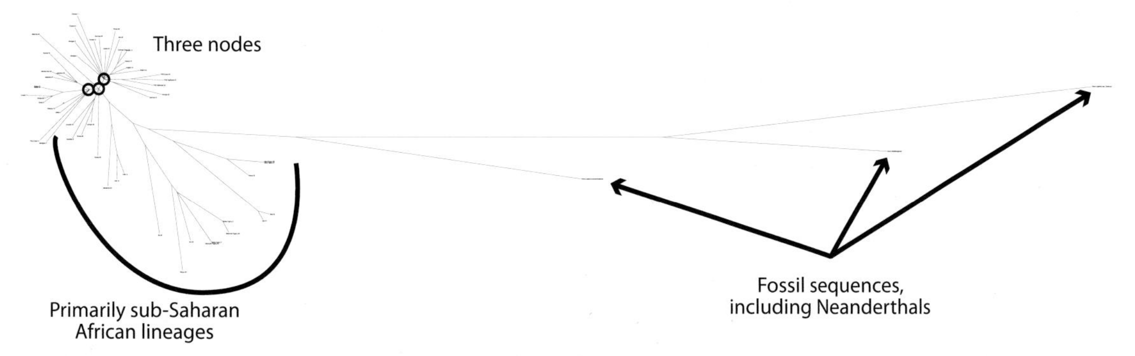

Figure 7.12. Fossil DNA sequences branch off primarily sub-Saharan African lineages. Fossil DNA sequences were the most divergent of all, reflected in the long branch lengths. However, they branched off of the primarily sub-Saharan African lineages, as if they were derived from African lineages. These fossil sequences might represent hyper-mutating sequences (which may have led to their extinction as a population), or they might represent degraded DNA. Adapted with permission from Figure 11C of N.T. Jeanson, "Mitochondrial DNA clocks imply linear speciation rates within 'kinds.' " *Answers Research Journal*, 2015, 8:273–304.

mtDNA faster than non-African people groups do, Neanderthal DNA might simply represent a hyper-mutating lineage — which eventually went extinct.

Alternatively, for technical reasons that I elaborate elsewhere,[76] the Neanderthal sequences might be too degraded to be reliable. In short, when I perform DNA sequence analyses in the lab, I tend to throw away DNA sequences that are older than a year. Despite storing them at -20° C, being 12 months removed from their normal cellular environment appears to do irreversible damage to DNA. How much more so when DNA sequences sit in fluctuating temperatures and environmental conditions for thousands of years. (My evolutionary colleagues disagree with my assessment regarding DNA degradation — which is why I still made predictions for Neanderthal DNA under the evolutionary model.)

Regardless of the actual explanation for Neanderthal sequences, the way to investigate these hypotheses is clear. It's the same method we would use to investigate any hypothesis. If someone thinks that they have an explanation for Neanderthal mtDNA, I would ask them what testable predictions their hypothesis makes. The hypothesis of an ancient timescale fails to make accurate predictions. My explanation of DNA degradation stems from the successful match between the predictions of the YEC timescale and mtDNA differences among modern humans. This may seem mundane. But can any other explanation do better?

❧ ❧ ❧ ❧

An alternative explanation for the results in the preceding section is that these successful YEC predictions simply represent an anomaly. After all, each of the mtDNA predictions were made using the mutation rate from a single species — humans. Though over 15 studies have attempted to measure the human mtDNA mutation rate, one species is not a statistically robust dataset. Furthermore, a 6,000-year timescale for the origin of humans flies in the face of the conventional scientific model.

Do similar results hold true in other species?

The breed-species comparisons (see chapters 5–6) suggest a natural place to start. As a general rule, creationists and evolutionists agree that species within vertebrate families share a common ancestor. These two origins views differ, however, on the timescale over which these speciation events occurred. Hence, by comparing mtDNA sequences among species within vertebrate families, we could test the general applicability of the human mtDNA results.

To date, only three vertebrate species possess a published mtDNA mutation rate — mice (*Mus musculus*),[77] chickens (*Gallus gallus*),[78] and Adélie penguins (*Pygoscelis adeliae*).[79] At current rates of mtDNA mutation,[80] neither the 6,000-year timescale nor the evolutionary timescale[81] captures mtDNA diversity among species within these families (Table 7.3).[82]

However, unlike the human mtDNA mutation rate studies, only a single study has been performed for each of these species. Over the last two decades of human mtDNA mutation rate research, the published papers often strongly contradicted one another. Only when each studied

	Published mutation rate (mutations per base pair per million years)	Converted mutation rate (mutations per mtDNA genome per year)	Predicted mutations (YEC timescale)	Predicted mutations (evolutionary timescale)	Actual mtDNA differences (average of species within the family)	Actual mtDNA differences (maximum pairwise species difference within the family)
Mus musculus (mouse)	0.23	0.0036	33	79,685	2,346	3,034
Gallus gallus (chicken)	0.04 to 1.12	0.00063 to 0.0176	6 to 158	46,517 to 1,302,489	2,007	2,567
Pygoscelis adeliae (Adélie penguin)	0.29 to 0.88	0.0046 to 0.014	55 to 168	156,774 to 475,728	1,570	1,763

Table 7.3. Under the assumption of constant rates of mitochondrial DNA mutation, neither the young-earth creationist model nor the evolutionary model accurately predicted mitochondrial DNA differences in vertebrate species.

was weighed by its statistical strength did a clear answer emerge. Similar statistical rigor might be required to elucidate the answer for these three vertebrate species.

Alternatively, the results in each of these species might remain unchanged. If so, then both timescales have a significant explanatory challenge — the evolutionary model would have to find a way to explain why so few mutations are seen today, whereas the 6,000-year model would have to explain why so many mutations are observed at present.

Could an alternate timescale explain these data? At a rate of 0.0036 mutations per mtDNA genome per year, we can calculate how many years would be required to produce the average number of mtDNA differences within the mouse family. Dividing 2,346 mtDNA differences by 0.0036 (mutations per year) and by 2 (because this is a divergence equation) yields a date of about 326,000 years ago. This is obviously in excess of the 6,000-year timescale. But it also severely conflicts with the paleontological data, on which the evolutionary timescale is based.

Similar calculations for chickens and penguins lead to the same conclusion. At a rate of 0.00063 to 0.0176 mutations per year (i.e., the chicken mutation rate), a total of 2,007 mutations could be produced via species divergence (i.e., my reason for including a factor of 2) in about 57,000 to 1.6 million years. At a rate of 0.0046 to 0.014 mutations per year (i.e., the penguin mutation rate), a total of 1,570 mutations could be produced via species divergence (i.e., my reason for including a factor of 2) in about 56,000 to 171,000 years. Again, these dates are not comfortable fits with either the creationist or the evolutionary timescales.

What could possibly explain these vertebrate results? Could natural selection reconcile the evolutionary predictions with reality? If so, what testable predictions does this model make? For example, the mtDNA mutation rate has not yet been measured in tens of thousands of other vertebrate species. What predictions does the evolutionary model make for these?

With respect to the 6,000-year timescale, a potential solution awaited another genetic discovery (see chapter 10).

⁂ ⁂ ⁂ ⁂

Vertebrates are not the only species possessing published mtDNA mutation rates. To date, rates are known for five invertebrate species and one fungal species — three roundworm species (*Caenorhabditis elegans*[83] (Figure 7.13), *Caenorhabditis briggsae*,[84] and *Pristionchus pacificus*[85] (Figure 7.14)), a fruit fly species (*Drosophila melanogaster*[86] (Figure 7.15)), a water flea species (*Daphnia pulex*[87] (Figure 7.16)), and baker's yeast (*Saccharomyces cerevisiae*[88] (Figure 7.17)).

Among the five genera to which the species above belong, only four[89] contain additional species possessing a published mtDNA genome sequence — *Caenorhabditis*, *Drosophila*, *Daphnia*, and *Saccharomyces*. In the genus *Caenorhabditis*, a total of three species have curated mtDNA sequences.[90] Ten additional *Drosophila* species have curated mtDNA sequences.[91] Two additional *Daphnia* species exist in the curated database.[92] In the genus *Saccharomyces*, eleven additional species possess curated mtDNA genome sequences.[93] However, for technical reasons that I explain in the endnotes,[94] only a handful of these 11 yeast species lend themselves to straightforward DNA analysis.

Our analysis of these mtDNA sequences employs the same methodology that we used for analyzing the vertebrate mtDNA sequences — with one exception. As we discovered in chapter 5 (see also the endnotes in chapter 5), modern creationists have a more nuanced view of speciation in invertebrates than they do of speciation in vertebrates. For invertebrates, modern creationists don't necessarily endorse common ancestry of species within the same family. But most would be comfortable with the common ancestry of species within the same genera. Thus, our calculations of mtDNA diversity within invertebrate genera allows us to compare the expectations of the YEC model side-by-side with the expectations of the evolutionary model.

Again, as we did for the human mutation rate calculations, the mutation rates in each of these four species must be converted from units of *mutations per generation* to *mutations per year*. Unlike humans, our knowledge of the generation times in these species is based primarily on the artificial environments of the laboratory. In the wild, each of the species can enter a state of dormancy. For example, dormancy in the roundworm *Caenorhabditis elegans* is termed the *daur* stage.[95] In the fruit fly[96] *Drosophila melanogaster* and in the water flea[97] *Daphnia pulex*, reproductive slowing is termed *diapause*. Baker's yeast can form spores.[98]

Practically, this means that generation times in the wild could be slower than the generation times in the laboratory.[99] How much slower depends on past environmental conditions. Since detailed histories of these conditions are not available, precise determination of historical generation times in the wild for these species is not possible. Nevertheless, the possibility of slower generation times can be incorporated into our calculations by predicting mutations for the full range of generation times — both for generation times observed in the lab and for generation

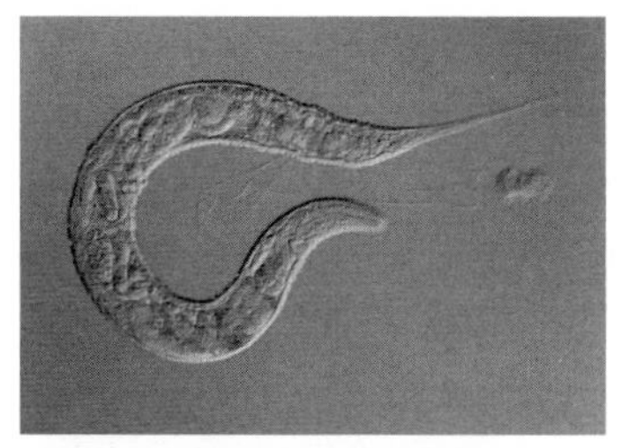

Figure 7.13. Image of the roundworm, *Caenorhabditis elegans.*

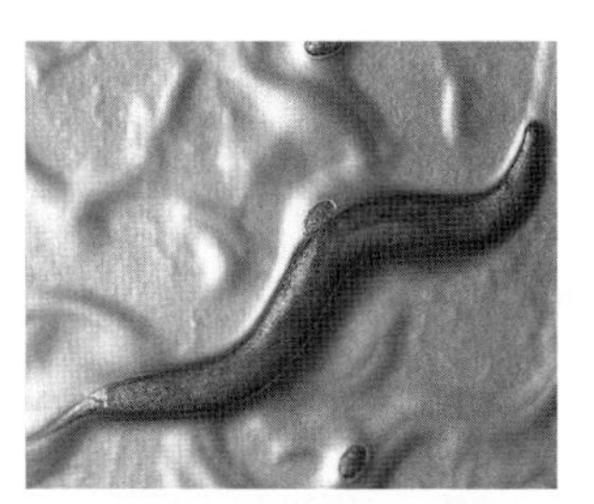

Figure 7.14. Image of the roundworm, *Pristionchus pacificus.*

Figure 7.15. Image of the fruit fly, *Drosophila melanogaster.*

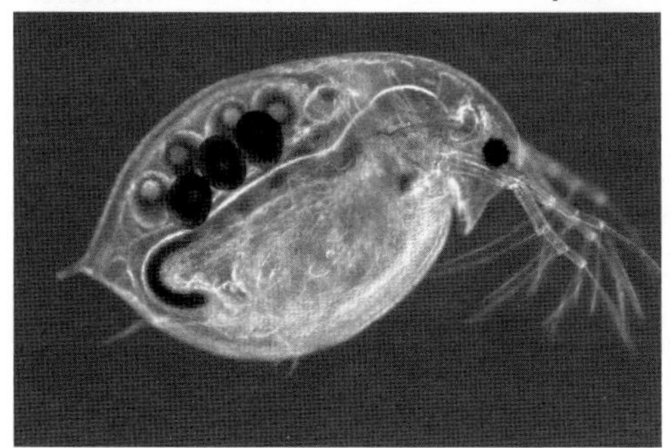

Figure 7.16. Image of the water flea, *Daphnia pulex.*

Figure 7.17. Image of baker's yeast (*Saccharomyces cerevisiae*). Image obtained with electron microscopy.

times estimated in the wild. I estimated generation times in the wild by dividing the laboratory generation times by a factor of 10.

Divergence predictions also require a timescale over which to make predictions. According to the evolutionary timescale, the three roundworm species began diverging around 18 million years ago.[100] Among the eleven fruit fly species, the evolutionary timescale spans around 50 million years.[101] For the three water flea species, the evolutionary timescale puts their divergence around 7.6 million years ago.[102] Evolutionists stretch the origin of the four yeast species (i.e., the only *Saccharomyces* species amenable to my analyses; see endnotes) over 15 million years.[103]

Multiplying each of these timescales by the appropriate mutation rate yields evolutionary predictions for mtDNA differences among the species within their respective genera. For example, differences among roundworm species can be predicted by multiplying the roundworm mutation rates (after converting to the appropriate units) by 18 million years and by a factor of 2 (since this is a divergence calculation). Similar calculations apply to the fruit fly group of species, the water flea group of species, and the yeast group of species.

In every case, the evolutionary timescale predicted mutations far in excess of mtDNA differences observed today (Figures 7.18, 7.19, 7.20, 7.21). Among roundworm species, a minimum of 45,000 mutations would have occurred in 18 million years (Figure 7.18).[104] Were generation times in the wild closer to generation times in the lab, several million mutations would have occurred in 18 million years. Neither of these predictions comes close to the less than 2,000 mtDNA differences that separate roundworm species today. Among fruit fly species, a minimum of 45,000 mutations would have occurred in 50 million years (Figure 7.19).[105] Again, if laboratory generation times approach generation times in the wild, millions of mutations would have been the result. Yet less than 2,000 mtDNA differences separate fruit fly species today. Among water flea species, at least 22,000 mutations would have occurred, but potentially up to several million (Figure 7.20).[106] Less than 5,000 mtDNA differences separate water flea species today. Among yeast species, at least 2 million mutations would have occurred over the 15-million-year timescale for the origin of yeast species (Figure 7.21).[107] Less than 20,000 mtDNA differences exist among the yeast species used in this analysis.

In other words, none of these predictions came close to the actual number of differences observed among these species (Figures 7.18– 7.21). In the best case, the evolutionary predictions for the minimum number of mutations in water flea species were nearly 5 times higher than the actual number of differences. In the other three cases, the predictions of the minimum number of mutations were around 20 to 100 times higher than the actual differences.

Like the situation we observed for the human-chimpanzee prediction, all of these calculations predicted a mutation number that exceeded the number of base pairs that were actually compared (Figures 7.18–7.21). Practically, this means that every single base pair in the mtDNA genomes of these creatures would have been mutated. In fact, some base pairs might have been mutated several times over.

Thus, in at least five independent measurements of the mutation rate — in humans, roundworms, fruit flies, water fleas, yeast — the evolutionary model failed to predict modern DNA differences. Furthermore, these species represented a very diverse swath of life. From a classification perspective, not only were separate *phyla* represented — Chordata (humans), Nematoda (roundworms), Arthropoda (fruit flies, water fleas) — but separate *kingdoms* (Animalia = humans, roundworms, fruit flies, water fleas; Fungi = yeast) were as well.

Consequently, the evolutionary failure becomes all the more difficult to dismiss as a statistical or biological anomaly. Instead, the results suggest a systematic problem with the evolutionary model.

In contrast, the YEC model successfully explained mtDNA differences among these animal and fungal species.[108] Just like the results for modern humans, I found similar congruence between prediction and fact. After 6,000 years of mtDNA mutation, a few thousand mtDNA differences should separate roundworm species from one another. This is exactly what we see today (Figure 7.22). The same agreement between prediction and fact held true in fruit flies, water fleas, and yeast (Figures 7.23, 7.24, 7.25).

Conservatively, these results do not necessarily imply that mtDNA differences are explicable over a 6,000-year timescale among species within a genus. Rather, several of these predictions captured modern diversity only under laboratory generation times. The *Drosophila* predictions are the strongest example of this fact — only the upper end of the prediction overlapped average

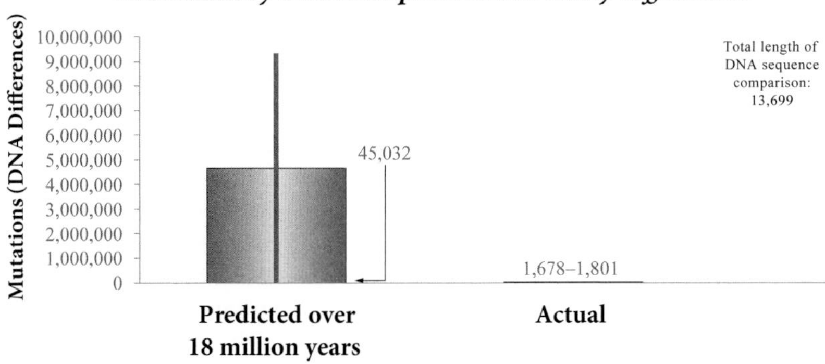

Figure 7.18. The evolutionary timescale predicts too many mitochondrial DNA differences among roundworm species. Single base pair mutations were predicted from the empirically determined roundworm mitochondrial DNA mutation rate. The height of the column represents the average prediction. The black line approximates the 95% confidence interval. In short, if roundworms consistently underwent only 9 generations per year, then only around 45,000 mitochondrial DNA mutations (the lower end of the black line) would have occurred over 18 million years. Instead, if roundworms consistently underwent 104 generations per year, then over 9 million mitochondrial DNA mutations would have occurred over 18 million years. Today, less than 2,000 single base pair mitochondrial DNA differences separate roundworm species in the genus *Caenorhabditis* (numbers above "Actual" column represent one standard deviation on either side of the average).

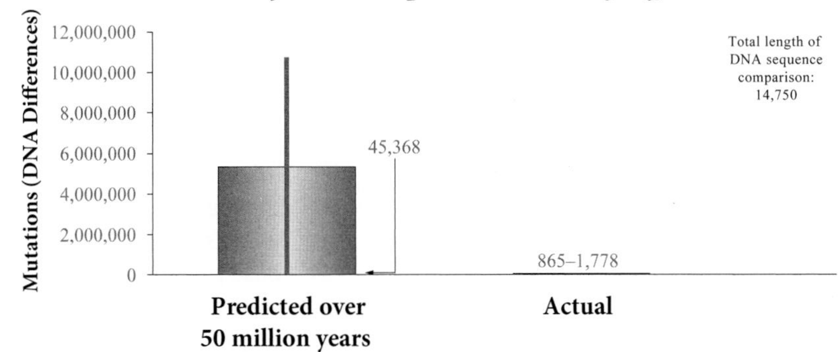

Figure 7.19. The evolutionary timescale predicts too many mitochondrial DNA differences among fruit fly species. Single base pair mutations were predicted from the empirically determined fruit fly mitochondrial DNA mutation rate. The height of the column represents the average prediction. The black line approximates the 95% confidence interval. In short, if fruit flies consistently underwent only 2 generations per year, then only around 45,000 mitochondrial DNA mutations (the lower end of the black line) would have occurred over 50 million years. Instead, if fruit flies consistently underwent 52 generations per year, then over 10 million mitochondrial DNA mutations would have occurred over 50 million years. Today, less than 2,000 mitochondrial DNA differences separate fruit fly species in the genus *Drosophila* (numbers above "Actual" column represent one standard deviation on either side of the average).

mtDNA diversity within the genus. Since laboratory generation times might be faster than what generation times have historically been in the wild, a more conservative conclusion would be that mtDNA differences are explicable over the 6,000-year timescale at, perhaps, the subgenus level.

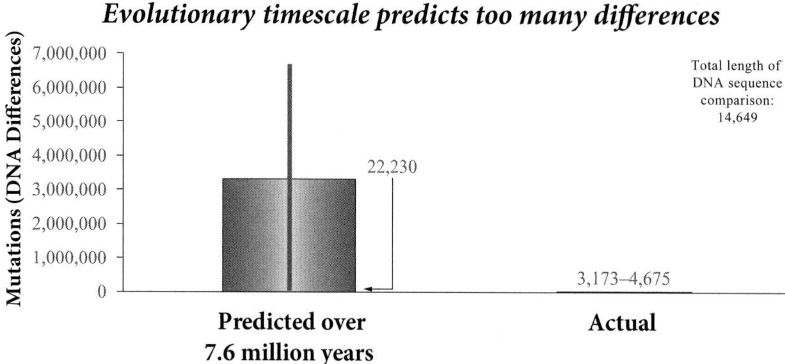

Figure 7.20. The evolutionary timescale predicts too many mitochondrial DNA differences among water flea species. Mutations (both single base pair and indels) were predicted from the empirically determined water flea mitochondrial DNA mutation rate. The height of the column represents the average prediction. The black line approximates the 95% confidence interval. In short, if water fleas consistently underwent only 2 generations per year, then only around 22,000 mitochondrial DNA mutations (the lower end of the black line) would have occurred over 7.6 million years. Instead, if water fleas consistently underwent 73 generations per year, then over 6 million mitochondrial DNA mutations would have occurred over 7.6 million years. Today, less than 5,000 mitochondrial DNA differences (both single base and indels) separate water flea species in the genus *Daphnia* (numbers above "Actual" column represent one standard deviation on either side of the average).

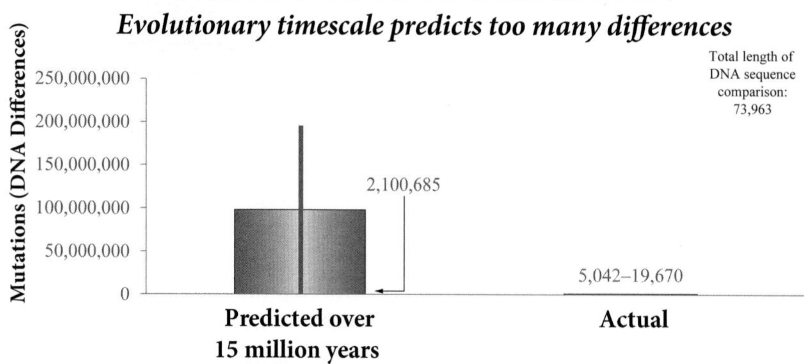

Figure 7.21. The evolutionary timescale predicts too many mitochondrial DNA differences among yeast species. Single base pair mutations were predicted from the empirically determined yeast mitochondrial DNA mutation rate. The height of the column represents the average prediction. The black line approximates the 95% confidence interval. In short, if yeast consistently underwent only around 526 generations per year (i.e., a generation time of 16 to 17 hours), then over 2 million mitochondrial DNA mutations (the lower end of the black line) would have occurred over 15 million years. Instead, if yeast consistently underwent 5260 generations per year (i.e., a generation time of 100 minutes), then almost 200 million mitochondrial DNA mutations would have occurred over 15 million years. Today, less than 20,000 single base pair mitochondrial DNA differences separate yeast species in the genus *Saccharomyces* (numbers above "Actual" column represent the range of pairwise mtDNA differences among various *Saccharomyces* species).

Technically, the data I just showed is not really a prediction. All I've done is taken published data, run them through an equation, and compared the results to other published data. The most appropriate term for this is a *retro*diction, rather than a *pre*diction.

Nevertheless, the agreement between my retrodictions and existing data was so strong that I'm confident that I can make bona fide predictions. Though these four groups of species represent a diverse sample of life, over a million invertebrate and fungal species still exist in which a mtDNA sequence has yet to be obtained, and in which the mtDNA mutation rate has yet to be measured.

Practically, the mtDNA sequence of a species is usually obtained and published before the mutation rate is measured. Consequently, the realm of predictions will be in the arena of mutation rates rather than in the arena of mtDNA differences. In other words, I'm confident that I can predict the mutation rate in the over 1 million invertebrate and fungal species in which the mtDNA mutation rate has yet to be measured.

In light of the successful retrodictions I made above, the predictive formula is easily derived. It's an algebraically rearranged divergence equation. In specific mathematical terms, I predict that the empirically measured mutation rate in fungal and invertebrate species will match the following equation:

mtDNA differences among species within a subgenus / 6,000 years / 2 =
mtDNA mutation rate per year.

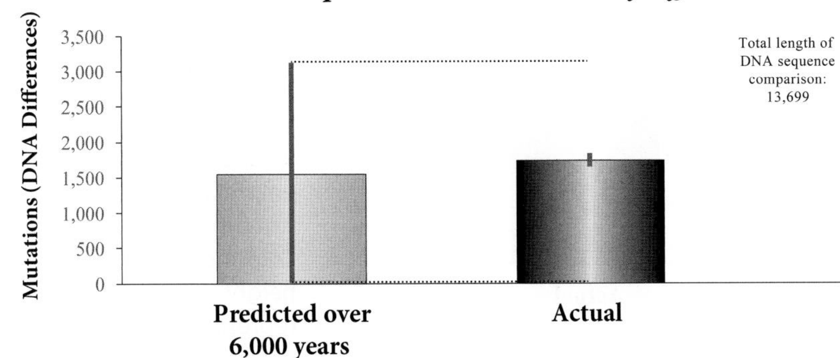

Figure 7.22. The 6,000-year timescale captures the mitochondrial DNA differences among roundworm species. Single base pair mutations were predicted from the empirically determined roundworm mitochondrial DNA mutation rate. The height of the column represents the average prediction. The black line in the "Predicted over 6,000 years" column approximates the 95% confidence interval; the black line in the "Actual" column represents one standard deviation on either side of the average. In short, if roundworms consistently underwent only 9 generations per year, then only around 15 mitochondrial DNA mutations (the lower end of the black line) would have occurred over 6,000 years. Instead, if roundworms consistently underwent 104 generations per year, then over 3,000 mitochondrial DNA mutations would have occurred over 6,000 years. Today, less than 2,000 single base pair mitochondrial DNA differences separate roundworm species in the genus *Caenorhabditis*.

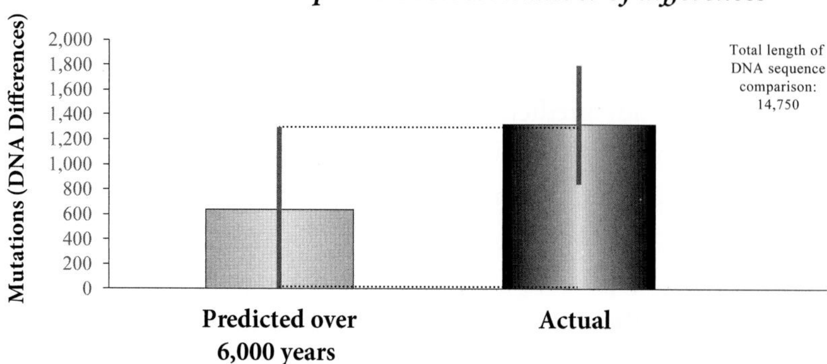

Figure 7.23. The 6,000-year timescale captures the mitochondrial DNA differences among fruit fly species. Single base pair mutations were predicted from the empirically determined fruit fly mitochondrial DNA mutation rate. The height of the column represents the average prediction. The black line in the "Predicted over 6,000 years" column approximates the 95% confidence interval; the black line in the "Actual" column represents one standard deviation on either side of the average. In short, if fruit flies consistently underwent only 2 generations per year, then only around 5 mitochondrial DNA mutations (the lower end of the black line) would have occurred over 6,000 years. Instead, if fruit flies consistently underwent 52 generations per year, then over 1,200 mitochondrial DNA mutations would have occurred over 6,000 years. Today, less than 2,000 single base pair mitochondrial DNA differences separate fruit fly species in the genus *Drosophila*.

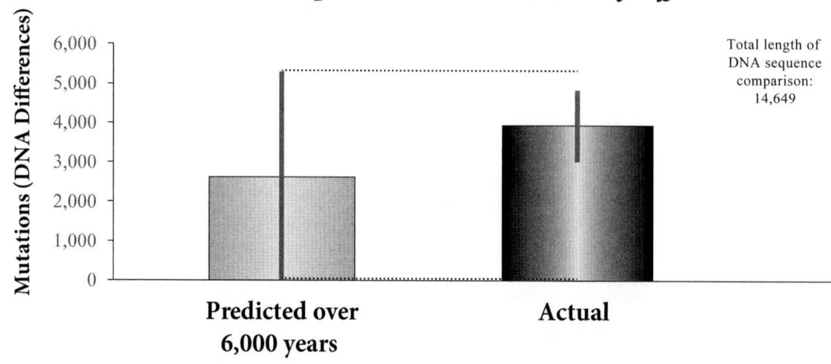

Figure 7.24. The 6,000-year timescale captures the mitochondrial DNA differences among water flea species. Mutations (both single base pair and indels) were predicted from the empirically determined water flea mitochondrial DNA mutation rate. The height of the column represents the average prediction. The black line in the "Predicted over 6,000 years" column approximates the 95% confidence interval; the black line in the "Actual" column represents one standard deviation on either side of the average. In short, if water fleas consistently underwent only 2 generations per year, then only around 18 mitochondrial DNA mutations (the lower end of the black line) would have occurred over 6,000 years. Instead, if water fleas consistently underwent 73 generations per year, then over 5,000 mitochondrial DNA mutations would have occurred over 6,000 years. Today, less than 5,000 mitochondrial DNA differences (both single base pair and indels) separate water flea species in the genus *Daphnia*.

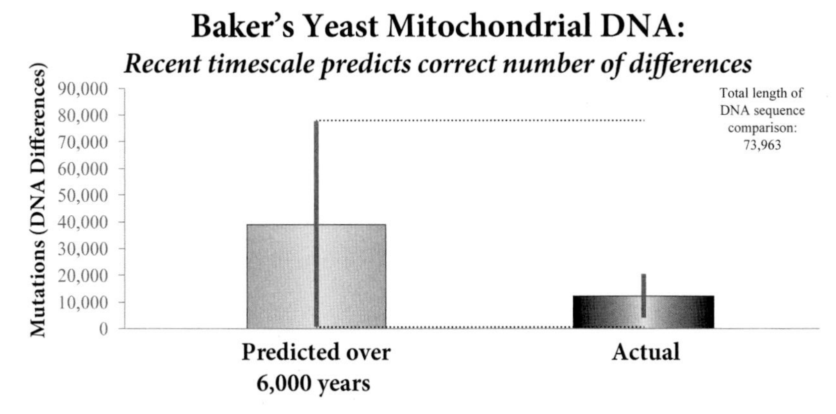

Figure 7.25. The 6,000-year timescale captures the mitochondrial DNA differences among yeast species. Single base pair mutations were predicted from the empirically determined yeast mitochondrial DNA mutation rate. The height of the column represents the average prediction. The black line in the "Predicted over 6,000 years" column approximates the 95% confidence interval; the black line in the "Actual" column represents the range of pairwise mtDNA differences among various *Saccharomyces* species. In short, if yeast consistently underwent only around 526 generations per year (i.e., a generation time of 16 to 17 hours), then only 840 mitochondrial DNA mutations (the lower end of the black line) would have occurred over 6,000 years. Instead, if yeast consistently underwent 5260 generations per year (i.e., a generation time of 100 minutes), then almost 80,000 mitochondrial DNA mutations would have occurred over 6,000 years. Today, less than 20,000 single base pair mitochondrial DNA differences separate yeast species in the genus *Saccharomyces*.

Naturally, this prediction is only as good as the terms that define it. For example, this equation contains a classification term — *subgenus*. We've already observed in chapter 6 that the classification term *species* is defined multiple ways. This fuzziness in terms extends to higher classification ranks.

In other words, the final and full classification description of each species or of groups of species has not yet been determined. As a specific example, in February of 2015, I downloaded the entire mammalian classification database from one of the major conservation organizations, the International Union for Conservation of Nature and Natural Resources (IUCN).[109] Two years later, I downloaded the entire mammal classification database again. In the 2017 version, more than 40 additional mammal species were listed. Furthermore, the 2017 database contained two more mammal families. The latter was not due to the sudden discovery of a whole new set of mammal species that belonged to a hitherto undiscovered family. Rather, it was due to the recognition that several known bat species had been reclassified. Formerly, these species had been included in a large bat family. Recently, most of the bat species in this large family were retained, while a handful were moved into two newly created families.[110] Even family-level classifications are still being defined.

If family-level classifications can change from year to year, then surely subgenus assignments can change as well. Furthermore, at least one of these bat species reassignments was based in part on mtDNA data. Thus, since my formula above deals explicitly with mtDNA, it contains a level of fuzziness on two counts. My prediction above is only as precise as the term *subgenus* will allow, and part of the precision of the term *subgenus* depends on mtDNA comparisons.

Despite these caveats, a clear picture emerged from all of these invertebrate and fungal mtDNA analyses. In each of these cases, the results were consistent. The evolutionary model failed to make accurate retrodictions, whereas the 6,000-year timescale explained modern diversity.

These results suggested that the observations that we made of human mtDNA were not an anomaly.

By the time that mtDNA sequence analyses were completed, two lines of biological evidence had accumulated *against* a millions-of-years timescale and *for* a recent timescale. Darwin's breed-species analogy suggested that the speciation process could happen quickly — much more quickly than commonly believed. However, because of the history of genetics, Darwin's argument was necessarily limited to visible traits. Genetics wasn't even a scientific term in 1859. Nevertheless, once mtDNA sequences were obtained for various species, the low levels of mtDNA diversity added genetic support to Darwin's analogy.[111]

However, this support went beyond simple numerical agreement. The mtDNA findings contained in this chapter called into question the entire foundation of the evolutionary timescale. For example, in the fields of geology and astronomy, the entire millions-of-years paradigm rests on the assumption that rates of change have been largely constant.[112] Yet, in the field of genetics, the assumption of constant rates of change (i.e., of mutational change) yields a 6,000-year timescale, not an ancient one. If constant-rate assumptions reject a millions-of-years timescale in the field of genetics, why should these assumptions be the only ones allowed in the fields of geology and astronomy?[113]

Thus, in jigsaw puzzle terms, the discovery of the mtDNA clock represented the discovery of a corner piece. It put a dramatic constraint — a 6,000-year timescale — on how the edge pieces (the DNA differences among species) could be related to one another. Furthermore, in other fields of science that were related to this question (e.g., geology), it constrained the types of explanations that could be invoked as well.

Actually, my use of a singular noun is incorrect. The discovery of the mtDNA clock in these six species (humans, fruit flies, water fleas, yeast, and two species of roundworms) represented the discovery of six different corner pieces. In other words, the discovery of the mtDNA clock represented the discovery of a *category* of corner pieces.

If similar results hold true in other species — especially in vertebrate species — the shape of the puzzle will eventually begin to emerge. For example, if millions of species show mutation rates consistent with the 6,000-year timescale, then this result will reveal the existence of millions of corner pieces. The sheer number of these pieces would imply that the edge pieces set the boundaries for multiple, independent, disconnected puzzles. How else could all these corner pieces fit into a unified whole? Perhaps the size of these individual puzzles will allow for species within a family to have a common ancestor. But connections beyond this would seem unlikely.

At this point, you might wonder how the discoveries in such a tiny genetic compartment — mtDNA — could have such massive ramifications. Furthermore, you might be skeptical that the mutation rates from just six species could turn the tables in so many scientific fields.

These concerns become less pressing upon four reflections. First, recall the conclusions we reached in chapters 2–3 on the question of the origin of species. We discovered that the most important scientific field in this investigation is genetics. Consequently, since mtDNA is a *genetic* dataset, this fact alone elevates its importance in the larger debate. In other words, the discoveries recorded in this chapter could not have come from a more relevant scientific discipline.

Second, recall from chapters 4–6 the central mechanism of evolution — descent with modification. Though Darwin had no knowledge of genetics, his scientific descendants have put his mechanism in concrete genetic terms. In modern evolutionary theory, DNA mutations are the driving force behind evolutionary change. To be sure, natural selection filters out certain DNA mutations and non-randomly shapes and sculpts the DNA sequences that we observe today. But mutations are at the heart of evolutionary change.

Yet, in this chapter, we observed that the mutation rate was not predicted by the evolutionary model. Instead, it was the 6,000-year timescale that made accurate retrodictions and even made *pre*dictions. In other words, in the branch of genetics that is most relevant to the evolutionary model, it's actually the creationists who have taken the lead.

Third, recall the common criticism that evolutionists have historically levied against creationists: evolutionists have said that the explanation "God did it" doesn't advance science and doesn't make testable predictions. Yet, as we've just observed, the 6,000-year timescale does, in fact, make testable and falsifiable predictions. By the evolutionists' own standard for what qualifies as science, this fact should bring the 6,000-year model into the discussion of the origin of species. A reversal this significant has very wide implications for the larger origins debate.

Fourth, recall some of the specific experiments from the history of genetics that we explored in chapters 2–3. The critical genetic discoveries were not made with massive experimental datasets that spanned hundreds of thousands of species. Rather, they were obtained with a few select organisms — Mendel's pea plants, the handful of species from Sutton and colleagues, Griffith's bacteria, Hershey and Chase's bacteriophages. In the years following each of these discoveries, the results specific to each of these organisms were generalized. But paradigm shifts in biological thought rested on a fulcrum of observations from a very small group of creatures.

Similarly, though the results in this chapter involved mtDNA mutation rates from only six species, they were consistent across a very diverse set of biology. Again, in classification terms, these six species belong to two separate kingdoms (Animalia, Fungi), several animal phyla (Chordata — humans, Nematoda — roundworms, Arthropoda — fruit flies, water fleas), and two major arthropod divisions (Insecta — fruit flies, Crustacea — water fleas). Aside from the puzzle of the three vertebrate species (mice, chickens, and penguins), these results suggested that the 6,000-year timescale would apply in general across life.

The discovery of the mtDNA clock was a category of corner piece that no one anticipated.

Chapter 8
A Preexisting Answer

When I assemble a rectangular jigsaw puzzle, the discovery of the first corner piece often has immediate ramifications. Sometimes it allows me to quickly connect and orient two rows of edge pieces. If my puzzle came without a box cover, this newfound orientation and shape for the puzzle would be a massive advance. It would also begin to suggest the identity of the remaining corner pieces.

In a similar way, the discovery of the first category of biological corner piece had several ramifications. For instance, in the previous chapter, direct measurements of mutation rates were key to uncovering the origin of DNA differences in the mitochondria. The success of this finding suggested that mutation rates would be useful to solving the puzzles in other genetic compartments, like the nucleus (Figure 8.1). Furthermore, these mtDNA results were obtained with a 6,000-year timescale. Naturally, these findings suggested a timescale over which nuclear DNA differences might be explained.

Actually, the mitochondrial results did much more than *suggest* answers to the question of nuclear DNA differences. The intersection between these two compartments had unanticipated implications that went far beyond either.

❧ ❧ ❧ ❧

Unlike mtDNA, nuclear DNA has been much more difficult to compare across species. For example, even among some species within the same genus (e.g., *Caenorhabditis*), the nuclear genome sequences are so different that rigorous comparisons are challenging to perform.[1]

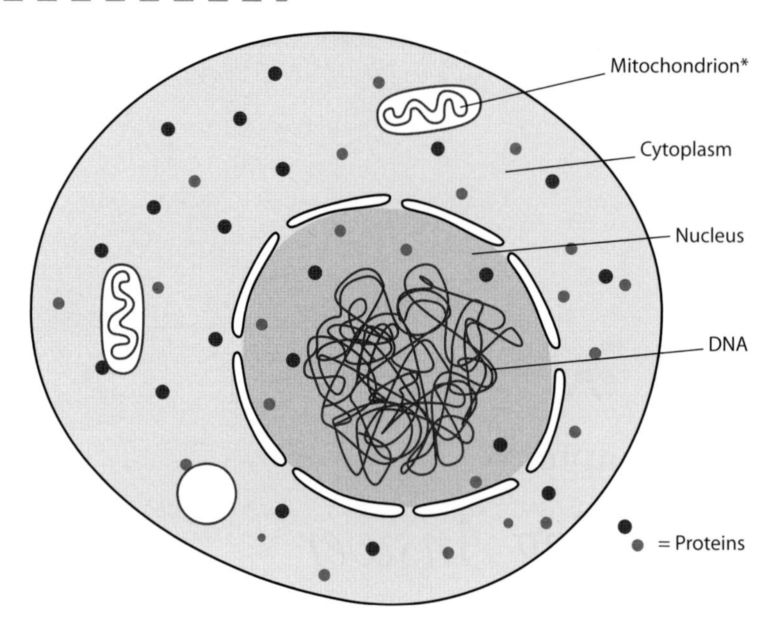

Figure 8.1. Two locations of DNA in the cell. The primary repository of DNA in the cell is the nucleus. *However, DNA is also found in another subcellular compartment, the mitochondria.

Nevertheless, at the level of genes, comparative genetic analyses have frequently been performed. And, similar to our observations of mtDNA, nuclear genes fall into nested hierarchical patterns.[2]

Twice in previous chapters we've engaged nested hierarchies. Twice we've come away without any new conclusions on the origin of species. Both at the anatomical level (chapter 5) and at the mtDNA level (chapter 7), we have discovered nested hierarchical patterns. And, at both levels, we have observed that both the evolutionary model and the creationist model expect nested hierarchies. For example, from the perspective of descent with modification, evolutionists have argued that a branching hierarchy is exactly what the process of evolution should produce. Similarly, creationists have pointed to the products of human design, which happen to fall in nested hierarchies. Creationists have argued that the matches between these designed hierarchies and the biological hierarchies is consistent with the hypothesis of design. In all of our previous attempts, nested hierarchies have failed to eliminate one of these two competing scientific explanations.

Similarly, in the realm of nuclear DNA comparisons, nested hierarchies cannot eliminate either explanation.

What, then, could nuclear DNA comparisons reveal about the origin of species?

In chapter 7, we observed that the *function* of DNA differences strongly discriminates between the creationist and evolutionary models. Evolutionists ascribe the origin of all DNA differences ultimately to mutations. Thus, between two species that share similar genes, they generally expect sequence differences between these genes to be functionally neutral.[3] In contrast, among species from separate families, creationists predict high levels of function for DNA sequence differences.[4]

These contrasting predictions lay the foundation for a head-to-head comparison of these two models in the realm of nuclear genes.

However, as we observed in chapter 3, only a small fraction of animal genomes — especially of mammal genomes — consist of protein-coding genes. Consequently, a head-to-head comparison of predictions for nuclear DNA gene sequences represents a test of only a small part of the total nuclear DNA. The bigger test of the creationist and evolutionary models is in the realm of the non-protein-coding nuclear DNA.

At the level of the entire genome, the evolutionary model has very specific expectations on DNA function. As a specific example, consider the evolutionary predictions for the overall level of function in the human genome:

> There exists a misconception among functional genomicists that the evolutionary process can produce a genome that is mostly functional. Actually, evolution can only produce a genome devoid of "junk" if and only if the effective population size is huge and the deleterious effects of increasing genome size are considerable...In the majority of known bacterial species, these two conditions are met; selection against excess genome is extremely efficient due to enormous effective population sizes, and the fact that replication time and, hence, generation time are correlated with genome size. In humans, there seems to be no selection against excess genomic baggage. Our effective population size is pitiful and neither the time it takes to replicate the genome nor generation time correlate with genome size.[5]

In other words, evolutionists expect the human nuclear genome to contain significant amounts of non-functional DNA sequences.

Practically, the actual level of function in the human genome has been hotly disputed. As we observed in chapter 3, the participants in the ENCODE project have directly tested the entire genome for function. The investigators for ENCODE have claimed that 80% of our DNA shows biochemical evidence for function. However, if this result holds true, it would contradict the expectations of the evolutionary model — expectations which the above quote details. Perhaps not surprisingly, some evolutionists have dismissed the ENCODE results out of hand, simply because they contradict the expectations of evolution.

Evolutionary disputes aside, other criticisms of ENCODE were prompted by purely technical considerations. For example, to unequivocally assign the label *functional* or *non-functional* to a sequence, comprehensive genetic knockout studies must be performed. While biochemical studies are useful first steps, they are not sufficient tests of DNA function. Currently, in humans, deliberate genetic knockout experiments are considered unethical. However, natural knockouts for certain nuclear genes exist in the global human population. About 6% of human genes have a natural knockout.[6] In other words, over 90% of human genes have yet to undergo the most fundamental genetic test for function. Since genes constitute less than 5% of the human genome, more than 90% of the human genome has never been rigorously tested for function. The ENCODE results are very preliminary.

Consequently, evolutionists would be justified in claiming that applying the label "functional" to 80% of the human genome is premature. They would also be justified in withholding judgment about the function of most nuclear DNA sequences.

But in practice, evolutionists have often swung the pendulum to the other extreme. For example, historically, many evolutionists have made claims about human-primate common ancestry based on the putative function (or lack thereof) of various non-protein-coding nuclear DNA sequences. One of their common arguments derives its strength from the following analogy: Suppose you are the proctor for an exam that consists of ten essay questions. If two students submit identical answers — the exact same words, sentences, and paragraphs — to these ten questions, you would be justified in suspecting that one student cheated off another. If you also found that the essay answers from one student contained several spelling mistakes, and if you discovered the exact same spelling mistakes in the other student's answers, your suspicions would grow. Similarly, evolutionists have argued that the existence of shared genetic *mistakes* between humans and the great apes indicates that humans plagiarized — they inherited — their nuclear DNA from a common ancestor with the great ape species.[7]

Specifically, as evidence of this genetic plagiarism, evolutionists have often cited the existence of shared *pseudogenes*. As their name implies, these DNA sequences resemble protein-coding genes, but they differ from actual protein-coding genes at several critical base pairs. These differences happen to be in a location that would disable the protein-coding function of these pseudogenes. Hence, the claim that pseudogenes represent sequences with "mistakes."

Yet, from a purely technical standpoint, can the label "mistake" be assigned to most sequences in the human genome? If more than 90% of the human genome has never been rigorously tested for function, can anyone assert that humans and chimpanzees share genetic errors? Is the label "pseudogene" even appropriate for certain human DNA sequences?

Curiously, pseudogenes have been tested for function via biochemical means. Naturally, since the ENCODE project used biochemical assays to test the entire human genome, pseudogenes would have been included in these tests. Consistent with the overall results of the ENCODE project, around 80% of human pseudogenes have at least one line of biochemical evidence for function.[8]

Again, these results are only biochemical experiments — not genetic knockouts. But they have set a trajectory that points toward pervasive, genome-wide nuclear DNA function.

Consistent with this trajectory, some of the most prominent examples of supposed genetic mistakes in the human genome have been overturned. For example, recall from chapters 2–3 that nuclear DNA is organized into chromosomes, and that these chromosomes are arranged in pairs. Among the great apes, 24 pairs of chromosomes exist in each species. In humans, only 23 pairs exist.

Evolutionists have long explained this discrepancy by postulating a chromosome fusion event (Figure 8.2) in the human lineage. In other words, evolutionists have claimed that an

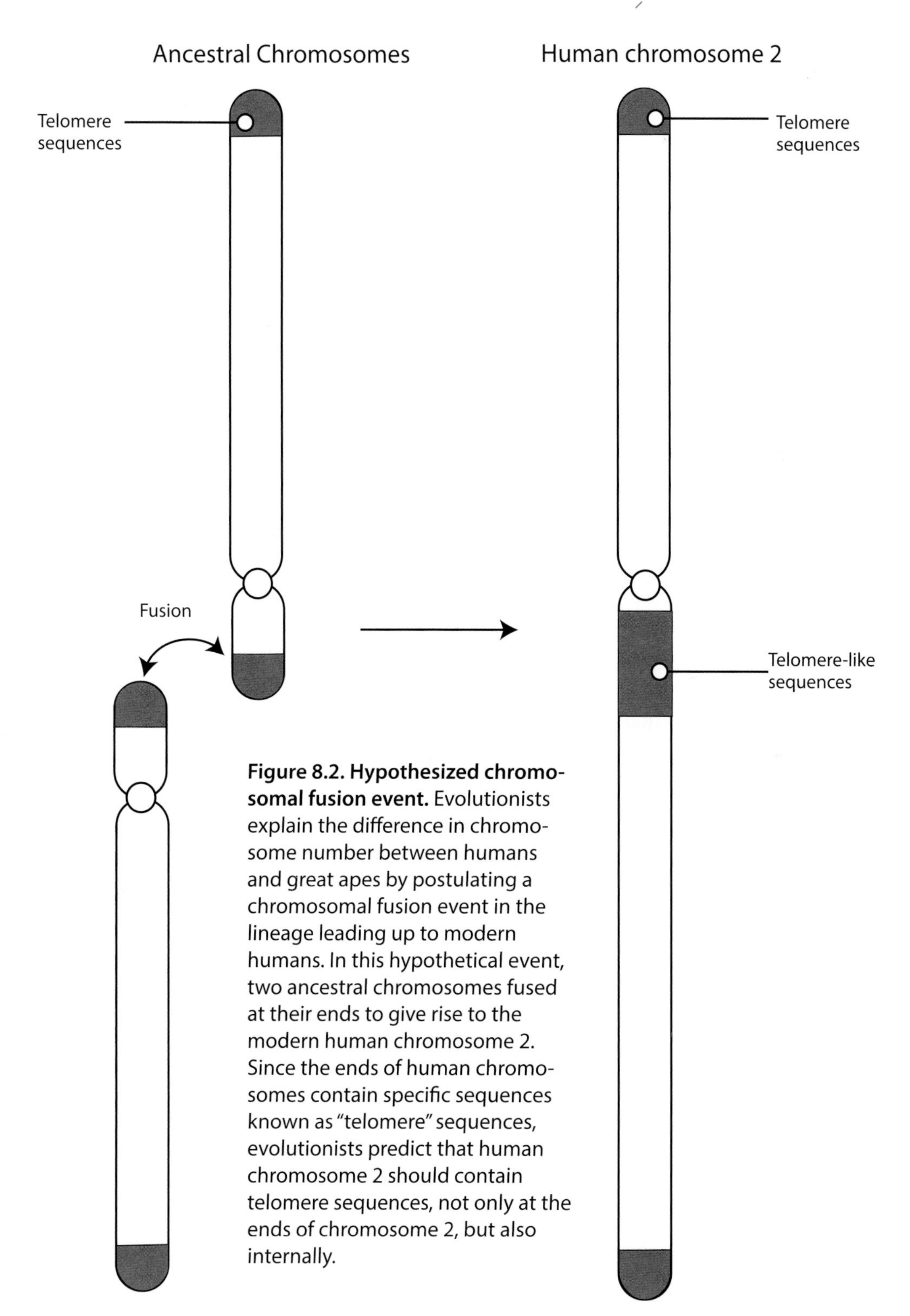

Figure 8.2. Hypothesized chromosomal fusion event. Evolutionists explain the difference in chromosome number between humans and great apes by postulating a chromosomal fusion event in the lineage leading up to modern humans. In this hypothetical event, two ancestral chromosomes fused at their ends to give rise to the modern human chromosome 2. Since the ends of human chromosomes contain specific sequences known as "telomere" sequences, evolutionists predict that human chromosome 2 should contain telomere sequences, not only at the ends of chromosome 2, but also internally.

accidental joining of two chromosomes — a genetic mistake — resulted in the reduced number of chromosome pairs in humans.

This hypothesis leads to expectations about function at the site where the proposed fusion occurred. At the site of the supposed fusion in human chromosome 2, evolutionists expect to find a genetic scar. They have even claimed that such a scar exists.

In contrast, recent investigations have demonstrated that the supposed fusion site does *not* bear the scar of an accidental chromosomal crash. Rather, the site sits in the middle of a functional gene, and the purported fusion sequence appears to participate in the regulation of this gene.[9]

At least one other classic example of a specific, supposedly non-functional human DNA sequence is following a similar path — an initial claim of non-function followed by experimental results that suggest the opposite conclusion.[10]

In essence, the evolutionary claim of shared genetic mistakes is simply another form of the anti-design argument that we explored in chapter 5. In the realm of anatomy and physiology, we observed how evolutionists have attempted to eliminate the design hypothesis by pointing to "vestigial" organs and other seemingly non-functional relics of our evolutionary past. We also observed that this line of reasoning was beset with two logical weaknesses. First, these arguments were, essentially, arguments from silence. Arguments from silence are weak because the absence of evidence is not necessarily evidence of absence. Second, the absence of evidence did not stem from exhaustive experimentation that failed to find any positive evidence for function. Rather, it stemmed from the absence of experimentation itself. The same critical weaknesses have beset the evolutionary argument for shared genetic "mistakes."

In summary, in the arena of nuclear DNA *patterns*, the patterns themselves cannot distinguish between the evolutionary model and the creationist/design models. Rather, the *function* of the nuclear DNA differences can distinguish between these two — and the trajectory of recent biochemical experiments is pointing toward high levels of function.

This trajectory would eventually have even bigger ramifications for the origin of species.

If the *function* of nuclear DNA differences is an experiment that is still in progress, what do the *number* of nuclear DNA differences among species reveal about the origin of species? Again, as we discovered in our examination of mtDNA in chapter 7, the fact of DNA mutation — in this case, *nuclear* DNA mutation — held the critical insights. And, again, like the situation we encountered for mtDNA, few species possess a directly measured nuclear mutation rate. And, yet again, among multicellular species, humans are the one multicellular species in which the nuclear DNA mutation rate has been studied the most intensely.[11] On average, in their nuclear genome, humans mutate around 78 base pairs per generation.[12]

Just like for mtDNA, the fact of nuclear DNA mutation suggests that nuclear DNA differences act like a clock. Again, to be consistent with the evolutionary practices in geology and astronomy, we should assume that the nuclear DNA clock ticks at a constant rate. If nothing else, observe that, in general, evolutionists already make this assumption when exploring nuclear DNA differences among species.

Unlike the situation that we encountered for mtDNA, the nuclear mutation rate has also been obtained for chimpanzees.[13] It is nearly identical to the human nuclear mutation rate — about 78 base pairs per generation.[14]

With the mutation rates from both of these species, we can make *retro*dictions for the nuclear DNA differences between humans and chimpanzees.

Traditionally, evolutionists have put the split time between these species at 4.5 to 7 million years ago.[15] Using this timescale, along with the measured human and chimpanzee nuclear mutation rates, we can predict how many nuclear differences should exist between humans and chimpanzees today.

Just like we did for mtDNA, before we can perform this calculation, we must convert the nuclear mutation rates from units of *mutations per generation* to units of *mutations per year*. The evolutionary community has already done this,[16] and the rates are around 4.25×10^{-10} mutations per base pair per year (chimpanzees = 4.52×10^{-10}; humans = 3.97×10^{-10}). Since only about 89% of the human and chimpanzee genomes appear to be alignable,[17] we have to use units of mutations *per base pair*, rather than mutations *per genome*. Nevertheless, we can still predict how many nuclear DNA differences should separate humans and chimpanzees *in the regions of nuclear DNA that can be aligned.*

Because we're making predictions for the split between two species, a divergence calculation is the most appropriate equation. Using these rates and a timescale of 4.5 to 7 million years, the

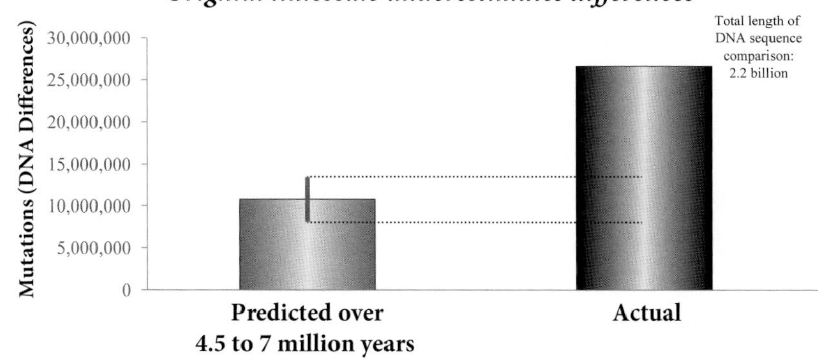

Figure 8.3. The traditional evolutionary timescale predicts too few nuclear DNA differences between humans and chimpanzees. Single base pair mutations were predicted from the published nuclear mutation rates for both species. The height of the column represents the average prediction. The black line represents the range of predictions over the range of split times (i.e., 4.5 to 7 million years). In short, if the human and chimpanzee lineages diverged 4.5 million years ago, then only around 8.4 million DNA differences would separate them today. Instead, if the human and chimpanzee lineages diverged 7 million years ago, then only around 13.1 million DNA differences would separate them. Today, over 26 million single base pair differences separate the human and chimpanzee nuclear DNA sequences.

divergence equation predicts that only around 11 million differences could have arisen (Figure 8.3).[18] In other words, of the DNA differences between humans and chimpanzees, only half can be explained by mutations.[19]

Thus, for human-chimpanzee comparisons on the evolutionary timescale, neither the mtDNA compartment (see chapter 7) nor the nuclear DNA compartment led to successful predictions.

Together, these failed predictions made the evolutionary explanatory challenge more acute. Consider the contrast in results between the mtDNA and nuclear DNA predictions. First, the two genetic compartments differed in the magnitude of their failed predictions. The mtDNA predictions (Figure 7.3) were over an order of magnitude higher than the actual number of differences; the nuclear DNA predictions (Figure 8.3) were only 50% different from the real result. Second, these failed predictions differed in the direction of the error. The evolutionary timescale vastly *over*predicted mtDNA differences (Figure 7.3), but *under*predicted nuclear DNA differences (Figure 8.3).

This contrast constrained the explanatory options for the evolutionary model. Consider the most likely evolutionary explanation for the mtDNA discrepancy. Given the massive number of predicted differences — differences that exceeded the length of the mtDNA genome — I anticipate that evolutionists will invoke natural selection to reconcile prediction with fact. Yet, in the realm of nuclear DNA, natural selection is excluded from the discussion, almost by definition. Since the nuclear DNA predictions *under*estimated the actual level of DNA differences, elimination of mutations via natural selection would only make this discrepancy worse — it would reduce the number of predicted differences even more. This presents a conflict for evolution. When does natural selection play a role? When does it not? Can the evolutionary model *predict* when natural selection sculpts the genome and when it doesn't? Or will natural selection always be an idea that is retrofitted to any result as needed — a "natural selection did it" type of explanation?

Consider another potential evolutionary explanation. In the realm of mtDNA, evolutionists have already discussed a phenomenon termed *time dependency*.[20] In other words, evolutionists have suggested that the mtDNA clock ticks at different rates at different points in history. Specifically, evolutionists have argued that mtDNA mutation rates have been *slower* in the distant past — an explanation which could, in theory, reconcile the erroneous predictions of the previous chapter with actual mtDNA differences. Yet, for nuclear DNA, a slower rate in the distant past would aggravate the magnitude of the underestimate. When does the molecular clock speed up and slow down? Can the evolutionary model *predict* when it accelerates and when it doesn't? Or will time dependency always be an idea that is retrofitted to any result as needed — a "time dependency did it" type of explanation?

Remarkably, the authors of the chimpanzee genome paper picked a third option as an explanation for the nuclear DNA data. They changed the timescale. Rather than put the split in the human and chimpanzee lineages at 4.5 to 7 million years ago, they suggested bumping it to 11 to

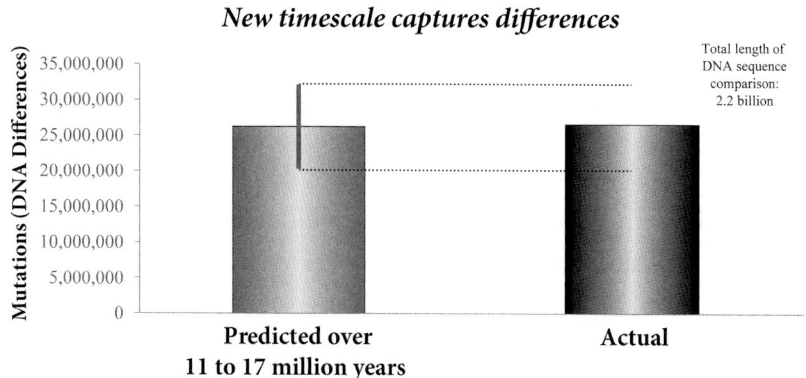

Figure 8.4. The revised evolutionary timescale correctly predicts the nuclear DNA differences between humans and chimpanzees. Single base pair mutations were predicted from the published nuclear mutation rates for both species. The height of the column represents the average prediction. The black line represents the range of predictions over the range of split times (i.e., 11 to 17 million years). In short, if the human and chimpanzee lineages diverged 11 million years ago, then around 20.6 million DNA differences would separate them today. Instead, if the human and chimpanzee lineages diverged 17 million years ago, then around 31.8 million DNA differences would separate them. Today, around 26.6 million single base pair differences separate the human and chimpanzee nuclear DNA sequences.

17 million years ago.[21] Over this timescale, the human and chimpanzee mutation rates accurately predict the differences between these species (Figure 8.4).[22]

Again, the mtDNA compartment complicates this explanation. Over the longer timescale, the minimum prediction of mtDNA mutations changes from 21,457 (prediction over 4.5 million years) to 52,450 (prediction over 11 million years).[23] The latter is three times as many base pairs as exists in the human or chimpanzee mtDNA sequence. How can evolutionists reconcile the results of these two compartments?

In addition, the implications of the revised timescale extend far beyond the question of the human-chimpanzee divergence.[24] Because the specific time points on the timescale of primate evolution are interconnected, the dates of the evolutionary divergences preceding the human-chimp

Diverging Lineages	Old Divergence Time (Millions of Years)	Divergence Time (Millions of Years) Predicted From Mutation Rates
Common Chimpanzee-Bonobo	2	4
Chimpanzee-Human	4.5 to 6	14
Chimpanzee-Gorilla	6 to 8	21
Gorilla-Orangutan	12 to 16	41
Orangutan-Rhesus Macaque	25 to 33	75

Table 8.1. The revised timescale for the human-chimpanzee split necessarily changes the timescale for other primate lineage splits.

split must also necessarily be bumped back. Using the average of the human and chimpanzee mutation rates (i.e., 4.25×10^{-10} mutations per base pair per year), and using the percent-difference among some of the primate species genomes,[25] we can recalculate when some of the major primate lineages diverged.[26] Since the human and chimpanzee mutation rates differed by only 12–14%, the assumption of rate constancy across evolutionary lineages seems fair. Under this assumption, each of the major divergence times is roughly doubled (Table 8.1). In fact, the split between the rhesus macaque and the orangutan lineages (i.e., 75 million years ago) is predicted to pre-date the first primate fossils (~55 million years ago[27]). Can this revised timescale survive?

Surprisingly, a fourth option exists for the human-chimpanzee nuclear DNA predictions. It's one that is invoked for the origin of modern humans.

Recall that evolutionists explain the origin of our own species on a timescale of about 200,000 years. Using the human nuclear mutation rate we just explored (i.e., 3.97×10^{-10} mutations per base pair per year), we can predict how many mutations should arise in the human population after 200,000 years. Since we're dealing with members of the same species, we'll use a coalescence calculation. Putting these numbers into the coalescence calculation yields a prediction of around 515,000 nuclear DNA differences.[28] Yet, on average, Africans (the first peoples to have evolved 200,000 years ago) differ from the rest of the people around the world at 4.31 *million* DNA base pairs (specifically, this number represents only the single nucleotide differences, not the indels or large insertions and deletions).[29] Has evolution made another erroneous prediction (Figure 8.5)? If the prediction for the 200,000-year timescale accounts for only 12% of the differences among modern humans, what explains the rest? Does the timescale of the origin of modern humans need to be bumped back as well?

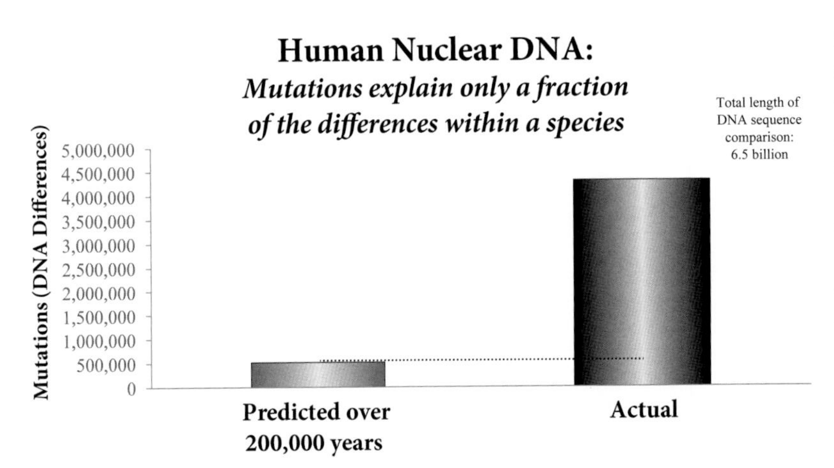

Figure 8.5. The evolutionary timescale predicts too few nuclear DNA differences among modern humans. Single base pair mutations were predicted from the published nuclear mutation rate for humans. If the modern human lineage arose 200,000 years ago, then a maximum of around 515,000 mutations would separate modern humans (population size considerations would lower this prediction). Today, over 4 million single base pair differences separate Africans from the rest of the world.

Let's ask a different question: Are all DNA differences the result of mutation? Unlike mtDNA, nuclear DNA mutations aren't the only source of nuclear DNA differences each generation.[30] In fact, they're one of the least significant sources. Recall from chapter 2 that DNA doesn't exist as a long stretched out ladder in the nucleus. Instead, it is wound around proteins, wrapped up, and (during cell division) concisely packaged into chromosomes (Color Plate 15). Thus, by following the behavior of chromosomes, we can follow the behavior of big chunks of the DNA double helix.

Recall a second discovery from chapter 2. From Mendel's experiments, we know that, each generation, both parents make a genetic contribution to their offspring. Observations subsequent to Mendel put these facts in chromosomal terms. Chromosomes come in pairs (Color Plates 11–12). One member of each chromosome pair comes from the paternal side; the other member of each pair comes from the maternal side.

In our own species, none of us have parents who are clones of each other. Today, around the globe, several *million* single base pair differences separate any two humans.[31] Therefore, each parent contributes chromosomes with unique genetic information. Since chromosomes represent tightly compacted chunks of DNA, we can extend our conclusions to the molecular level: each parent contributes a *different DNA sequence* than the other parent. For example, the DNA sequence of human chromosome 1 that comes from Dad differs from the DNA sequence of human chromosome 1 that comes from Mom (Color Plate 79).

Millions of DNA differences are the result in each generation.[32] I don't just mean DNA differences *between* individuals (though this is also true). I mean DNA differences *within* individuals. If you were to obtain the DNA sequence of both members of the chromosome 1 pair in your body, you'd discover that these sequences are different. The same conclusion would hold true for both members of the chromosome 2 pair. And the chromosome 3 pair. And so on.

In other words, in every person alive today, pairs of each chromosome exist. The sequence of one member of a pair in an individual is different from the sequence of the other member of the pair in the same individual. In technical terms, geneticists refer to this state as *heterozygous* (Color Plate 79).* If the sequence on one member of a pair was the same as the other member of the pair, this state would be referred to as *homozygous* (Color Plate 79).**

To clarify, the term *heterozygosity* can be applied to a chromosomal pair at multiple levels of comparison. It can refer to a state in which every single base pair on one member of the chromosome pair is different from every single base pair on the other member of the chromosome pair. It can also refer to a state in which all base pairs but one are identical between two members of a chromosome pair. Heterozygosity simply refers to the state of being genetically different. The context of each use determines how many base pairs are involved.

---

* For a helpful memory tool, note the "hetero" — meaning "different" — prefix.

** Again, as a helpful memory tool, note the "homo" — meaning "same," as in *homo*genized milk — prefix.

In short, the fact that we each have two different parents results in nuclear DNA heterozygosity at millions of sites. The exact number of sites depends on the genealogical connection between our parents. If the connection is too distant to measure, the number of heterozygous sites will be high. When close relatives marry and bear children, heterozygosity levels drop.

Figure 8.6. Image of the wall cress, *Arabidopsis thaliana.*

In contrast, each generation, mutations contribute very little to heterozygosity. At a rate of just 78 base pairs each generation, mutations are swamped by millions of inherited differences. Per generation, the process of inheritance — rather than mutation — is the primary contributor to within-individual DNA differences.

Experiments in plants have revealed a new twist on this fact.[33] Unlike humans, close plant relatives and siblings can be deliberately inbred for research purposes. Recently, inbreeding experiments were performed on two lines (the *Col* and *Ler* lines) of plants from a species in the mustard family: *Arabidopsis thaliana* is a wall cress, a type of green weed (Figure 8.6).

At the genetic level, the initial plant cross was more akin to a normal human marriage. Significant nuclear DNA differences existed between the two strains — about 0.35% of the genome differed. Similarly, in humans, the millions of single base pair differences (i.e., excluding the differences due to indels and large insertions and deletions) between any two people represent about 0.1% of the genome.* Conversely, hybridization between the two plant strains produced offspring with high levels of heterozygosity (Color Plate 80).

However, subsequent experiments were deliberately designed to inbreed the progeny. Following the initial cross between plant strains, siblings in each generation were self-fertilized. Effectively, this cut the heterozygosity in half every generation. For example, the progeny (i.e., the F_1 generation) of the Col and Ler cross were self-fertilized to produce the next generation (i.e., the F_2 generation) (Color Plate 80). These F_2 plants were also self-fertilized to produce the following generation (i.e., the F_3 generation) (Color Plate 80). This process was repeated with the F_3 plants to produce the F_4 generation (Color Plate 80). Each of these inbreeding events reduced heterozygosity (Color Plate 80), similar to the way in which marriage between human siblings reduces heterozygosity.

The investigators then measured the mutation rate in each generation. Surprisingly, as heterozygosity decreased, the mutation rate also decreased (Figure 8.7). In the most heterozygous plants (i.e., the immediate offspring — the F_1 progeny — of the Col-Ler cross), the mutation

* Using rounded numbers: 4.3 million single-nucleotide differences / 3.2 billion base pair (haploid) genome = 0.1%.

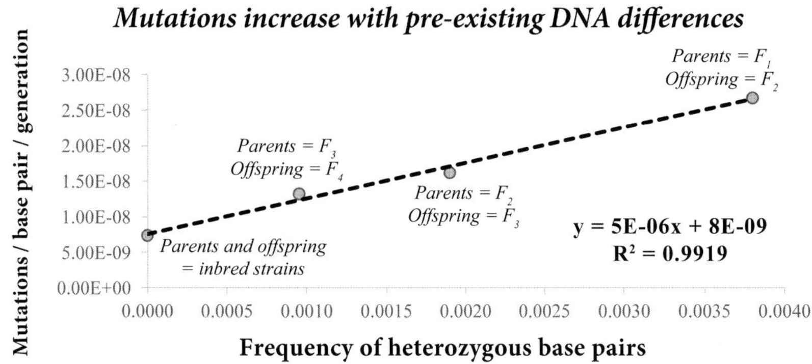

Figure 8.7. Positive relationship in plants (*Arabidopsis thaliana*) between nuclear DNA mutation rate and nuclear heterozygosity. Different lines of *A. thaliana* were crossed and then inbred to produce varying levels of heterozygosity. The mutation rates in these plants were measured and plotted against the heterozygosity levels. The parents and offspring involved in each cross are indicated next to each plotted point. Adapted with permission from Figure 9 of N.T. Jeanson and J. Lisle, "On the Origin of Eukaryotic Species' Genotypic and Phenotypic Diversity: Genetic Clocks, Population Growth Curves, and Comparative Nuclear Genome Analyses Suggest Created Heterozygosity in Combination with Natural Processes as a Major Mechanism," *Answers Research Journal*, 2016, 9:81–122.

rate was highest.* In contrast, in the F_3 cross (which produced the F_4 offspring), the mutation rate was lower. For every two units of decrease in heterozygosity, a single unit of decrease occurred in the mutation rate.

Regardless of the specific quantity in the drop, the fact that a drop occurred was the most striking element of this discovery. The more heterozygosity that existed in the *parents*, the higher the number of mutations in the *offspring*. When parents had low levels of heterozygosity, offspring gained fewer mutations. In other words, it seemed as if preexisting nuclear DNA differences were required to produce more nuclear DNA differences via mutation. It was almost a catch-22.

If mutations were the sole cause of all nuclear DNA diversity, you might predict the opposite relationship to hold true. Or perhaps no relationship at all.

Conversely, this relationship appeared to be true in more than this plant species. In fact, it looked like more than a plant-specific phenomenon. For example, in humans, the average heterozygosity among individuals around the globe is known. The average nuclear mutation rate is known as well.[34] Using the discoveries in plants, we can predict what the values of one parameter should be if we know the value of the other. This relationship fits exactly what we see in humans (Figure 8.8).

It also fits exactly what we see in chimpanzees (Figure 8.8). Like humans, the average heterozygosity among chimpanzee individuals is known, as is the average nuclear mutation rate. The relationship between the two is exactly what the plant graph predicts (Figure 8.8).

---

* The mutation rate was measured by comparing the DNA sequences of the F_2 offspring to the DNA sequences of the F_1 plants.

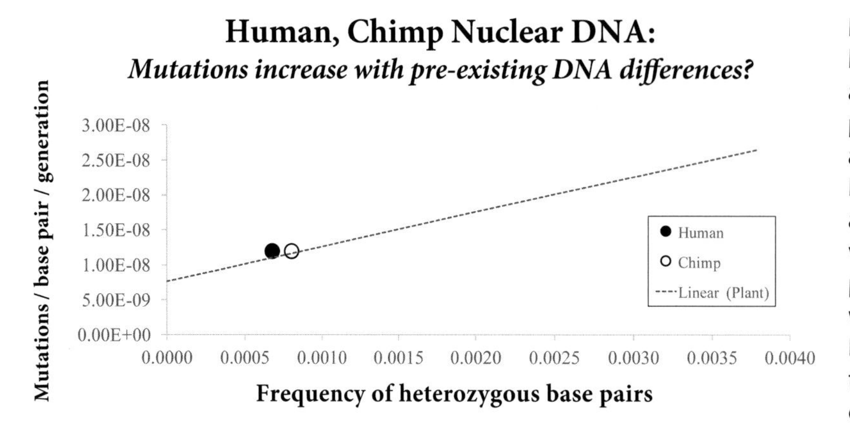

Figure 8.8. Relationship between mutation rate and heterozygosity in plants predicts human and chimpanzee values. Human mutation rate and heterozygosity value were taken from published literature, as were chimpanzee values. Dotted line represents the relationship discovered in plants (*Arabidopsis thaliana*). Modified with permission from Figures 10 and 13 of N.T. Jeanson and J. Lisle, "On the Origin of Eukaryotic Species' Genotypic and Phenotypic Diversity: Genetic Clocks, Population Growth Curves, and Comparative Nuclear Genome Analyses Suggest Created Heterozygosity in Combination with Natural Processes as a Major Mechanism," *Answers Research Journal,* 2016, 9:81–122.

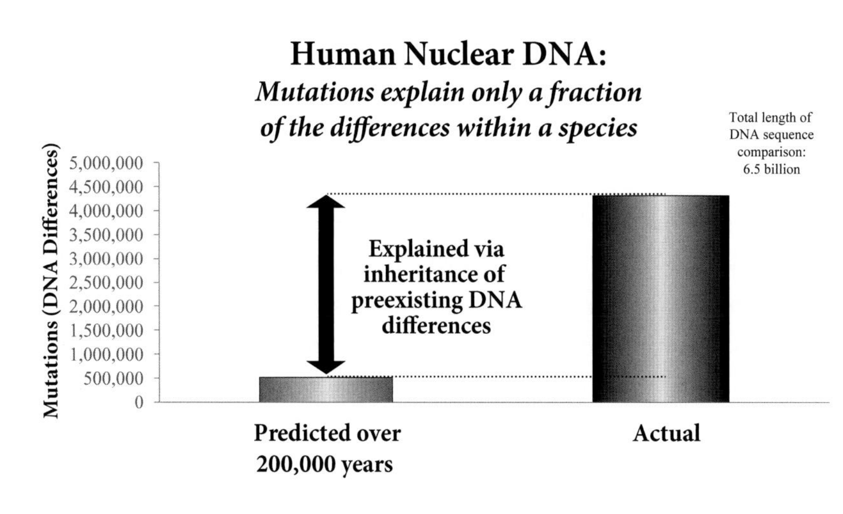

Figure 8.9. The evolutionary timescale reconciles predictions with reality by invoking inherited DNA differences. If the modern human lineage arose 200,000 years ago, then a maximum of around 515,000 mutations would separate modern humans (population size considerations would lower this prediction). To explain why over 4 million differences separate modern human individuals from one another, evolutionists invoke the accumulation of mutations prior to the origin of modern humans. In other words, from the time that the human lineage split from the chimpanzee lineage, to the time (200,000 years ago) that the modern human lineage arose, evolutionists say that mutations would have accumulated and would have therefore contributed to the nuclear DNA diversity that we observe today.

Thus, across diverse species, preexisting nuclear DNA differences looked like they might be required to produce more nuclear DNA differences via mutation.

Naturally, when you extrapolate these results backward in time, they raise an important question. In these species, what was the original nuclear DNA state? Did the ancestors of modern humans and chimpanzees have preexisting DNA differences (i.e., preexisting heterozygosity) in their nuclear genomes?

For the origin of modern humans, evolutionists answer the latter question with *yes*. Over the 200,000-year timescale, they explain the discrepancy between predicted mutational differences and actual differences with preexisting DNA differences (Figure 8.9). In other words, they say that from the time of the split in the human and chimpanzee lineages (i.e., 11–17 million years ago) to the origin of modern humans (i.e., 200,000 years ago), the ancestors to modern humans were mutating. When modern human evolved 200,000 years ago, they inherited the accumulated mutations, and then added a couple hundred thousand more over the last 200,000 years.

Actually, evolutionists have modified this narrative even more. Even if we put the human-chimpanzee split at 4.5 million years ago, a constant rate of mutation would produce almost 6 million DNA differences.[35] Evolutionists bring this number down to 4.31 million[36] by invoking changing population sizes.*

If preexisting differences are used to explain the discrepancy between prediction and fact for humans, why not for the population present at the human-chimpanzee split? In fact, why not for any evolutionary split? When can preexisting differences be invoked, and when should they not be invoked? What testable predictions does the hypothesis of preexisting differences make?

With respect to this explanation, here is one arena in which the evolutionary model does not have trouble reconciling the mtDNA and nuclear DNA results. Unlike nuclear DNA, mtDNA is not organized into pairs of chromosomes. Thus, based solely on the biology of these two compartments, it is possible to invoke separate explanations for DNA differences in these two compartments — and to do so without pitting the results against each other.

However, this does not relieve the evolutionary model of all challenges — especially when considering both genetic compartments together. Again, the mutational predictions for *human-human* nuclear DNA differences underestimated actual results. The mutational predictions for human-human *mtDNA* differences overestimated actual results (see chapter 7, Figure 7.5). Almost by definition, for human-human nuclear DNA differences, the evolutionary model will *not* be invoking natural selection, time-dependent mutational slowdown, a shortened timescale, or any other such device as might be invoked for the mtDNA results. This raises again the question we've been considering all along in this chapter: Can the evolutionary model explain mtDNA and nuclear DNA differences in a scientifically consistent manner? If they can, will their explanation make testable predictions?

* Though beyond the scope of this book, the theory behind this is straightforward. In small populations, DNA base pair differences are lost to chance — more easily than in large populations. The argument is statistical in nature. For further discussion, see any standard population genetics textbook.

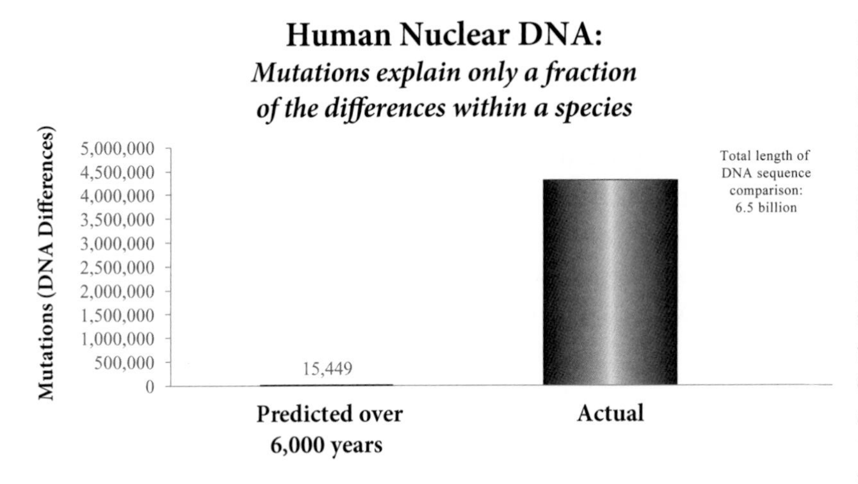

Figure 8.10. Based on mutations alone, the 6,000-year timescale predicts too few nuclear DNA differences among modern humans. Single base pair mutations were predicted from the published nuclear mutation rate for humans. If the human lineage arose 6,000 years ago, then a maximum of around 15,000 mutations would separate modern humans (population size considerations would lower this prediction). Today, over 4 million single base pair differences separate Africans from the rest of the world.

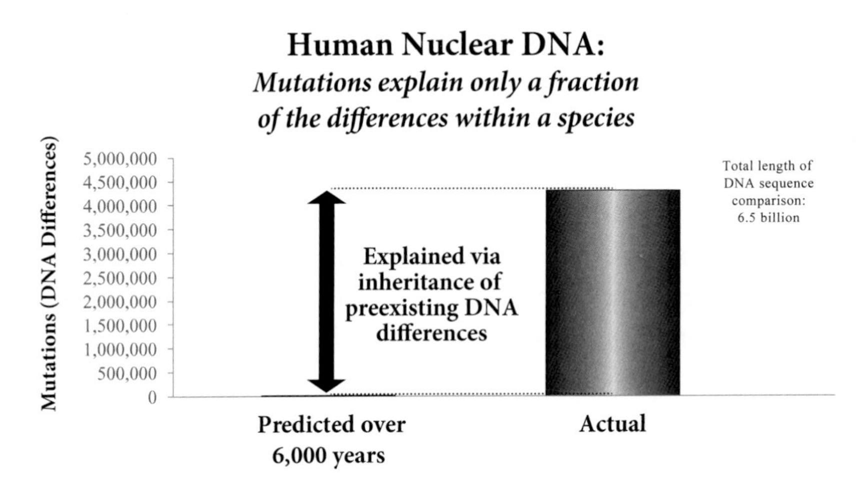

Figure 8.11. The 6,000-year timescale invokes the same explanations that evolutionists do for nuclear DNA differences. If the human lineage arose 6,000 years ago, then a maximum of around 15,000 single base pair mutations would separate modern humans (population size considerations would lower this prediction). To explain why over 4 million single base pair differences separate modern human individuals from one another, young-earth creationists invoke the creation of heterozygosity within Adam and Eve. Transmission of these heterozygous differences to subsequent generations would be the major contributor to the nuclear DNA diversity that we observe today.

Given the trajectory of the nuclear DNA predictions that we've explored, it should be no surprise that a constant rate of mutation (3.97×10^{-10} mutations per base pair per year) over 6,000 years[37] captures only a small fraction (0.4%) of DNA differences (i.e., the single nucleotide differences, not the indels or large insertions and deletions) that exist among humans today (Figure 8.10). But if evolutionists can invoke preexisting differences, why can't creationists (Figure 8.11)? If evolutionists explain 88% of our differences as preexisting, is it too much of a stretch to bump the number up to 99.6%?

Consider the scientific coherence of this explanation. Since mtDNA does not exist in chromosome pairs like nuclear DNA does, it is possible to invoke separate explanations for DNA differences in the mitochondria (i.e., mutations) versus DNA differences in the nucleus (i.e., mostly preexisting, with some mutations) — and to do so without pitting the results against each other. The distinctive biology of each of these compartments allows for this.

Furthermore, nothing else about these explanations would create a contradiction. For example, I'm not invoking natural selection to explain either compartment. My treatment of natural selection across both compartments is consistent. In addition, I'm not invoking separate timescales for each compartment; rather, both would be explicable over the 6,000-year timescale. Also, I'm not invoking changing mutation rates in one compartment and constant mutation rates in another. In fact, the relationship between nuclear DNA heterozygosity (i.e., preexisting DNA differences) and nuclear mutation rates (see Figure 8.8) would seem to prohibit any consideration of faster mutation rates in the nucleus.

Finally, I'm not invoking separate mechanisms for different species. This is most easily seen in the response to a potential objection to what I've just shown. Naturally, when a creationist invokes preexisting nuclear DNA differences, someone might object, "Preexisting *from what*?" In other words, since creationists don't invoke an unending chain of ancestors back to the first cell, from where did these preexisting differences arise?

A creationist would respond by saying that these differences were created. For humans, creationists would invoke this at the origin (i.e., creation) of Adam and Eve. For vertebrate species, based on the breed-species analogy (see chapters 5–6), creationists would invoke this event for the family-specific ancestor of species within vertebrate families. For invertebrates, they might invoke it for the subgenera-specific or genus-specific ancestors of invertebrate subgenera or genera.

This explanation is underscored by the nuclear DNA predictions on the YEC timescale across a variety of species. For example, since chimpanzees are vertebrates, creationists would expect their ancestry to be shared among other members of the family (Pongidae) to which they belong. Among the categories of great apes within Pongidae (chimpanzees/bonobos, gorillas, orangutans), the most divergent from orangutans are chimpanzees.[38] Using the chimpanzee mutation rate that we invoked earlier (i.e., 4.52×10^{-10} mutations per base pair per year), only about 0.02% of the DNA differences between chimpanzees and orangutans were explicable via mutation over 4,500 years* (Figure 8.12). In other words, the vast majority (99%) of nuclear DNA differences among species within the great ape family would be due to preexisting DNA differences (Figure 8.13).[39]

* Because the great apes are land-dependent, air-breathing creatures, YEC scientists would explain their origins from a common ancestral pair on board the Ark of Noah (see chapter 7), and they place the date of the Flood about 4,500 years ago. The first Pongidae members would have been created around 6,000 years ago (i.e., less than 2,000 years before the Flood).

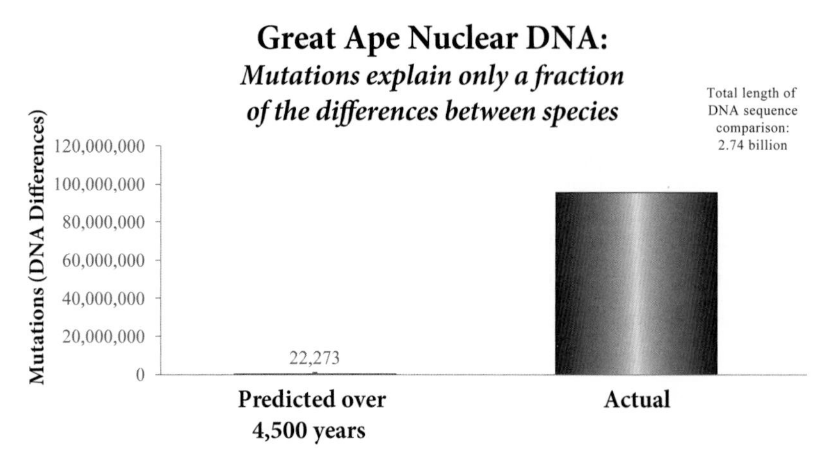

Figure 8.12. Based on mutations alone, the 6,000-year timescale predicts too few nuclear DNA differences among great apes (family Pongidae). Single base pair mutations were predicted from the published nuclear mutation rate for chimpanzees. If the great ape family began speciating after the Flood of Noah (about 4,500 years ago), then a maximum of around 22,000 mutations would separate modern species from one another. Today, over 95 million single base pair differences separate the most divergent great ape lineages from one another.

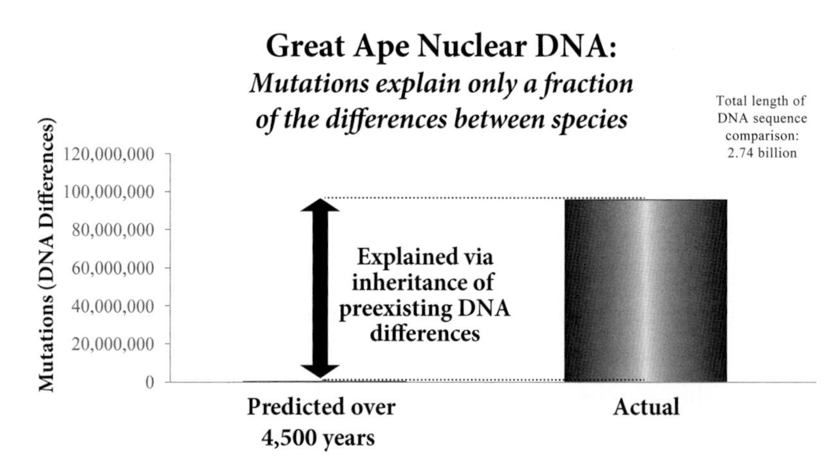

Figure 8.13. The 6,000-year timescale invokes the same explanation for the origin of human-human nuclear DNA differences and of animal-animal nuclear DNA differences. If the great ape family began speciating after the Flood of Noah (about 4,500 years ago), then a maximum of around 22,000 single base pair mutations would separate modern species from one another. To explain why over 95 million single base pair differences separate the most divergent great ape lineages from one another, young-earth creationists invoke the creation of heterozygosity within the first (created) members of the great ape family. Transmission of these heterozygous differences to subsequent generations (which would later become new species) would be the major contributor to the nuclear DNA diversity that we observe today.

Nuclear DNA mutation rates have been measured in other vertebrate species. In both mouse species[40] (Figure 8.14) and bird species[41] (Figure 8.15), current rates of nuclear DNA mutation over 4,500 years* were insufficient to explain the nuclear DNA differences even within a *genus*. Since the family-wide nuclear DNA differences are likely even higher than the differences within

* Again, because mice and this particular species of bird are both land-dependent, air-breathing creatures, YEC scientists would explain their origins from their respective common ancestors on board the Ark of Noah (see chapter 7), and they place the date of the Flood about 4,500 years ago. The first mouse and bird ancestors would have been created around 6,000 years ago (i.e., less than 2,000 years before the Flood).

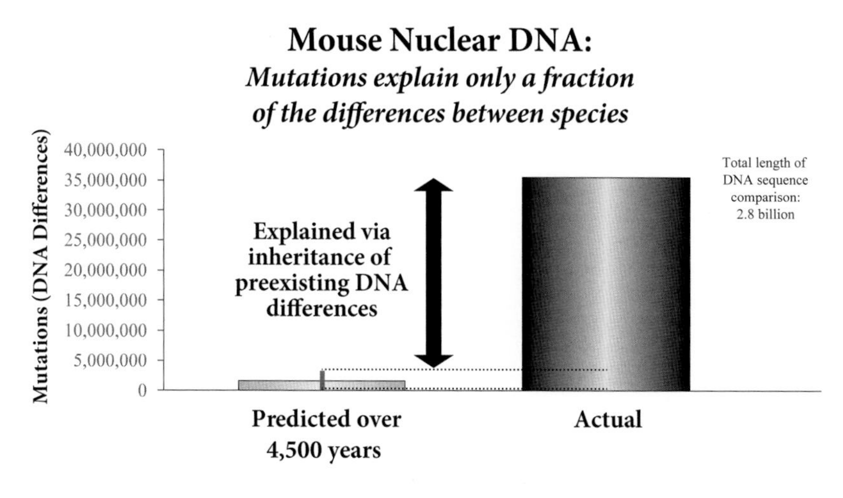

Figure 8.14. The 6,000-year timescale invokes a consistent explanation for the origin of mammalian nuclear DNA differences. If the mouse family (Muridae) began speciating after the Flood of Noah (about 4,500 years ago), then less than 3 million single base pair mutations would separate modern species from one another. To explain why over 35 million single base pair differences separate species within the same genus (let alone species within the same family), young-earth creationists invoke the creation of heterozygosity within the first (created) members of the Murid family. Transmission of these heterozygous differences to subsequent generations (which would later become new species) would be the major contributor to the nuclear DNA diversity that we observe today.

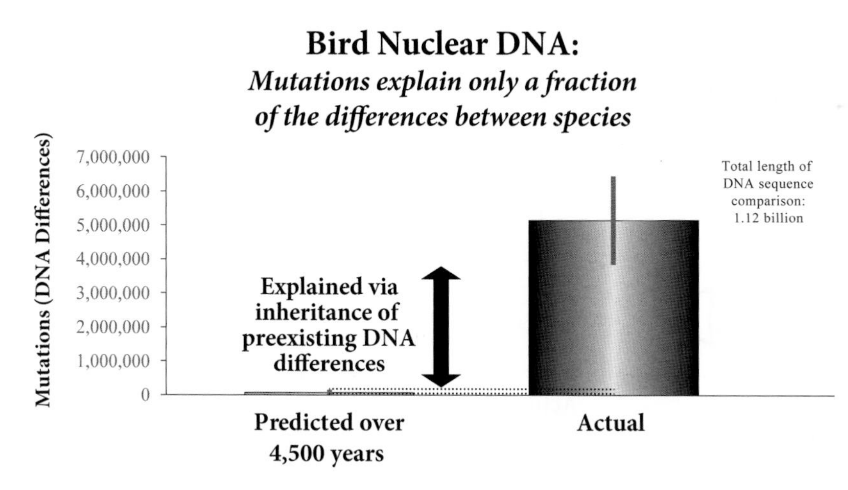

Figure 8.15. The 6,000-year timescale invokes a consistent explanation for the origin of vertebrate nuclear DNA differences. If the Old World flycatcher family (Muscicapidae—a bird family) began speciating after the Flood of Noah (about 4,500 years ago), then less than 60,000 single base pair mutations would separate modern species from one another. To explain why over 3.9 million single base pair differences (black bar in the "Actual" column represents a range of differences) separate species within the same genus (let alone species within the same family), young-earth creationists invoke the creation of heterozygosity within the first (created) members of the Muscicapidae family. Transmission of these heterozygous differences to subsequent generations (which would later become new species) would be the major contributor to the nuclear DNA diversity that we observe today.

a genus, these discrepancies would be magnified even more. In other words, across vertebrate classes, the vast majority of nuclear DNA differences among species within a family are due to preexisting DNA differences.

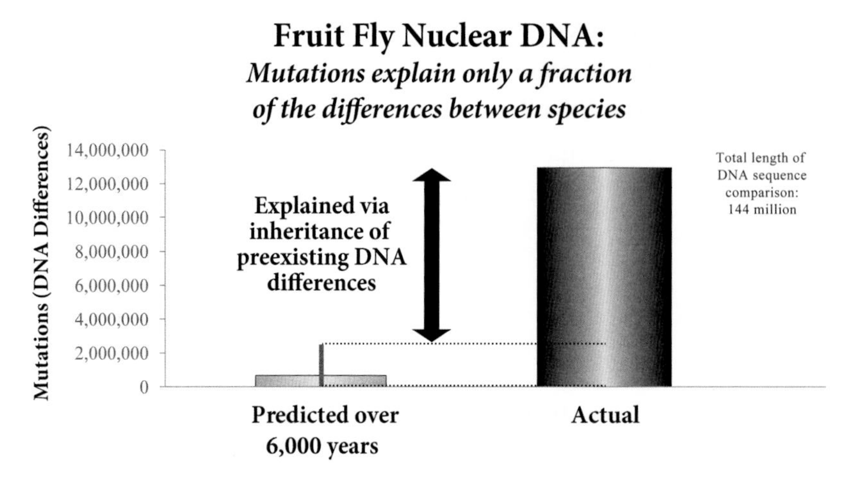

Figure 8.16. The 6,000-year timescale invokes a consistent explanation for the origin of animal nuclear DNA differences. If the first members of the fruit fly genus (*Drosophila*) arose 6,000 years ago, then less than 2.4 million single base pair mutations would separate modern species from one another. To explain why nearly 13 million single base pair differences separate some of the most divergent species within this genus, young-earth creationists invoke the creation of heterozygosity within the first (created) members of the *Drosophila* genus. Transmission of these heterozygous differences to subsequent generations (which would later become new species) would be the major contributor to the nuclear DNA diversity that we observe today.

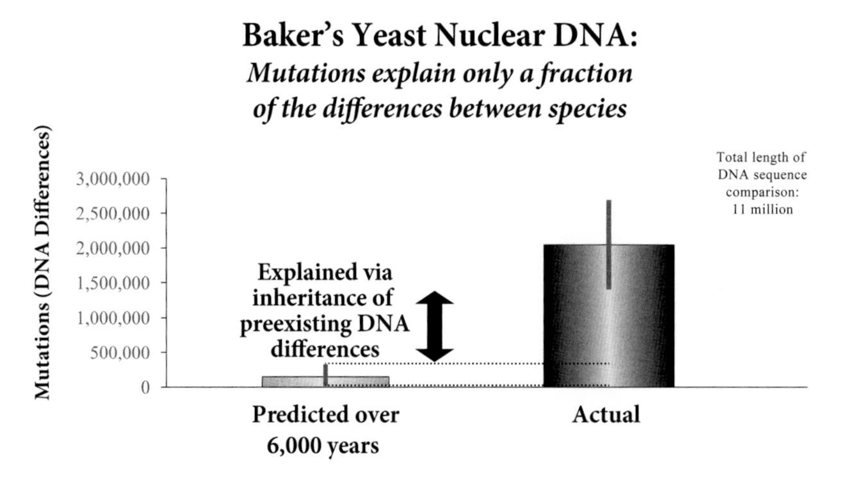

Figure 8.17. The 6,000-year timescale invokes a consistent explanation across kingdoms for the origin of nuclear DNA differences. If the first members of the yeast genus (*Saccharomyces*) arose 6,000 years ago, then less than 300,000 single base pair mutations would separate modern species from one another. To explain why over 1.4 million single base pair differences (black bar at the top of the "Actual" column represents a range of differences) separate some of the most divergent species within this genus, young-earth creationists invoke the creation of heterozygosity within the first (created) members of the *Saccharomyces* genus. Transmission of these heterozygous differences to subsequent generations (which would later become new species) would be the major contributor to the nuclear DNA diversity that we observe today.

Among the invertebrate and fungal genera that we examined in the previous chapter, two were amenable to nuclear DNA predictions.[42] However, in these two genera, some of the species used in the mtDNA analysis were not easily analyzable at the nuclear DNA level. Nevertheless, for those species that could be analyzed, the nuclear DNA mutational predictions produced the same results as the vertebrate predictions did. The vast majority of fruit fly[43] nuclear DNA differences (Figure 8.16) and the majority of yeast[44] nuclear DNA differences (Figure 8.17) were due to preexisting DNA differences.

In short, as an explanation within the YEC model, preexisting nuclear DNA differences were applicable across animal and fungal kingdoms. Even though the fraction of differences that were preexisting might vary from species to species, the requirement for preexisting variety spanned very diverse taxa.

Scientific coherence is a necessary — but not sufficient — requirement for an explanation to be considered truly *scientific*. The mark of a scientific model is the testable predictions that it makes. Up to this point, I have strongly challenged the evolutionary model along these lines. Not only have I pitted the evolutionary explanation for mtDNA differences against the evolutionary explanation for nuclear DNA differences, I have also asked what testable predictions the model makes.

As we have observed for the human and chimpanzee nuclear DNA predictions, evolutionary divergence time does not *predict* mutation rates. The human-chimpanzee divergence time was revised in light of the mutation rates in these species. Furthermore, evolutionists explain the vast majority of human-human nuclear DNA differences, not by mutation, but by preexisting differences.

With respect to other species, mutation rates and divergence times run into additional problems. For example, among yeast species, the current mutation rate over the 15 mil-

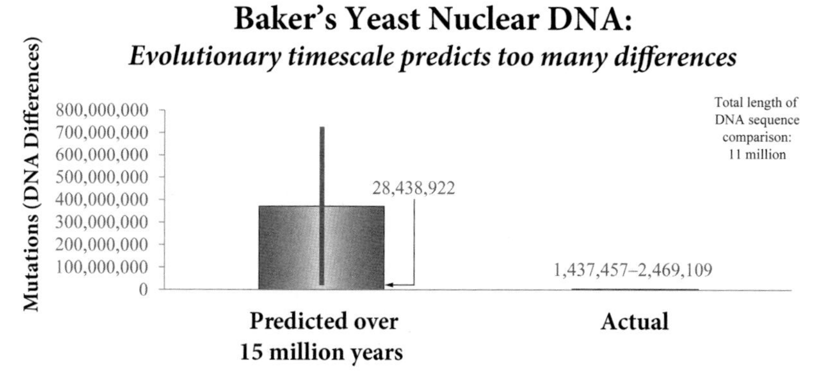

Figure 8.18. The evolutionary timescale predicts too many nuclear DNA differences among yeast species. Mutations were predicted from the empirically determined yeast nuclear DNA single base pair mutation rate. The height of the column represents the average prediction. The black line approximates the 95% confidence interval. In short, if yeast consistently underwent only around 526 generations per year (i.e., a generation time of 16 to 17 hours), then over 28 million nuclear DNA mutations (the lower end of the black line) would have occurred over 15 million years. Instead, if yeast consistently underwent 5260 generations per year (i.e., a generation time of 100 minutes), then over 715 million nuclear DNA mutations would have occurred over 15 million years. Today, less than 3 million nuclear DNA single base pair differences (numbers at the top of the "Actual" column represent a range of differences) separate yeast species in the genus *Saccharomyces*.

lion-year evolutionary time of divergence predicts[45] far too many mutations among yeast species (Figure 8.18). Just like we observed for mtDNA, the number of predicted mutations actually exceeded the yeast genome size. This result raises again the questions of what role natural selection plays, when it plays its role, and how much of a role it plays in each compartment. If nothing else, it demonstrates that evolutionary divergence times do not consistently predict mutation rates.

If divergence times do not predict the mutation rate, what does? If evolutionary divergence times do not make accurate predictions in genetics, should they be accepted as scientific?

At this point, you might ask the same questions of the YEC model. Though the model is internally consistent and does not pit genetic compartments against one another, it would be right to ask what testable predictions the YEC model makes in the arena of nuclear DNA. For example, can the YEC model predict the nuclear DNA mutation rate across a diversity of species? Or, to start smaller, can it predict the mutation rate in at least one species — humans?

In short, the answer is . . . *sort of.*

The YEC model can *retro*dict the human mutation rate. This retrodiction follows from the YEC population history that we discussed previously. In summary, from a YEC perspective, humanity has undergone two to three population bottlenecks — one at the initial creation of Adam and Eve, and one at the time of the Flood where only eight people survived on the Ark. A third might be the Tower of Babel incident, where the confusion of languages resulted in populations of unknown (but probably small) size. Likely, each of these three bottlenecks was followed by rapid — if not exponential — population growth.

Under the model of preexisting nuclear DNA differences, these bottlenecks would have been *population* bottlenecks, not genetic bottlenecks. In other words, population bottlenecks tend to eliminate large amounts of genetic variety only when the population size is kept small *for many generations.* A single-generation population bottleneck followed by exponential population growth (e.g., the kind of event that describes the bottlenecks at the creation of Adam and Eve, at the Flood, and at the Tower of Babel) would result in the loss of far fewer varieties.[46]

On the flip side, a post-Flood population explosion would have left a strong stamp of a different sort on the nuclear genome. In an exponentially growing population, new mutations add DNA differences to the overall population. However, as the population grows larger, each new mutation in each individual has a very small statistical chance of spreading to the rest of the individuals in the population. Unless the other individuals suddenly die out or unless they fail to reproduce, the DNA sequences and differences present in the rest of the population will swamp the signal of the new mutation. In other words, rapid population expansion results in a large number of mutations that are rare in frequency.

Today, human nuclear DNA differences are classified into "common" and "rare" categories — depending on their statistical frequency in the global human population.[47] Evolutionists have already noted that the amount of rare differences in the human population fits a model of recent population expansion.[48] In other words, the distribution of human DNA differences fits the YEC model of (1) a vast number of DNA differences that were preexisting differences in Adam and Eve; followed by (2) transmission of these differences over several generations to the eight people who boarded the Ark; followed by (3) rapid population expansion post-Flood where new mutations were almost automatically rare in frequency. To put it more bluntly, the recent population expansion *predicts* (technically, it *retro*dicts) the human mutation rate.[49]

However, both the YEC model and the evolutionary model account for this fact. Therefore, when comparing these models based on their testable predictions, this successful retrodiction doesn't distinguish between the models.

Nevertheless, the YEC anthropology that I just outlined makes another prediction. Since my model traces the origin of nearly all the "common" variants back to Adam and Eve, my model suggests that the history of civilization can be read off of the nuclear DNA differences among the peoples of the globe — and on a timescale consistent with the YEC model.

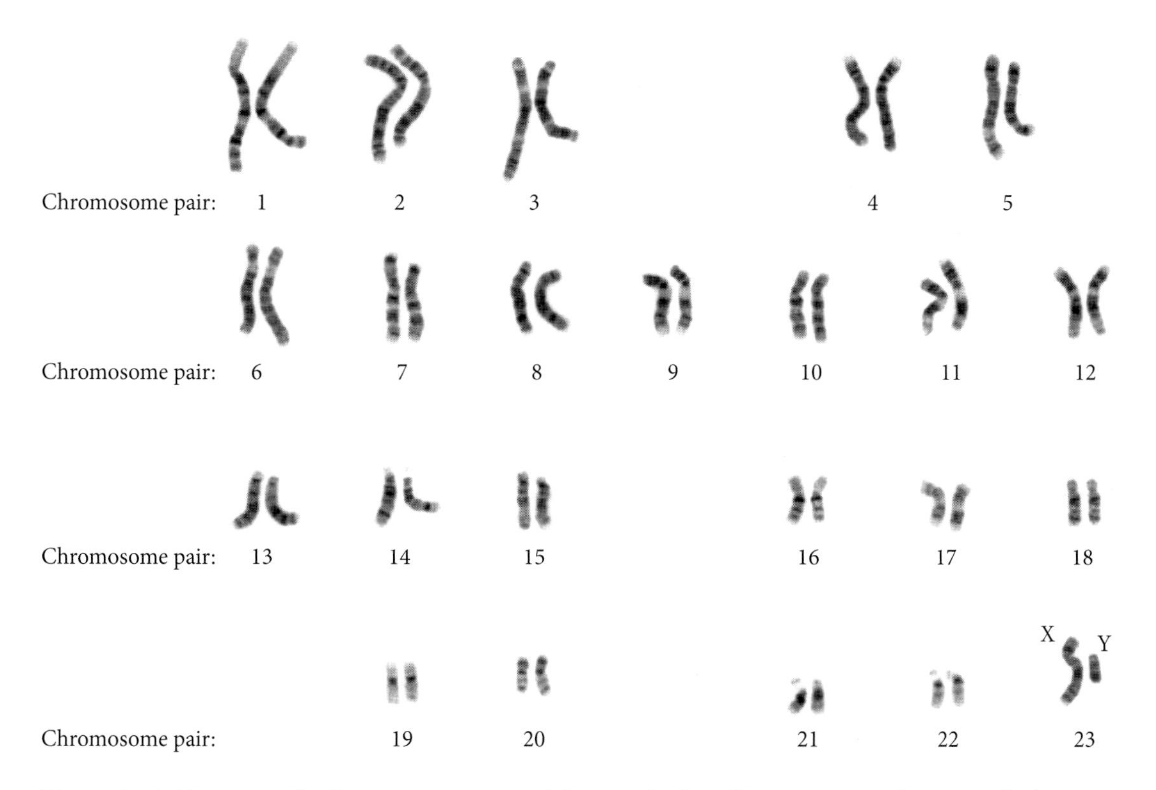

Figure 8.19. Human male chromosome spread from a single cell. In non-reproductive cells, human chromosomes exist in 23 pairs. The twenty-third pair distinguishes males and females. Females have two X chromosomes; males have an X chromosome and a Y chromosome.

For example, in the last few hundred years, European colonization and the Trans-Atlantic slave trade have resulted in major geographic movements of peoples around the globe. These movements will leave a signature in the genetics of each of these peoples.

Specifically, in the nuclear DNA compartment, two types of signatures will be present. As we observed earlier in this chapter, humans have 23 pairs of chromosomes. Twenty-two of the 23 pairs consist of two chromosomes of similar size (Figure 8.19). In females, the 23rd pair follows this rule; it consists of two X chromosomes. However, in males, the 23rd pair consists of two chromosomes of mismatched size — an X chromosome and a Y chromosome (Figure 8.19). Since males are the only possessors of a Y chromosome, and for technical reasons that will be explained more fully in the next chapter,* the Y chromosome sequence is a direct record of humanity's *paternal* ancestry.

Because of its uniparental inheritance, the Y chromosome shares a number of characteristics with the other uniparentally inherited genetic compartment, the mtDNA genome. In both compartments, all modern DNA differences are the result of mutations to the sequence that was present in the first ancestors. Both compartments go back to a single ancestral sequence: the mtDNA sequence to Eve's mtDNA sequence (Adam would have possessed mtDNA, but since males do not pass on mtDNA sequences, he would not have passed it on to offspring), and the Y chromosome sequence to Adam's Y chromosome sequence (the 23rd pair in Eve would have been X-X; in Adam, it would have been X-Y).

Thus, both the mtDNA and Y chromosome sequences have the potential to act as strict, absolute molecular clocks. The rate of mutation in each of these compartments will determine how precisely each can measure time. But the fact of a clock is true in both compartments. Therefore, the timing of the origin and migration of various people groups can be interrogated with these genetic tools.

The Y chromosome differences among modern humans represent, in theory, the first type of nuclear DNA signature of the history of civilization.

Currently, I'm exploring whether the timing of the Trans-Atlantic slave trade has left a genetic signature in the Y chromosome. If so, I can theoretically test the predictions of the 6,000-year model for the origin of humanity.

The second type of nuclear DNA signature arises from the rest of the chromosome pairs. It's more complicated than the Y chromosome signature, and it involves processes that I discuss at greater length in chapter 9. Hence, I will postpone discussion of this signature until the next chapter.

---

* Unlike the rest of the chromosome pairs, the X and Y chromosomes largely do not recombine with each other (the process of recombination is explored in detail in chapter 9).

With respect to the rest of the species around the globe, the YEC model does not yet have a predictive formula for the nuclear DNA mutation rate. Too few results have been obtained to see a general pattern of what precise percentage of nuclear DNA differences are preexisting, and what percentage are due to mutation. In this respect, the YEC model and evolutionary models find themselves in a similar position on the question of nuclear DNA differences.

However, in a more qualitative sense, the results that have been obtained thus far for the YEC model suggest that the majority — if not the vast majority — of nuclear DNA differences in species were preexisting. Since the YEC model would attribute the first appearance of these preexisting differences to the act of creation, these differences must have been *designed* into the genomes of these species. In other words, creationists predict that the majority — if not the vast majority — of nuclear DNA differences among species within families (for vertebrates) or within subgenera/genera (invertebrates and fungi) are functional participants in the biology of each species, not functionally neutral.

Can the evolutionary model make a similar prediction? In other words, though neither model appears to have a predictive explanation for nuclear DNA mutation rates, the YEC model can predict levels of function in the nuclear DNA genome. Can the evolutionary model match this scientific standard?

To clarify, the prediction I made above for animals naturally has an exception for humans. Creationists have long explained human origins apart from primate ancestry. Nevertheless, when we explored the YEC explanation for the origin of nuclear DNA differences among modern humans, we found that the vast majority (99%) were due to preexisting differences. In other words, the YEC model predicts that around 99% of the DNA differences among humans play a functional role in our biology — similar to the high level of function that we predicted for animal nuclear DNA sequences.

This is an experimental prediction that is being evaluated right now. At the beginning of this chapter, we observed that every single base pair in the human genome has already been tested biochemically for function — with 80% of human base pairs showing biochemical evidence for function. In roundworms, though the exact amount of genome-wide biochemical evidence for function is still forthcoming, the genome-wide evidence for *pseudogene* activity has been published. Nearly 70% of roundworm pseudogenes show biochemical evidence for function.[50] In humans and fruit flies, 79% and 68% (respectively) of pseudogenes manifest the same evidence.[51] In fact, in fruit flies, 89% of the *entire genome* shows preliminary evidence for function.[52]

In other words, experimental results are accumulating across diverse species that set a trajectory for pervasive, genome-wide function. Given this trajectory, I would argue that the YEC explanation for nuclear DNA differences — namely, that the vast majority were preexisting — looks promising.

❧ ❧ ❧ ❧

At the beginning of this chapter, I made a claim about nuclear DNA differences in terms of the jigsaw puzzle analogy that we've employed in this book. I implied that discoveries in the field of nuclear DNA differences represented another category of corner piece. Do they?

I also said that the mtDNA corner pieces naturally suggested a shape for some of the remaining corner pieces. Have they done so?

In actual jigsaw puzzles, corner pieces severely constrain how the puzzle can be put together. When connected to edge pieces, the constraints become even tighter. In a rectangular jigsaw puzzle, if two corner pieces are connected to each other via a row of edge pieces, the puzzle acquires its first hard edge. Furthermore, if these two linked corner pieces are also each connected at a 90-degree angle to an additional row of edge pieces, the puzzle begins to take shape, and the number of possible arrangements in the center of the puzzle are reduced.

Together, the mtDNA and nuclear DNA results of the previous chapter and of this chapter severely constrained the possible explanations within the evolutionary model. In fact, pitting the results from the two compartments against one another raised the question of whether the evolutionary model would be able to connect these pieces of evidence into a coherent whole.

With respect to the YEC model, no such difficulty was found. Rather, the mtDNA and nuclear DNA results harmonized coherently. In short, it appears that mtDNA differences are due to a constant rate of mutation over 6,000 years, whereas the majority — if not vast majority — of nuclear DNA differences are due to preexisting DNA differences that were placed in the ancestors of modern species 6,000 years ago. This model for the origin of nuclear DNA differences even led to testable predictions on nuclear DNA function.

Under the YEC view, the puzzle of the origin of species is beginning to take shape.

However, given the current inability of the YEC model to predict the nuclear mutation rate, you might pause at my assertion. Though the evolutionary model fares no better on this question, you might question the strength of the evidence for my claim about preexisting DNA varieties.

I myself might pause at the evidence that I've given in this chapter. If these were all the data that we possessed in support of the model of preexisting nuclear DNA differences, we might suspend judgment indefinitely.

However, recall our conclusion to chapter 6. We observed that Darwin's breed-species analogy was weak evidence for the recent origin of species — but weak only in isolation. Once we explored the discovery of the mtDNA clock (chapter 7), the breed-species analogy became an independent line of evidence for a conclusion that was growing in strength. Similarly, in this chapter, the evidences for preexisting nuclear DNA differences were not the end of the discussion on this question. They were the start.

Chapter 9
From DNA to Visible Traits

In a typical rectangular jigsaw puzzle, each new successful link between corner pieces accelerates my assembly of the puzzle. For example, once I've connected two corner pieces via a row of edge pieces, setting the rest of the edges becomes easier. Furthermore, since each corner piece has two sides that connect to edges, in theory I can frame three sides of the puzzle with just two corner pieces. I might even be able to get a sense for the identity of the remaining corner pieces.

Similarly, the discovery of the mtDNA and nuclear DNA corner pieces quickly suggested the identity of the remaining corner pieces. Together, they framed the question of the origin of species within a robust 6,000-year timescale.

Since the timescale of speciation is linked to the mechanism of speciation (see chapter 6), the new discoveries on the former (see chapters 6–8) had immediate implications for the latter.

At first blush, the link between the YEC timescale of speciation and the mechanism of speciation might seem like a challenge to overcome, not the start of an eye-opening discovery. For example, consider the numerical implications of a 6,000-year timescale in vertebrates. From chapters 6–8, we concluded that species within vertebrate families have a common ancestor. Today, over 70,000 vertebrate species exist, yet they belong to just over 1,100 families.[1] How could tens of thousands of new species form in just a few thousand years?[2]

This question is different from the one we explored in chapter 6. In chapter 6, we asked whether the measured rates of speciation are compatible with a timescale of a few thousand years, and we discovered that no one has measured current rates of speciation. In other words, no one has been able to disprove or confirm the fast rates of speciation implied by the 6,000-year timescale. In this chapter, a different question greets us. Even if someone were to measure the rate of speciation in the present, and even if this rate were consistent with a 6,000-year timescale, this discovery would leave a major question unanswered. *How* could species have formed so quickly?

The discoveries of the previous chapter provide the first clue to the *how* of speciation. Let's combine the findings of chapter 8 with the early genetic discoveries that we explored in chapter 2. In chapter 8, I argued that preexisting nuclear heterozygosity plausibly explains the origin of the vast majority of nuclear DNA differences that divide species today. In other words, I claimed that the ancestors to modern species were created with heterozygous chromosome pairs. In chapter 2, we followed Walter Sutton's observations of chromosome behavior. Sutton calculated the number of possible combinations that could theoretically result from pairs of heterozygous chromosomes. We observed that enormous numbers of chromosome combinations are possible from a pair of heterozygous parents (see Tables 2.1–2.3 in chapter 2).

In other words, given a set of heterozygous ancestors, a massive potential for chromosome diversity exists. Since chromosomes encode visible traits, these results imply that heterozygous ancestors also contain massive potential for diversity in visible traits.

Recent genetic discoveries in humans have cast a new light on this conclusion. To see how, let's perform a thought experiment on the first humans. In previous chapters, I argued that the first humans were Adam and Eve, a couple that was created about 6,000 years ago with heterozygous DNA. Let's trace some of their chromosomes through time and compare the results to actual human chromosomal data.

To make the question even more tangible, let's focus on the potential for diversity in a single chromosome pair — say, the chromosome 1 pair (Color Plate 82). If both Adam and Eve were heterozygous, then Adam would have had DNA differences between the two chromosomes of his chromosome 1 pair. Eve would also have had DNA differences between the two chromosomes of her chromosome 1 pair (Color Plate 82).

Now let's put some rounded numbers on this scenario. Today, we know that millions of DNA differences separate any two humans. These are distributed among 46 chromosomes. For sake of simplicity, let's assume that 1.84 million of Adam's base pairs were heterozygous. (This number is a reasonable approximation of current levels of heterozygosity in the human population.) Let's also assume that each chromosome was the same size. (This is not true, but it simplifies the math.) Under this scenario, each chromosome pair would have possessed about 40,000 differences (i.e., 1.84 million / 46 chromosomes = 40,000). The same numbers would have applied to Eve.

Now let's follow these four versions of chromosome 1 over subsequent generations (Color Plate 82). In any one offspring, Adam could have passed on only one of his two versions. Eve could have passed on only one of her two versions. Consequently, in their offspring, only four combinations of chromosome 1 pairs were possible.

As these offspring matured and reproduced, only four additional combinations of chromosome 1 pairs were possible in their offspring. In other words, given the two combinations in the parents, the four combinations in their offspring, and the four additional combinations in the following generation, only 10 combinations of chromosomes were possible in total. This limited number of possible pair combinations for chromosome 1 would have remained unchanged for generations.

If we extrapolate this result to the other chromosome pairs, a similar scenario holds true. If we all came from two people, a limited number of versions of each chromosome pair would have existed. Even if Adam and Eve were heterozygous in each chromosome pair, the total number of versions for each chromosome would have been small.

To clarify, as per Sutton's calculations that we explored in chapter 2, an enormous number of chromosome combinations would have existed — when we consider the number of possible ways to combine *46 chromosomes* in one individual. But for each chromosome pair, the total number of combinations in each pair would have been small.

Mutations would have added a slight nuance to this prediction. To make the math simpler, let's say that the average per-generation mutation rate is 92 mutations rather than 78 mutations. This way, with 46 pairs of chromosomes, each individual member of each pair receives about 2 mutations per generation. Over time, each member of each chromosome pair will grow slightly different from the other member of the pair.

However, over 6,000 years, these differences would have been trivial. As we observed in chapter 8, about 15,400 mutations would have occurred between two human lineages over the last 6,000 years. Again, to make the math simple, let's reduce this number to 13,800 — or about 300 mutations per individual member of each chromosome pair (13,800 mutations divided by 46 individual chromosomes equals 300 mutations per chromosome). Recall that, in our calculations above, we postulated that Adam and Eve would have *started* with about 40,000 heterozygous sites per chromosome. An additional 300 mutations represents less than 1% of this total (300 divided by 40,000 equals 0.0075 or 0.75%).

In short, you might expect the creationist model to predict that members of each chromosome pair would come in one of only four versions. Mutations might increase the diversity slightly, but around 99% of the DNA differences between chromosomes would be due to inheritance. And these differences should be easily classifiable into one of the four starting chromosomal versions in the original parents.

Today, global human chromosome diversity is astounding. For example, if you examine chromosome 1 in human populations around the globe, it comes in all sorts of versions. When you compare Africans, Asians, Native Americans, and Europeans, the versions appear to have been scrambled[3] (Color Plate 82) — consistent with the striking visible differences among the populations. How could two ancestors have produced this variety?

Actually, nuclear DNA inheritance happens in a much more nuanced way than what I've just described. When discussing inheritance of chromosomes, I've treated each individual chromosome as a single unit of inheritance (Color Plate 82). In other words, I've acted as if chromosomes come in one of only four versions — the two paternal versions and the two maternal versions — which can be combined in a very limited number of ways. Consequently, when we traced this simple pattern down through multiple generations, we found that each chromosome *pair* had very little variability (Color Plate 82). If chromosomes are blocks of sequence that change only via mutation, then little diversity results (Color Plate 82).

This simplistic view fails to capture the complicated way in which chromosomes are actually inherited. To make the illustrations easier for the moment, let's explore chromosome behavior in just one parent — the father. Again, we'll use the rounded numbers that we employed above. For sake of simplicity, we'll again assume that each chromosome is equal in size — about 141 million base pairs in length (6,484,000,000 total base pairs / 46 chromosomes = roughly 141,000,000 base pairs). We'll also assume that each member of each chromosome pair differs from the other member of the pair by about 40,000 base pairs.

Under the scenario I've just described, specialized genetic processes dramatically increase the potential for genetic diversity. When a father produces sperm cells, chromosomes aren't copied in his cells as unchanging entities. Parts and chunks of chromosome pairs are swapped with one another to produce unique chromosomes in each sperm cell (Color Plate 83). When chunks of chromosomes are swapped, the process is termed *recombination*. For example, a chunk of the first member of the chromosome 1 pair might be swapped with a segment of the second member of the chromosome 1 pair. When tiny sections or even individual base pairs are swapped, the process is termed *gene conversion*[4] (Color Plate 83).

On average, in a single human sperm, about one recombination event happens per chromosome per generation.[5] In contrast, a total of about 12 gene conversion events happen per generation per sperm.[6] In other words, on average, each chromosome will experience one recombination event, while every other chromosome (12 / 23 = 0.5) will experience a gene conversion event. If average behavior is consistent from generation to generation, then every chromosome will experience at least one type of chromosome diversification event each generation.

The results of these two processes are starkly different from the result of the process of mutation. The biggest effect is in the sequence of the offspring. For example, recombination and gene conversion would occur, not just in the hypothetical father that we examined. It would also occur in the mother (Color Plate 83). Thus, the offspring of these two parents would have a unique chromosome set. On average, no chromosome in the child would exactly match any of the chromosomes in either parent. Recombination and gene conversion would prevent exact matches from occurring (Color Plate 83). However, at any DNA *position* in the child, you could trace the base pair to an exact chromosome in the parents. Without mutation, the specific base pairs at each position would be preserved each generation. Only the *combination* of base pairs along the length of any individual chromosome would differ (Color Plate 83).

In contrast, mutation increases DNA diversity by changing the *identity* of individual base pairs. In fact, this is how mutations are defined. Rates of mutation are measured by comparing the DNA sequences of offspring to the DNA sequences of parents. Where the DNA base pair (or sequence of base pairs) in a child cannot be matched to any of the chromosomes of the parents,* a mutation is scored.

Over time, the combined processes of recombination and gene conversion produce a unique outcome. Again, let's go back to our hypothetical father. In each member of each chromosome pair, the DNA sequence along a particular chromosome occurs in a defined linear order. For example, at position 1,000 of the first chromosome in the chromosome 1 pair, the base pair might be an A-T pair. At position 100,000 a G-C pair might exist. In the second chromosome of the chromosome 1 pair, position 1,000 might be a G-C pair, and position 100,000 might be an A-T pair. In technical terms, we'd say that, on the first chromosome of the chromosome 1 pair, the A-T at position 1,000 and the G-C at position 100,000 were *linked*. The same term applies to the two positions on the second member of the chromosome 1 pair. In fact, on a single chromosome, every DNA position is linked to every other DNA position on the chromosome.

Recombination and gene conversion disrupt this linkage. For example, after recombination, the first member of the chromosome 1 pair might have an A-T pair at both position 1,000 and position 100,000. If so, then the second member of the chromosome 1 pair would have a G-C pair at both positions. Effectively, the result represents a disruption of the original linkage.

When this process occurs over many generations, the linkage that existed in the original father becomes more and more disrupted. In fact, over 6,000 years, these processes would significantly disrupt the original linkage that would have existed in Adam and Eve. Depending on the generation times, anywhere from around 6,200 (if generation times were 50 years) to 39,600 (if generation times were 15 years) linkage-breaking events could have occurred in a single lineage in the last 6,000 years.[7]

Let's take these discoveries and revisit Sutton's calculations. Since Sutton published his table, further discoveries on chromosome behavior suggest that the parent-of-origin for each chromosome makes a minor contribution to the function of the chromosome. In other words, calculations on the theoretical combinatorial diversity from a given set of chromosomes are best done using formulas for combinations rather than for permutations.

Using the numbers that we just derived, Adam and Eve's chromosomes would have each undergone hundreds of lineage-breaking events (i.e., 6,200 events / 46 chromosomes = 135 events

* Due to single base pair mutations, deletions, insertions, translocations, etc. (see chapter 8).

per chromosome; 39,600 events / 46 chromosomes = 861 events per chromosome; average of 498 events per chromosome).[8] In other words, the versions of each chromosome that originally existed in Adam and Eve have become hundreds of different versions. Furthermore, since these numbers represent the situation for a single lineage, and since multiple lineages have descended from Adam and Eve, a massive number of versions of each chromosome would exist today.

For simplicity, let's conservatively say that chromosome 1 now exists in 100 different versions. Let's also say that these same numbers apply to every other one of the remaining 22 pairs of chromosomes in humans.[9] In light of the fact of recombination and gene conversion, how many different chromosome combinations would be possible in the 6,000 years from Adam and Eve until the present? *A nearly incalculable amount*—over 10^{85} (i.e., the number 10 with 85 zeros after it).[10]

Now apply these calculations to other species. In chapter 8, I invoked preexisting heterozygosity for more than humans. I argued that preexisting heterozygosity explains the vast majority of nuclear DNA differences across animal and fungal kingdoms as well. Since species in these kingdoms undergo recombination (and probably gene conversion as well), *a nearly incalculable potential for chromosome diversity also exists in these species.*

In other words, given a set of heterozygous ancestors, a fantastic potential for chromosome diversity exists. Since chromosomes encode visible traits, these results imply that heterozygous ancestors also contain a fantastic potential for visible diversity.

On an individual-to-individual basis, the numbers that we just explored easily explain genetic and visible diversity. But the origin of species is not a question of individuals. One unusual offspring does not a species make. To be sure, unusual or distinct individuals are the prerequisite for new species. But, by definition, the origin of new species involves the origin of *populations.*

From chapter 2, recall Mendel's experiments on populations of pea plants. Consider the fate of the individuals that he obtained from the self-fertilization of doubly heterozygous parents (Color Plate 10). One of the most visibly distinct offspring (i.e., distinct as compared to parents, which had smooth and yellow seeds) was the double recessive offspring — a pea plant with green and wrinkled seeds. Achieving this success was fairly simple. About 1 in every 16 offspring ended up this way.

Yet this result would have been temporary had not self-fertilization been maintained in subsequent generations. For example, consider the potential breeding partners for this double recessive individual (Color Plate 10). Among the other types of offspring, seven were double dominant in at least one trait. If the double recessive individual were crossed to one of these seven, at least one of the recessive traits would have disappeared in the next generation. Of the remaining eight potential types of breeding partners,* four had one dominant allele in both traits. If the double

* I'm ignoring the other double recessive individuals in my tally of potential breeding partners--since breeding with one of these other double recessive individuals would basically be equivalent to a self-fertilization event (with respect to the traits that we're following).

recessive individual were crossed to one of these individuals, the disappearance of one of the recessive traits in the offspring was also a highly probable result (i.e., 50% chance). Thus, if the double recessive individuals were ever to grow into a population, at least 70% of its potential breeding partners were poor choices. In one generation, the recessive traits could have been re-hidden. Only under circumstances where specific crosses were prevented (e.g., to dominant individuals) or promoted (e.g., to other double recessive individuals; or to self) could a *population* of double recessive individuals arise.

In sum, due to the recessive nature of some traits, heterozygous parents have tremendous potential for producing diverse offspring. But translating recessive individuals into recessive populations requires very specific parental pairings in subsequent generations.

In Mendel's experiments, he deliberately supervised and directed breeding and self-fertilization events. However, in the wild, no breeders exist. No human picks and chooses mates for each individual. How, then, can populations — new species — form? In other words, what mechanism isolates distinct individuals and limits their breeding partners, such that they eventually form a distinct population?

The answer goes back to Darwin. In the wild, individuals are engaged in a competition for resources. As populations grow larger, resources on a per-individual basis are reduced. In some cases, individuals with superior traits outcompete the inferior individuals. The strongest survive and pass on — via DNA sequences — the code for these superior traits to the next generation, while the weaker individuals die off. In other cases, a subpopulation simply breaks off and migrates to a new area where the competition is less. This struggle for survival naturally isolates groups from one another.

Of the two mechanisms that I just listed, the second should be especially familiar. In chapter 4, we concluded that the geographic distribution of species around the globe is the result of *migration*. Geography naturally isolates and subdivides populations from one another.

In other words, the mechanisms that I just listed aren't even controversial. They are established fact. Migration is a very plausible mechanism for isolation of populations.

Plausibility aside, let's move the discussion from *qualitative* considerations to *quantitative* arguments. Can migration in the wild isolate populations quickly enough to produce tens of thousands of new species in 6,000 years?

Again, let's take a cue from Darwin. As we observed in chapters 5–6, the process of breeding has produced a large *number* of breeds in a very short amount of time. In fact, in our most domesticated families, far more breeds exist than species. So if we're looking in the wild for a natural substitute for the human element of the breeding process, we don't need a very efficient one.

Let's now consider this breed-species comparison in quantitative genetic terms. For example, how many genetic differences have human breeders manipulated to produce over 850 breeds of equids? Thanks to modern DNA sequencing, we know exactly how many nuclear DNA differences separate horse breeds from one another — 5 to 6 million.[11] We also know how many

DNA differences separate equid species from one another — around 10 to 28 million.[12] In other words, with fewer DNA differences at their disposal, human breeders have created much more morphological variety on the farm than exists in the wild. If human breeders can do so much with so little genetic potential, and if much more genetic potential exists in the wild, then a natural "substitute" for humans is almost unnecessary — natural processes need to accomplish so little.

With these numbers in hand, let's consider a potential mechanism by which species might arise in the wild. As a representative example, let's consider the origin of all seven living equid species. In light of the discussion of chapter 8, our starting point is a pair of highly heterozygous equid ancestors. Since I argued that the vast majority of nuclear DNA differences among species were preexisting, let's start with an equid ancestor that possesses 28 million heterozygous DNA base pairs. In addition, since modern equid species have 32–64 chromosomes, let's spread these 28 million heterozygous sites across 32 chromosomes.

According to (1) basic Mendelian genetics, (2) the fact of recombination, and (3) an updated version of Sutton's calculations, we know that the potential for chromosome diversity in the offspring of these ancestors is almost incalculable. In other words, we know the next generation will immediately be genetically diverse. Furthermore, since chromosomes encode visible traits, we can logically conclude that this genetic diversity will result in visible diversity.

By this stage, we've advanced the process of speciation to the production of distinct individuals. However, to produce seven distinct species, these distinct individuals will need to be *isolated* from their parent populations.

Today, the seven living species of equids are distributed across Africa and Asia (for zebra geography, see Figure 4.3 in chapter 4). We know from our discussion of chapter 4 that these species reside in their current locations because they *migrated* there. Migration isolates populations.

Migration also aids in cementing the distinctiveness of each isolated population. Two population-level processes easily accomplish this. First, when a small population breaks off from a large population, inbreeding occurs by definition. If your population consists of only you and your close relatives, your mate options are severely limited. Even if the breakaway population includes more than close relatives, the mate options of the next generation will be limited to other offspring in the breakaway population. Their offspring will also be forced to mate with one another, and inbreeding will occur.

When inbreeding occurs, heterozygosity is reduced. Since, by definition, both parents are closely related, they both have similar DNA sequences. Thus, in their offspring, the chance is high that, at a particular DNA position, both chromosomes of these offspring will end up with the same DNA sequence. By definition, when an individual has the same DNA sequence at a particular DNA position in both chromosomes, this individual is homozygous at this position. Inbreeding quickly moves a population towards homozygosity.

Second, in small populations, some DNA variety is lost simply due to chance. For example, in each human generation, only a subset of individuals reproduce. For instance, some people remain unmarried their entire lives. Others marry, but then discover that they are infertile. Still others have just one child, or maybe two.

Consider the fate of the DNA varieties present in the single (i.e., unmarried and childless) and in the infertile. With no offspring, their DNA variants go extinct. Unless other members of the population share their DNA varieties, the fact that they don't reproduce effectively eliminates these variants from the population.

Now consider also the families with just one or two children. The offspring will receive their genetic information from both parents. In Mom and Dad, potentially four different versions of DNA exist, which can be recombined into a large number of different combinations. But, by chance, some variants might get passed on to both children, while other variants fail to be passed on to either. The latter will also go extinct apart from their presence in other members of the population. Loss of DNA variants eventually reduces the heterozygosity of the entire population.[13]

The process of migration involves both mechanisms of heterozygosity reduction. If the breakaway population is founded by a small number of individuals, chance loss of variants can occur. Inbreeding will happen by definition. In other words, unless a breakaway population is founded by a large splinter group from a parent population, migration will reduce heterozygosity over time.

The more homozygous a population becomes, the less it will resemble the original (and very heterozygous) parent population. The less it resembles the parent population, the more likely it is to be recognized as a new species.

To produce a new, visibly distinct population, a particular combination of homozygous DNA positions is not needed. *New* species are not defined by their ability to achieve a specific morphological shape or target. Taxonomists are not waiting with clipboards, rejecting new populations that fail to meet their favorite heights, weights, and colors. Rather, a breakaway population is scored as a new species based on its genetic and morphological relationship to existing populations. In other words, for a new zebra species to form, it would not have to possess either six stripes or seven — with all other numbers being inadmissible. No specific number of stripes are necessary for speciation to occur. Instead, in order to be called a new species, a breakaway zebra population would just have to be different enough from the existing zebra populations.

In short, under the model that I just described, formation of new species is easy. As long as subpopulations historically broke away from the original ancestral populations, and as long as these breakaway populations moved toward homozygosity at some of the original heterozygous DNA loci such that isolated, distinct populations appear, then new species would have formed.

Since highly heterozygous ancestors appear to be the rule for animals and fungi alike (see chapter 8), then around the globe the formation of new species is a comparatively simple process.

⁂ ⁂ ⁂ ⁂

In our discussion of the process of speciation, I've left out one of the most important steps — one with strong temporal implications. After distinct individuals are born and then isolated via migration, they must produce more offspring in order to be recognized as a population. Can this process of population regrowth occur fast enough to produce tens of thousands of species in 6,000 years?

Naturally, the rate of population growth varies from species to species.[14] In general, species that consist of individuals of small physical size take less time to reproduce than species that consist of individuals of large physical size. However, in all cases, population growth happens much quicker than we might intuitively assume. Because population growth is an exponential mathematical function, the phrase "breeding like rabbits" applies to more than fluffy pets.

To appreciate this, we need just a few parameters from each species. Let's use placental mammals (i.e., not marsupials like kangaroos, or monotremes like echidnas) as an example. After conception, growth in the womb will take time. For example, in zebras, gestation time is around a year.[15]

After birth, individuals aren't typically ready to reproduce. They must be weaned. Then they must mature before being sexually competent. In zebras, sexual maturity takes two and one-half to four years.[16]

Furthermore, once they reproduce, species vary in their litter sizes — another factor which determines the speed at which populations can grow. The lifespans of the parents also vary; since tallies of population size at any point in time include living parents, this parameter is also critical to our predictions. Taken together, all of these parameters can be used to calculate how fast a population can grow from, say, just two individuals.

Currently, we have reliable data on these parameters for around 300–400 mammal species. These species belong to 23 mammal orders,[17] spanning small rodents to elephants and whales.[18] If we make predictions for just 200 years of growth, many of the rodent species produce population sizes that exceed the calculating capacity of Microsoft Excel — in other words, a population that is greater than 10^{300} (that's the number 10 with 300 zeroes after it).[19] The slowest growers — elephants, dugongs, dolphins, and whales — produce a minimum of 2,000 individuals, with some producing up to 300,000 individuals. Even the great ape species produce population sizes of several million.

When we calculate population sizes after 1,200 years of growth, the results are even more striking. Around half of the species produce population sizes in excess of the calculating capacity of Microsoft Excel. Over 60% of the remaining species produce at least 10^{100} individuals — that is, the number 10 with one hundred zeros after it. Even the slowest growers — elephants, dugongs, dolphins, and whales — produce more than 10^{18} individuals. This number represents a *quintillion* individuals. Populations can recover extremely rapidly.

Obviously, population regrowth would not constrain the 6,000-year timescale.

But, just to be thorough, let's walk through the appropriate population regrowth numbers for a species-rich mammal family. Today, the family Muridae — a mouse family — has 712 documented species. If we assume that all these mouse species formed within the last 4,500 years,* then an average of 1 new mouse species would have formed every 6 years. The mammalian population growth dataset has 36 of these mouse species present. In just 5 years, these species could form anywhere from 2 thousand to 3 quintillion** individuals (i.e., different mouse species have different reproductive parameters).[20] Even if mouse populations had to grow, crash, and regrow from two individuals repeatedly for the last 4,500 years, population growth curves demonstrate the mathematical plausibility of this scenario.

In summary, the three elements of the speciation process — formation of genetic distinctiveness, isolation of the distinct individuals, and growth of a new population — are all very plausible under the model that I just outlined. However, they need not occur in the order that I've presented them.[21] For example, subpopulations might first separate geographically. That is, a chance environmental change could isolate members of an originally united population. Then each population might acquire genetic and visible distinctiveness — due to inbreeding, to the statistics of small populations, etc. Regardless of the specific order in each case, these three elements are sufficient to produce tens of thousands of vertebrate species from 1,100 pairs of vertebrate ancestors in just a few thousand years.

As an aside, these same three steps apply to the formation of new, distinct populations *within* a species — for example, to the formation of the various human ethnolinguistic groups.[22] This model makes testable predictions — the predictions to which I alluded at the end of chapter 8. In other words, the movements from heterozygous human ancestors to more homozygous modern humans would have left a stamp on our nuclear genomes — especially in the non-X-Y chromosome pairs. The derivation is technical (see Color Plate 81). But my model predicts that the history of civilization should be readable off of our nuclear genomic differences.

The 6,000-year model of speciation that I've just presented represents a sharp departure from the evolutionary model of speciation. Though the three steps of speciation are the same under both models, the practical outworking of these steps looks very different under the evolutionary model.

Let's set these two explanations side-by-side to highlight the major points of difference (Color Plate 84). For instance, according to evolution, the ultimate cause of genetic change is mutation. When mutations accumulate in a genome, recombination and gene conversion can reshuffle their original arrangement to produce new combinations. Consequently, in some generations the immediate cause of genetic distinctiveness might not be a mutational event. But, under the evolutionary model, mutations are the ultimate cause.[23]

* Because mice are land-dependent, air-breathing creatures, YEC scientists would explain their origins from ancestors on board the Ark of Noah (see chapter 7), and they place the date of the Flood about 4,500 years ago.

** Three quintillion is the number 3 with 18 zeroes after it.

The ramifications of this fact are clear. Today, millions of DNA differences separate species. Yet, each generation, mutations generate a trivial amount of variety. To generate millions of DNA differences via mutation, enormous amounts of time are required. Since the accumulation of these mutations must eventually produce the visible variety that we see among species, the first element of the speciation process — formation of genetic (and visible) distinctiveness — could not have occurred in just a few thousand years.

In contrast, if the ancestors of modern species had millions of heterozygous DNA sites from the start (as under the 6,000-year model), genetic (and visible) distinctiveness could have happened in a single generation.

Under the evolutionary model, the second step of the speciation process can occur via at least three major mechanisms. As an illustration, let's consider the fate of a new mutation. Furthermore, let's say that this mutation produces obvious visible consequences. At a minimum, let's say it produces definite functional consequences.

When the mutation first occurs, it likely occurs in a single individual. Consequently, at this stage of the speciation process, "subpopulation" is hardly an appropriate term for mutant. Nearly every other member of the population will lack the new trait. The mutant will be swamped by non-mutants.

One way to make a distinct mutant *population* is for the offspring of the mutant individual to eventually replace the offspring of the non-mutant individuals. For this to occur, the mutant would have to produce numerous offspring, while the non-mutants would have to bear few offspring. Then the mutant progeny would have to dominate the next round of reproduction, leading to even more mutant offspring. As this cycle repeated itself, and as each previous generation died out, the mutant trait would increase in frequency in the population.

Alternatively, the mutants could become distinct via the process of natural selection. Let's say that, after a few generations of reproduction, the environment changed, such that the mutant trait was advantageous and promoted survival. If the non-mutant members of the population died, then the mutant survivors would be the only ones to produce offspring. As the population recovered, it would consist almost exclusively (depending on the dominant/recessive nature of the mutation) of mutant individuals.

A third way to produce a similar effect is via migration. For example, if the mutant bore several mutant offspring, and if this group of mutants migrated away from the original population, they could found a new population. Once again, as this new population recovered, it would consist almost exclusively (depending on the dominant/recessive nature of the mutation) of mutant individuals. Effectively, the major difference between this scenario and the natural selection scenario and the replacement scenario is the fate of the non-mutants.

How long might these processes take? In the first scenario, multiple generations would be required to spread the mutation throughout the population. Reproductive replacement is time-intensive. It's even more time intensive when it has to be repeated multiple times over for multiple mutations. Millions of mutations would require enormous amounts of time.

In the second scenario, natural selection probably wouldn't operate on multiple traits simultaneously. Instead, a single trait would likely confer a survival advantage. Other mutations around the critical one might be selected passively, but the direct actions of selection tend to be specific. Consequently, the accumulation of millions of mutations via natural selection would also be time-intensive.

Similarly, if migration isolated a single trait — or perhaps a single mutation — at a time, then very long periods of time would be required to isolate millions of them.

In contrast, under the model of preexisting genetic diversity, the second element of the speciation process can occur much more quickly. If new varieties appear each generation — if millions of DNA differences (i.e., new combinations of heterozygous and homozygous DNA positions) can arise in a single generation, then the isolation of genetic and visible distinctiveness is comparatively easy. No need to *sequentially* isolate or select individual DNA variants for generation after generation. Effectively, the model of preexisting heterozygosity eliminates the most time-intensive step of the speciation process under the evolutionary model.

Again, the formation of a *specific* new species is not the goal; rather, the goal is the formation of *new* species, period.

To clarify, the model of preexisting genetic diversity invokes multiple mechanisms as this second step — natural selection, migration, genetic drift, etc. —though Color Plate 84 simulates only two of these mechanisms (due to space constraints).[24]

In summary, under the evolutionary model, the first two steps of the speciation process involve long periods of time. As compared to the total number of differences among species, mutations are infrequent and accumulate slowly. Consequently, the formation of new populations from these individual mutants is gradual, especially if each mutant is isolated sequentially.

In contrast, both the first and second elements of the speciation process are rapid under the model that I've proposed.

The third element of the speciation process occurs very quickly under both the evolutionary model and the preexisting genetic diversity model. On this point, my model doesn't deviate as sharply from the evolution model. But, in combination with the first two elements, the ramifications of my model are starkly different from the evolutionary ramifications.

Thus, my model radically rewrites the traditional view of speciation (Color Plate 84). Despite this departure, the 6,000-year model of speciation that I've just presented is both necessary and sufficient to explain the origin of species. It's necessary to explain the origin of mtDNA differences and nuclear DNA differences. Furthermore, the combination of heterozygous ancestors and of the operation of normal population genetic process is sufficient to explain the origin of modern species within a family.

In fact, the ease with which species can form turns the tables on the objection with which we began this chapter. In light of what we've just discussed, we don't need to wonder about the plausibility of the origin of tens of thousands of species in just a few thousand years. Instead, under the model that I've outlined, we've observed that morphological changes happen every generation. These offspring can grow into large populations much more quickly than anyone appreciates. The isolation of these populations is easy. Together, these steps can occur very rapidly. Consequently, instead of asking if thousands of years is enough for the process of speciation, we should ask why this timescale hasn't led to more species.

What's even more remarkable about the data that I've described in this chapter is their source. The mathematical behavior of populations was solved on a theoretical level long before the modern genetic era. And it was solved by evolutionists. Yet, when combined with modern genetic data, it perfectly merges with the model that I've proposed.

In jigsaw puzzle terms, the genetics of populations (a.k.a. the field of population genetics) clicks perfectly into the shape we predicted from our observations in the preceding two chapters. The orientation of the puzzle provided by these two chapters naturally joined the observations we made in this chapter. In other words, the genetics of populations represents the third category of corner piece, linking mtDNA and nuclear DNA clock data.

In light of this success, the last category of corner piece seemed as if it would just fall into place.

Chapter 10
On the Origin of New Species

Since we began our quest to solve the puzzle of the origin of species, we've discovered many of the crucial pieces — the edge pieces and the corners. Specifically, species appear to have arisen recently, within the last few thousand years. Most of their traits seem to have been present in coded, genetically heterozygous form in their ancestors. In the descendants of these ancestors, shifts from heterozygosity to homozygosity would have been easy. These shifts would have revealed traits that were previously hidden. For example, the genes controlling zebra stripes likely come in dominant and recessive forms. As offspring arose with homozygous forms of one or the other,* distinct traits would have appeared — striped and unstriped individuals. Population subdivision via migration would have isolated these distinct traits and promoted the rise of new species, such as zebras and asses. These scientific pieces suggest a picture of species' origins that is slowly coming together.

Like actual jigsaw puzzles, the identification of these three categories of biological corner pieces (mtDNA clocks, nuclear DNA clocks, population genetics) makes assembly of the rest of the puzzle much easier. In actual rectangular jigsaw puzzles, if three corner pieces connect to four rows of edge pieces, you can virtually *predict* how the last remaining corner piece will look. Similarly, the discoveries of the preceding three chapters nearly predicted the final discovery.

* I.e., homozygous dominant or homozygous recessive.

❧ ❧ ❧ ❧

Up to this point in this book, whenever we've discussed patterns in biology, we've almost always found the discussion to end unresolved. For example, in chapters 5, 7, and 8, we discovered patterns of nested hierarchies at anatomical and molecular levels. Each time, we observed that both the evolutionary model and the creation model expect nested hierarchies to characterize life. Because nested hierarchies cannot eliminate either the evolutionary hypothesis or the creationist hypothesis, nested hierarchies alone cannot resolve the question of the origin of species.

Biological patterns have not borne much fruit in our discussions thus far.

Are there any biological patterns that could productively inform our investigation? In theory, a subset of species and subset of genetic data suggest themselves as candidates. For example, though creationists and evolutionists disagree on ancestry above the level of family, they agree that vertebrate species within a family share a common ancestor. Furthermore, they agree that the mtDNA differences among these species arose as a result of mutation over time. Though they disagree on the rate at which mutation has occurred over long periods of time, they both agree that mtDNA differences act as a clock.

As per chapter 7, the fact of a clock might appear to be a premature conclusion. Recall that neither side has solved the discrepancy between their predictions and actual mtDNA differences among vertebrate species. Nevertheless, both sides in the origins debate would concede that a clock exists — however fast or slowly it ticks.

When the ticks of this clock (i.e., mtDNA differences) are visualized, both sides agree that the resultant tree suggests a timeline of speciation. For example, the mtDNA differences among equid species[1] (see Table 7.1 in chapter 7) can be visualized as a tree[2] (Figure 10.1A). Similar to what we observed for the mtDNA trees in chapter 7, the lengths of the horizontal lines in the equid tree are proportional to the amount of DNA differences separating any two nodes or any two tips of these branches. In fact, these trees can be used to display the exact amount of DNA differences between these points (Figure 10.1B). In this case, the number above each branch represents the fraction of the total sequence that differs between the two points. Thus, to calculate the total DNA difference between any two species, you simply start with one species and trace the horizontal lines connecting it to another species, adding each fractional number. (The vertical length of each line has no relationship to the amount of DNA differences separating any two nodes or any two tips of these branches.) When this sum is multiplied by the total sequence length, the result is the total number of mtDNA differences that separate these species.

Creationists and evolutionists agree that these branch points approximate speciation events. They must — since mtDNA trees are used, in part, to define species. Thus, for vertebrate species within a family, both creationists and evolutionists agree that mtDNA trees represent the history of speciation within the family.

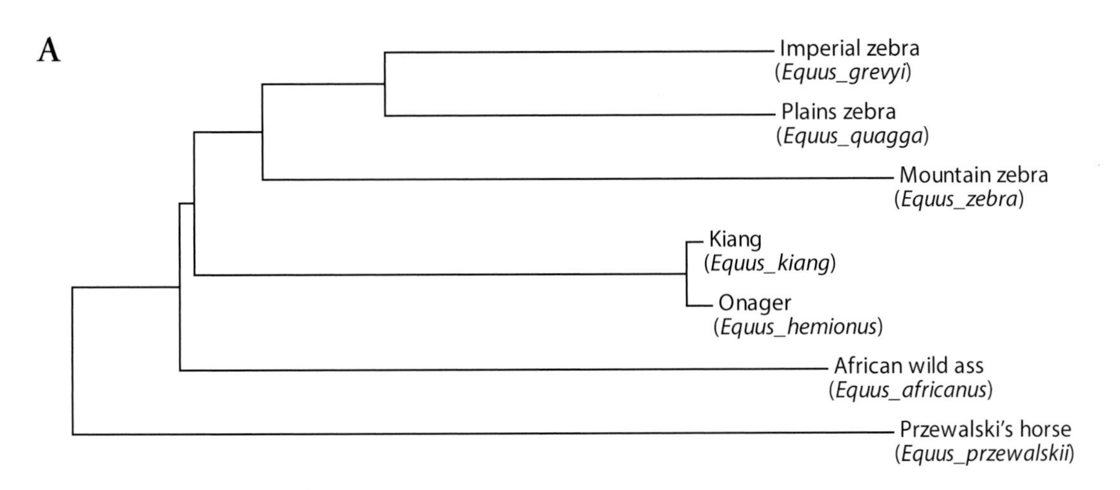

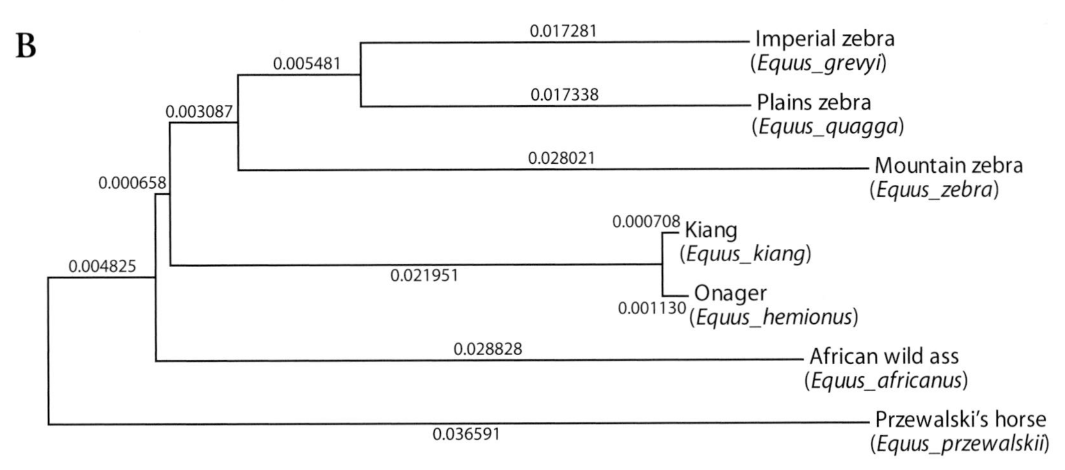

Figure 10.1. Mitochondrial DNA trees for equid species.

(A) Mitochondrial DNA differences were visualized as a branching tree. The horizontal length of each line is proportional to the number of mitochondrial DNA differences that separate one species from another.

(B) Same as (A) but with branch lengths displayed. The branch length above or below each line represents the fraction of total DNA sequence that separates a species from the nearest branch point. To find the total difference between any two species, simply add the horizontal length of each line connecting them. For example, the Onager is 0.001130 units away from its nearest branch point, and the Kiang is 0.000708 units away from its nearest branch point. Added together, a total of 0.001838 units separate the two species. Since the total length of compared DNA sequence is 16,345 base pairs, a total of 30 base pairs (i.e., 0.001838 * 16,345 = 30) divides the two species. (Note that, due to the specifics of the algorithm that was used to create this tree, the differences predicted by branch lengths do not always exactly match the DNA differences obtained by direct pairwise comparison. The reason for this is beyond the scope of this book.)

In light of these facts, we can now attempt to interrogate the history of each vertebrate species. In other words, we can now try to ask and answer the question, *In what pattern did these species arise?*

To clarify, let's identify what we're *not* asking. Because the YEC model and evolutionary model differ on the question of *absolute* timescales, we're not probing the absolute dates of species' divergences. We're also not exploring the pattern in which now-extinct species arose. Creationists and evolutionists have a sharp disagreement on the interpretation of the fossil record, especially on the absolute dates assigned to these fossils. Instead, we're attempting to document the *relative* pattern of speciation that gave rise to *living* species.

To answer this question, we'll have to make some simplifying assumptions. Though our goal is not an absolute timescale, we still must have a definite *start* and *stop* to the speciation process. The *stop* is easy — it's the present. In terms of the tree, the present is the current mtDNA sequence in each species, represented as the tip of each branch. By extension, to look back in time on the tree, we would need to move from the tips to the nodes (i.e., the intersection points) among the branches (i.e., in Figure 10.1, we would need to move from right to left in the diagram).

But where do we put the start of the speciation process? Since we're limiting our focus to living species within vertebrate families, we cannot assess what happened before the first splitting event took place. In other words, when using the mtDNA trees to explore patterns of speciation, we are necessarily limited to interrogating speciation patterns *only after speciation has begun.*

In the diagram of equid speciation (Figure 10.1A–B), I've implied that the speciation process began with the split between the Przewalski's horse lineage and the lineage leading to the rest of the equid species. Is this true? Recall from chapter 7 that, in the human mtDNA tree, I suggested that African lineages might mutate their mtDNA at faster rates than non-African lineages. Evolutionists also have a history of invoking different rates of change in different lineages. If rates vary among lineages, then the tree would have to be rearranged. For example, the root might need to be placed between two of the zebra lineages.

How can we be sure that we're picking the actual start of the speciation process?

In short, we can't — until we measure the mtDNA mutation rates in all of the species we're examining. However, we can make a simplifying assumption and then conduct a preliminary analysis. I've already made this simplifying assumption in the equid mtDNA tree. The assumption invokes a technique termed *rooting on the midpoint.* This method takes the two most divergent branches and sets the root of the tree as the halfway point between the two tips.[3] Effectively, this strategy helps minimize variety in branch lengths, and it seems to have done so for the equid tree (Figure 10.1A–B).

Nevertheless, some families exist in which the branch length differences are so great that rooting on the midpoint still fails to smooth out this variety. For simplicity, we can limit our focus to those families without wild variation in branch lengths.

With respect to the equid mtDNA tree, we can indirectly test the tree orientation via another method. Recall from chapter 9 that, under the hypothesis that most nuclear DNA differences are preexisting, the nuclear DNA differences among species can be used to create a *relative* hierarchy of branching events (see Color Plate 81). The nuclear DNA sequences are known for all equid species.[4] The differences between each species and the domestic horse sequence (a good surrogate sequence for the Przewalski's horse sequence) are also known.[5] The biggest differences are between the horse sequences and all six of the ass and zebra sequences. Each ass and zebra species differs from the domestic horse genome by 24.1 million to 27.8 million base pairs.[6] Conversely, ass and zebra species are closer to one another than any of these six species are to the horse.[7] For example, zebra and ass genomes differ from the domestic donkey genome by only 11 million base pairs or less.[8] In addition, zebra species tend to be closer to one another than they are to ass species, and vice-versa.[9] In short, based on methodology that I describe in Color Plate 81, these nuclear DNA differences depict the following relative branching order: first, the horse lineage split from the ass/zebra lineage. Then the ass and zebra lineages split from one another.

Thus, the order of events implied by the nuclear DNA differences matches the major splitting events depicted by the midpoint-rooted mtDNA tree.[10] This suggests that the midpoint rooting technique is a good first approximation.

With these simplifying assumptions in mind, we can try to answer the question, "What was the pattern of speciation among living species *once speciation had begun*?"

The first answer to this question was not found with equid species. With only seven species in the entire family, Equidae represents a dataset with low statistical potential for making discoveries. By definition, few total speciation events happened in this family (i.e., in the context of living species), making detection of speciation patterns more difficult.

In general, most mammal families have few species. If you calculate the average number of species per family, this point isn't obvious. The average number of mammal species per family is 35.[11] But the average is sensitive to extreme outliers, and the few mammal families that are species-rich skew the average. The mode — the most frequent value — reveals this fact. In mammals, the mode of the number of species per family is just 1 species. In light of these extremes, the median[12] is a better reflection of "average" behavior, and the median number of species per mammal family is 8.[13] If a pattern of speciation was to be found, most mammal families offered few datasets from which statistically robust conclusions could be drawn.

Nevertheless, a few species-rich mammal families do exist, and they offered insights into speciation patterns that other families could not. Ten mammal families have one hundred or more living species representatives — a statistically powerful dataset[14] (Table 10.1). Unfortunately, only one of these families (Bovidae) has complete mtDNA genome sequence information published for 75% or more of its species, and only two families (Bovidae, Cercopithecidae) have

Order	Family	Common Name	Number of species per family	Theoretical number of species with mitochondrial DNA sequences available	% of species within the family	Actual number of species with mitochondrial DNA sequences available
Rodentia	*Muridae*	*mice, rats*	*712*	*45*	*6*	*40*
Rodentia	*Cricetidae*	*lemmings, hamsters*	*683*	*25*	*4*	*24*
Chiroptera	*Vespertilionidae*	*common bats*	*394*	*24*	*6*	*24*
Eulipotyphla	*Soricidae*	*shrews*	*387*	*24*	*6*	*24*
Rodentia	*Sciuridae*	*squirrels, chipmunks*	*280*	*29*	*10*	*29*
Chiroptera	Pteropodidae		183	7	4	
Chiroptera	Phyllostomidae		179	12	7	
Cetartiodactyla	**Bovidae**	**cattle, antelopes**	**138**	**117**	**85**	**107**
Primates	**Cercopithecidae**	**Old World monkeys**	**134**	**65**	**49**	**62**
Chiroptera	Molossidae		103	0	0	
Didelphimorphia	Didelphidae		96	5	5	
Rodentia	Echimyidae		87	2	2	
Chiroptera	Hipposideridae		82	1	1	
Chiroptera	Rhinolophidae		74	8	11	
Dasyuromorphia	Dasyuridae		72	5	7	
Carnivora	**Mustelidae**	**weasels, ferrets**	**63**	**25**	**40**	**25**
Diprotodontia	Macropodidae		63	4	6	
Rodentia	Nesomyidae		63	0	0	
Lagomorpha	Leporidae		62	11	18	
Rodentia	Ctenomyidae		60	2	3	
Rodentia	Heteromyidae		59	0	0	
Cetartiodactyla	**Cervidae**	**deer**	**54**	**46**	**85**	**34**
Chiroptera	Emballonuridae		53	0	0	
Rodentia	Dipodidae		51	6	12	
Eulipotyphla	Talpidae		42	11	26	
Primates	Pitheciidae		42	6	14	

Table 10.1 Mitochondrial DNA sequence representation in mammal families. Only families with 25 species or more are shown. The theoretical number of species with mitochondrial DNA sequences available represents a raw count; the actual number represents the sequences remaining after sequences from domestic animals, subspecies, and other individuals were filtered out.

Order	Family	Common Name	Number of species per family	Theoretical number of species with mitochondrial DNA sequences available	% of species within the family	Actual number of species with mitochondrial DNA sequences available
Primates	Callitrichidae		41	0	0	
Carnivora	**Felidae**	**big & small cats**	**38**	**39**	**103**	**35**
Cetartiodactyla	**Delphinidae**	**oceanic dolphins**	**36**	**19**	**53**	**19**
Rodentia	Geomyidae		36	0	0	
Carnivora	**Canidae**	**wolves, foxes**	**35**	**17**	**49**	**13**
Carnivora	Herpestidae		35	1	3	
Afrosoricida	Tenrecidae		34	1	3	
Carnivora	Viverridae		33	5	15	
Primates	Cheirogaleidae		32	2	6	
Lagomorpha	Ochotonidae		29	3	10	
Rodentia	Gliridae		28	1	4	
Diprotodontia	Phalangeridae		26	2	8	
Primates	Lepilemuridae		26	3	12	
Primates	Atelidae		25	4	16	

complete mtDNA genome sequence information published for 45% or more of their species (Table 10.1). If any mammal families could reveal unique insights into the pattern of speciation, these families would.

What did mtDNA comparisons among the species within the family Bovidae reveal about the pattern of speciation within Bovidae? Like equids, just a few thousand mtDNA differences (at most) separate bovid species from one another. Again, like equids, when we visualize these differences in tree form, we see a similar hierarchical structure.[15] Some bovid species are genetically close; others are more distant from one another (Figure 10.2).

Since mtDNA differences act like a clock, and since branch points represent speciation events, we can get a sense for when each speciation event happened by moving on the tree from left to right. To anchor this to an absolute timescale, we would

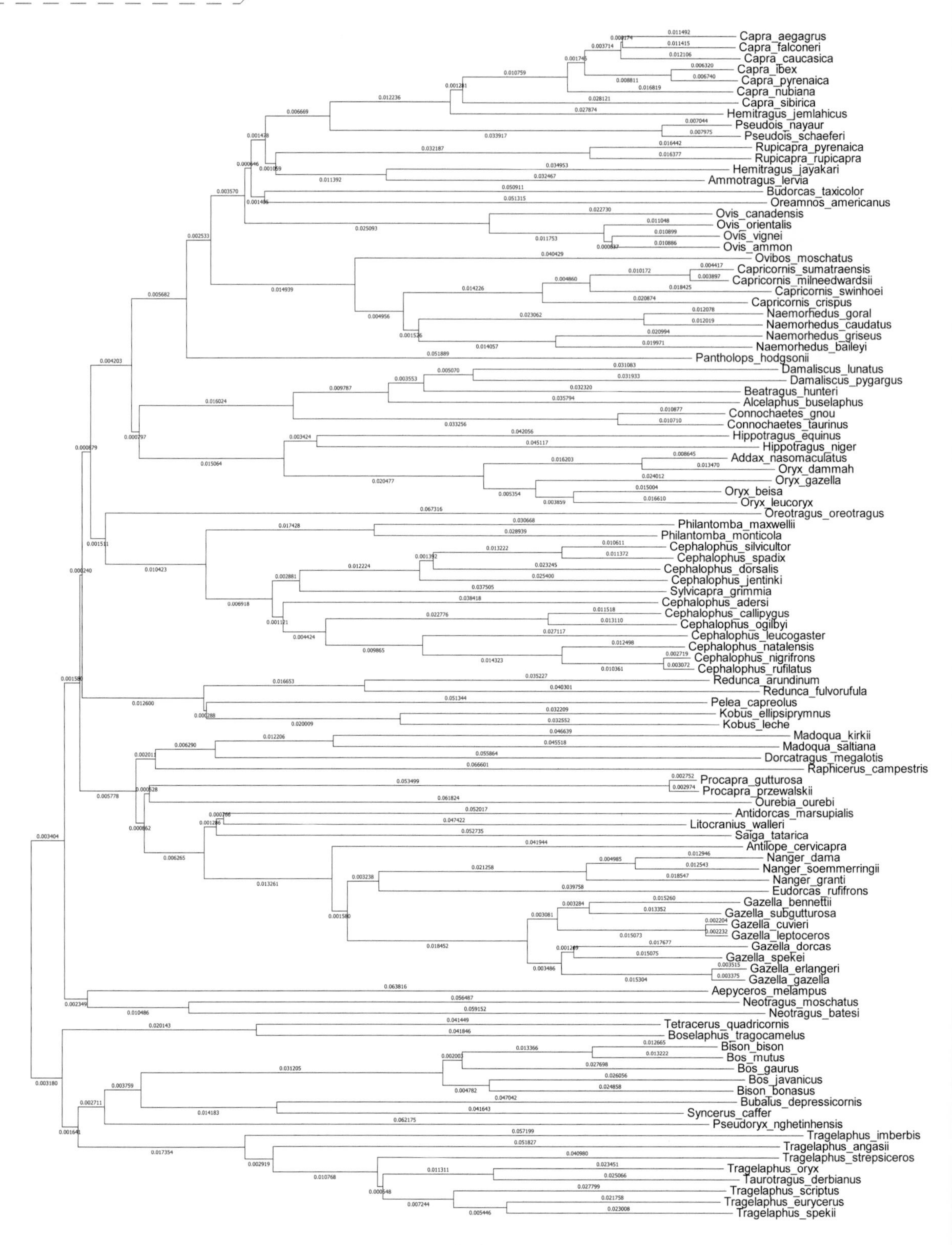
Capra_aegagrus
Capra_falconeri
Capra_caucasica
Capra_ibex
Capra_pyrenaica
Capra_nubiana
Capra_sibirica
Hemitragus_jemlahicus
Pseudois_nayaur
Pseudois_schaeferi
Rupicapra_pyrenaica
Rupicapra_rupicapra
Hemitragus_jayakari
Ammotragus_lervia
Budorcas_taxicolor
Oreamnos_americanus
Ovis_canadensis
Ovis_orientalis
Ovis_vignei
Ovis_ammon
Ovibos_moschatus
Capricornis_sumatraensis
Capricornis_milneedwardsii
Capricornis_swinhoei
Capricornis_crispus
Naemorhedus_goral
Naemorhedus_caudatus
Naemorhedus_griseus
Naemorhedus_baileyi
Pantholops_hodgsonii
Damaliscus_lunatus
Damaliscus_pygargus
Beatragus_hunteri
Alcelaphus_buselaphus
Connochaetes_gnou
Connochaetes_taurinus
Hippotragus_equinus
Hippotragus_niger
Addax_nasomaculatus
Oryx_dammah
Oryx_gazella
Oryx_beisa
Oryx_leucoryx
Oreotragus_oreotragus
Philantomba_maxwellii
Philantomba_monticola
Cephalophus_silvicultor
Cephalophus_spadix
Cephalophus_dorsalis
Cephalophus_jentinki
Sylvicapra_grimmia
Cephalophus_adersi
Cephalophus_callipygus
Cephalophus_ogilbyi
Cephalophus_leucogaster
Cephalophus_natalensis
Cephalophus_nigrifrons
Cephalophus_rufilatus
Redunca_arundinum
Redunca_fulvorufula
Pelea_capreolus
Kobus_ellipsiprymnus
Kobus_leche
Madoqua_kirkii
Madoqua_saltiana
Dorcatragus_megalotis
Raphicerus_campestris
Procapra_gutturosa
Procapra_przewalskii
Ourebia_ourebi
Antidorcas_marsupialis
Litocranius_walleri
Saiga_tatarica
Antilope_cervicapra
Nanger_dama
Nanger_soemmerringii
Nanger_granti
Eudorcas_rufifrons
Gazella_bennettii
Gazella_subgutturosa
Gazella_cuvieri
Gazella_leptoceros
Gazella_dorcas
Gazella_spekei
Gazella_erlangeri
Gazella_gazella
Aepyceros_melampus
Neotragus_moschatus
Neotragus_batesi
Tetracerus_quadricornis
Boselaphus_tragocamelus
Bison_bison
Bos_mutus
Bos_gaurus
Bos_javanicus
Bison_bonasus
Bubalus_depressicornis
Syncerus_caffer
Pseudoryx_nghetinhensis
Tragelaphus_imberbis
Tragelaphus_angasii
Tragelaphus_strepsiceros
Tragelaphus_oryx
Taurotragus_derbianus
Tragelaphus_scriptus
Tragelaphus_eurycerus
Tragelaphus_spekii

Figure 10.2 (previous page). Mitochondrial DNA tree for cattle, sheep, and antelope species (family Bovidae). Mitochondrial DNA differences were visualized as a branching tree. The horizontal length of each line is proportional to the number of mitochondrial DNA differences that separate one species from another. The number above each branch depicts the fraction of the total sequence represented by the branch.

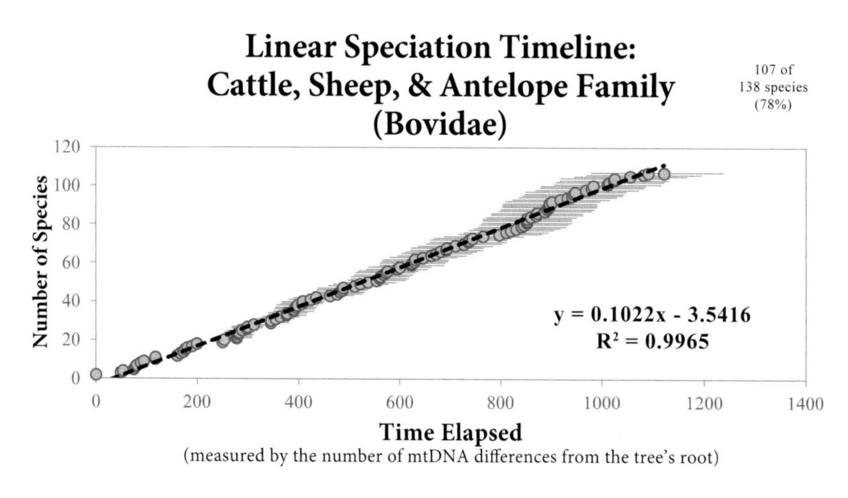

Figure 10.3. The pattern of speciation in the family Bovidae matches a linear function. Splitting events on the Bovidae mtDNA tree were scored as speciation events, and the timing of these events was obtained from the branch length values of the Bovidae mtDNA tree. Each circle represents a single speciation (splitting) event. The total number of species represented in this graph (i.e., those with publically available mtDNA genome sequences) is shown in the top right as a percentage of the total number of species within the family. The black dashed line represents the linear regression function (shown in the equation on the center-right) for these events. The light horizontal gray lines on either side of the circles approximate the statistical uncertainty of the exact timing of these splitting events. The uncertainty arises from the fact that the branches that terminate on each species are not all the exact same distance from the root of the entire tree. This can be seen in the horizontal jaggedness of the branches of the Bovidae tree in Figure 10.2.

need to adopt a particular origins model. Instead, since we're interested simply in the relative pattern of speciation, the relative branching order is sufficient.

Again, the branch lengths are easily obtained electronically. Using these branch lengths (i.e., mtDNA differences) as a surrogate for time, we can create a timeline[16] of sorts for speciation within the family Bovidae (Figure 10.3). Remarkably, it showed a very strong match to a linear mathematical function (correlation coefficient ["R^2" value] of 0.9965; see Figure 10.3).

Similar results followed from analysis of the Old World monkey family (Cercopithecidae). The differences among species can be visualized as a tree,[17] with branch lengths depicted above each branch (Figure 10.4). Since mtDNA differences act like a clock, and since branch points represent speciation events, we can get a sense for when each speciation event happened by moving on the tree from left to right, and we can create a timeline with the branch lengths. Again, plotting these speciation events[18] revealed a pattern of speciation that showed a very

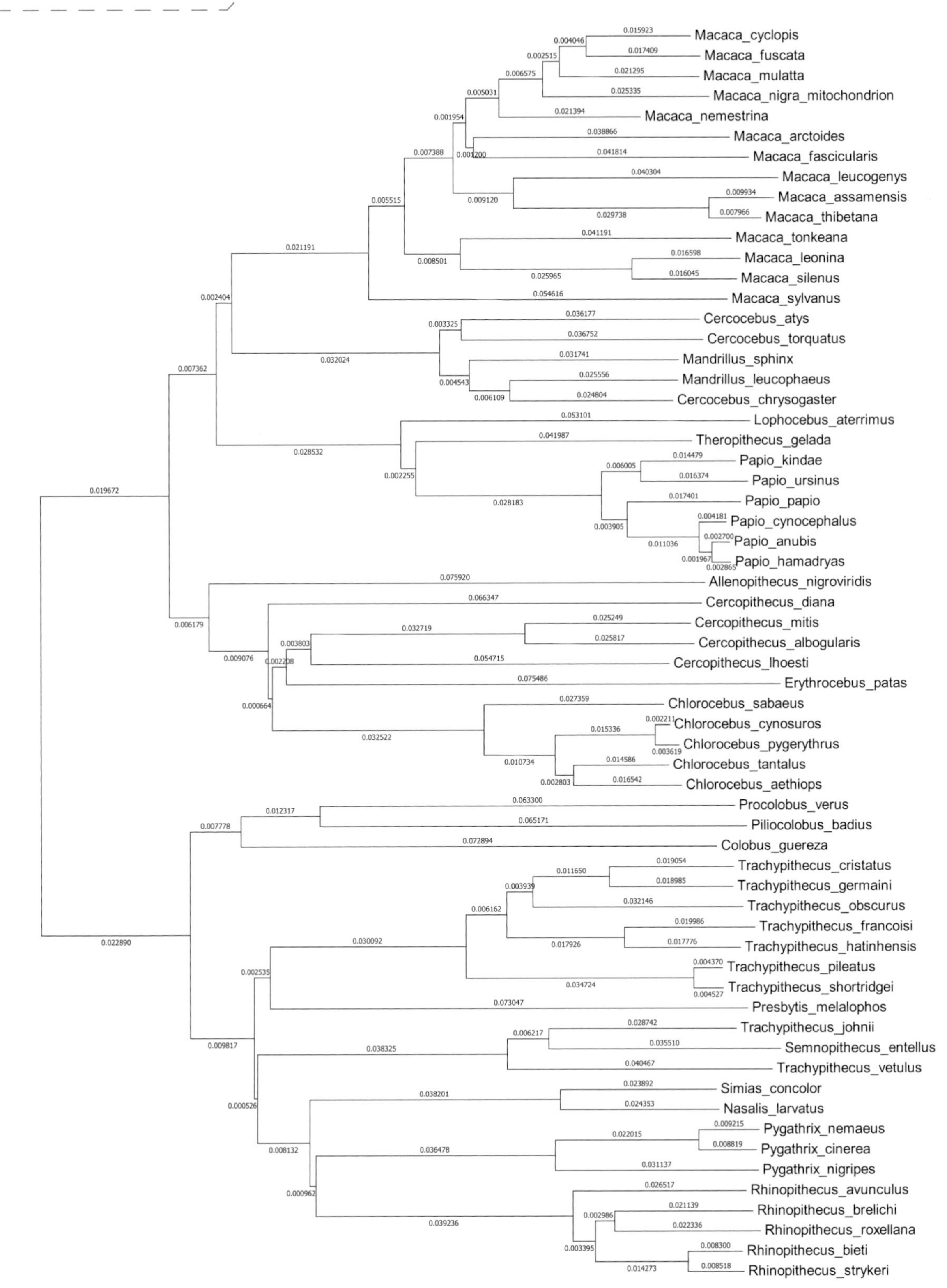
Macaca_cyclopis
Macaca_fuscata
Macaca_mulatta
Macaca_nigra_mitochondrion
Macaca_nemestrina
Macaca_arctoides
Macaca_fascicularis
Macaca_leucogenys
Macaca_assamensis
Macaca_thibetana
Macaca_tonkeana
Macaca_leonina
Macaca_silenus
Macaca_sylvanus
Cercocebus_atys
Cercocebus_torquatus
Mandrillus_sphinx
Mandrillus_leucophaeus
Cercocebus_chrysogaster
Lophocebus_aterrimus
Theropithecus_gelada
Papio_kindae
Papio_ursinus
Papio_papio
Papio_cynocephalus
Papio_anubis
Papio_hamadryas
Allenopithecus_nigroviridis
Cercopithecus_diana
Cercopithecus_mitis
Cercopithecus_albogularis
Cercopithecus_lhoesti
Erythrocebus_patas
Chlorocebus_sabaeus
Chlorocebus_cynosuros
Chlorocebus_pygerythrus
Chlorocebus_tantalus
Chlorocebus_aethiops
Procolobus_verus
Piliocolobus_badius
Colobus_guereza
Trachypithecus_cristatus
Trachypithecus_germaini
Trachypithecus_obscurus
Trachypithecus_francoisi
Trachypithecus_hatinhensis
Trachypithecus_pileatus
Trachypithecus_shortridgei
Presbytis_melalophos
Trachypithecus_johnii
Semnopithecus_entellus
Trachypithecus_vetulus
Simias_concolor
Nasalis_larvatus
Pygathrix_nemaeus
Pygathrix_cinerea
Pygathrix_nigripes
Rhinopithecus_avunculus
Rhinopithecus_brelichi
Rhinopithecus_roxellana
Rhinopithecus_bieti
Rhinopithecus_strykeri

Figure 10.4 (previous page). Mitochondrial DNA tree for Old World monkey species (family Cercopithecidae). Mitochondrial DNA differences were visualized as a branching tree. The horizontal length of each line is proportional to the number of mitochondrial DNA differences that separate one species from another. The number above each branch depicts the fraction of the total sequence represented by the branch.

strong match to a linear mathematical function (correlation coefficient ["R^2" value] of 0.95; see Figure 10.5).

In fact, in each mammal family that I examined, I found strong evidence for a linear pattern of speciation. Among the remaining mammal families with higher numbers of species and with a mtDNA genome representation of around 40% of these species (Table 10.1), the same result ensued. The weasel (Mustelidae),[19] deer (Cervidae),[20] cat (Felidae),[21] dolphin (Delphinidae),[22] and dog (Canidae)[23] families all displayed a linear pattern of speciation with strong statistical support — correlation coefficients of 0.9 or higher (Figures 10.6-10.10).

The consistency among these results was surprising. These datasets sampled four mammal orders (Artiodactyla — Bovidae, Cervidae; Primates — Cercopithecidae; Carnivora — Mustelidae, Felidae, Canidae; Cetacea — Delphinidae), and they touched very diverse biology — herbivores,

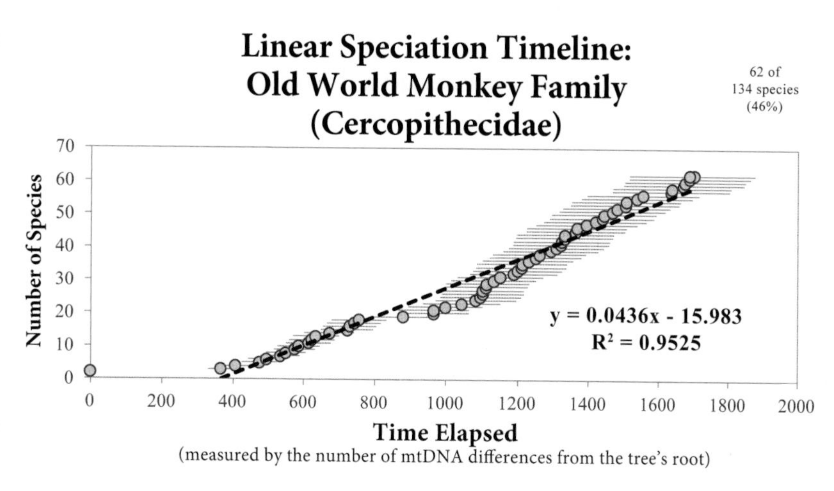

Figure 10.5. The pattern of speciation in the family Cercopithecidae matches a linear function. Splitting events on the Cercopithecidae mtDNA tree were scored as speciation events, and the timing of these events was obtained from the branch length values of the Cercopithecidae mtDNA tree. Each circle represents a single speciation (splitting) event. The total number of species represented in this graph (i.e., those with publically available mtDNA genome sequences) is shown in the top right as a percentage of the total number of species within the family. The black dashed line represents the linear regression function (shown in the equation on the center-right) for these events. The light horizontal gray lines on either side of the circles approximate the statistical uncertainty of the exact timing of these splitting events. The uncertainty arises from the fact that the branches that terminate on each species are not all the exact same distance from the root of the entire tree. This can be seen in the horizontal jaggedness of the branches of the Cercopithecidae tree in Figure 10.4.

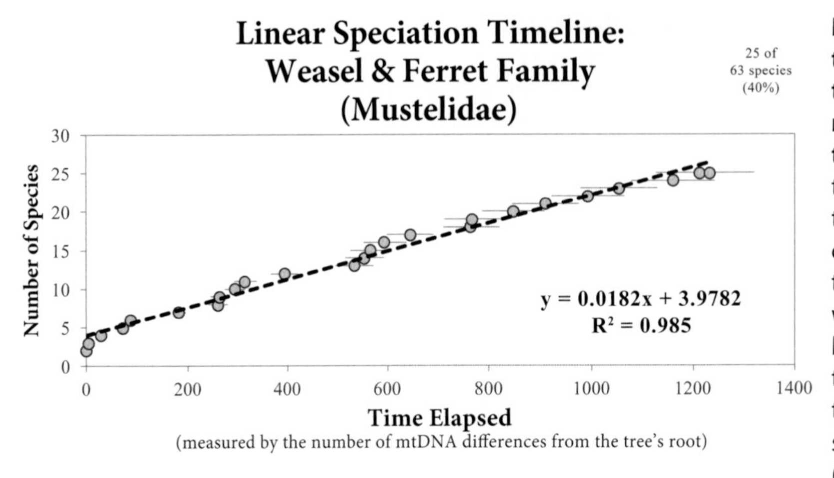

Figure 10.6. The pattern of speciation in the family Mustelidae matches a linear function. Splitting events on the Mustelidae mtDNA tree were scored as speciation events, and the timing of these events was obtained from the branch length values of the Mustelidae mtDNA tree. Each circle represents a single speciation (splitting) event. The total number of species represented in this graph (i.e., those with publically available mtDNA genome sequences) is shown in the top right as a percentage of the total number of species within the family. The black dashed line represents the linear regression function (shown in the equation on the center-right) for these events. The light horizontal gray lines on either side of the circles approximate the statistical uncertainty of the exact timing of these splitting events. The uncertainty arises from the fact that the branches that terminate on each species are not all the exact same distance from the root of the entire tree.

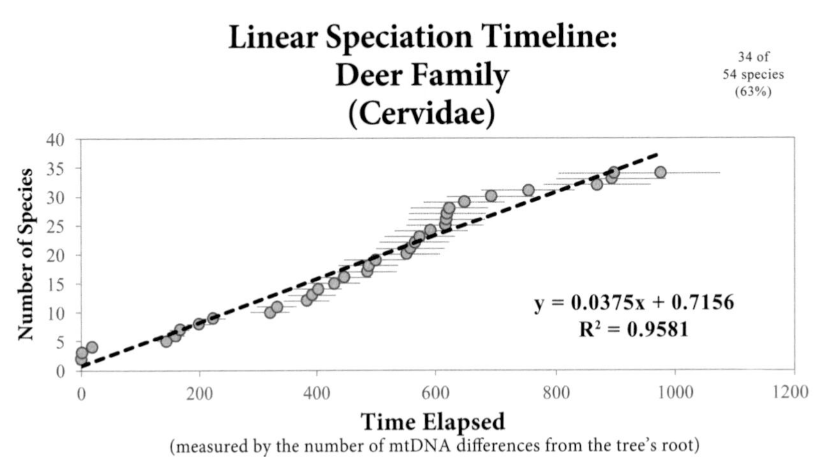

Figure 10.7. The pattern of speciation in the family Cervidae matches a linear function. Splitting events on the Cervidae mtDNA tree were scored as speciation events, and the timing of these events was obtained from the branch length values of the Cervidae mtDNA tree. Each circle represents a single speciation (splitting) event. The total number of species represented in this graph (i.e., those with publically available mtDNA genome sequences) is shown in the top right as a percentage of the total number of species within the family. The black dashed line represents the linear regression function (shown in the equation on the center-right) for these events. The light horizontal gray lines on either side of the circles approximate the statistical uncertainty of the exact timing of these splitting events. The uncertainty arises from the fact that the branches that terminate on each species are not all the exact same distance from the root of the entire tree.

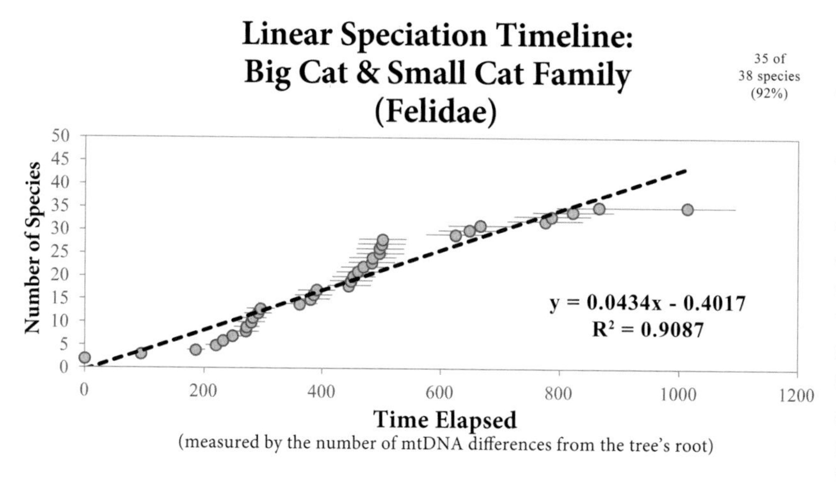

Figure 10.8. The pattern of speciation in the family Felidae matches a linear function. Splitting events on the Felidae mtDNA tree were scored as speciation events, and the timing of these events was obtained from the branch length values of the Felidae mtDNA tree. Each circle represents a single speciation (splitting) event. The total number of species represented in this graph (i.e., those with publically available mtDNA genome sequences) is shown in the top right as a percentage of the total number of species within the family. The black dashed line represents the linear regression function (shown in the equation on the center-right) for these events. The light horizontal gray lines on either side of the circles approximate the statistical uncertainty of the exact timing of these splitting events. The uncertainty arises from the fact that the branches that terminate on each species are not all the exact same distance from the root of the entire tree.

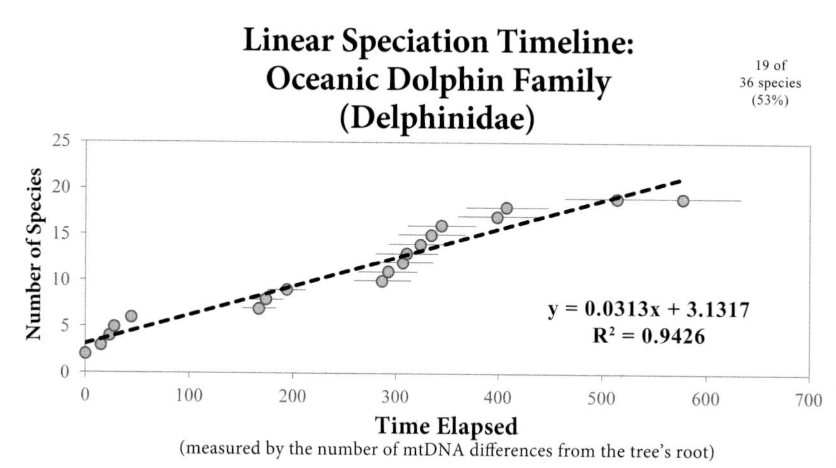

Figure 10.9. The pattern of speciation in the family Delphinidae matches a linear function. Splitting events on the Delphinidae mtDNA tree were scored as speciation events, and the timing of these events was obtained from the branch length values of the Delphinidae mtDNA tree. Each circle represents a single speciation (splitting) event. The total number of species represented in this graph (i.e., those with publically available mtDNA genome sequences) is shown in the top right as a percentage of the total number of species within the family. The black dashed line represents the linear regression function (shown in the equation on the center-right) for these events. The light horizontal gray lines on either side of the circles approximate the statistical uncertainty of the exact timing of these splitting events. The uncertainty arises from the fact that the branches that terminate on each species are not all the exact same distance from the root of the entire tree.

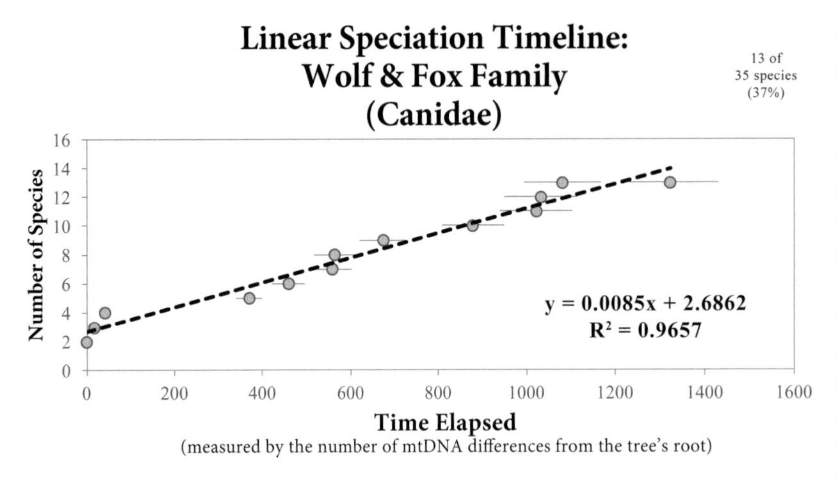

Figure 10.10. The pattern of speciation in the family Canidae matches a linear function. Splitting events on the Canidae mtDNA tree were scored as speciation events, and the timing of these events was obtained from the branch length values of the Canidae mtDNA tree. Each circle represents a single speciation (splitting) event. The total number of species represented in this graph (i.e., those with publically available mtDNA genome sequences) is shown in the top right as a percentage of the total number of species within the family. The black dashed line represents the linear regression function (shown in the equation on the center-right) for these events. The light horizontal gray lines on either side of the circles approximate the statistical uncertainty of the exact timing of these splitting events. The uncertainty arises from the fact that the branches that terminate on each species are not all the exact same distance from the root of the entire tree.

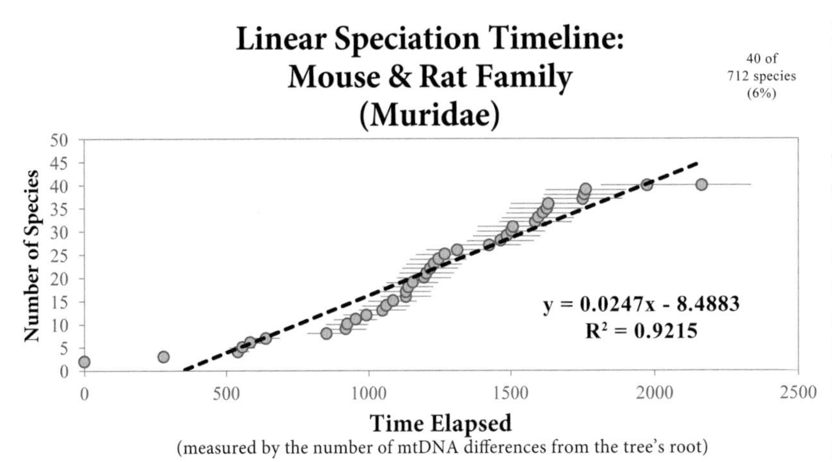

Figure 10.11. The pattern of speciation in the family Muridae matches a linear function. Splitting events on the Muridae mtDNA tree were scored as speciation events, and the timing of these events was obtained from the branch length values of the Muridae mtDNA tree. Each circle represents a single speciation (splitting) event. The total number of species represented in this graph (i.e., those with publically available mtDNA genome sequences) is shown in the top right as a percentage of the total number of species within the family. The black dashed line represents the linear regression function (shown in the equation on the center-right) for these events. The light horizontal gray lines on either side of the circles approximate the statistical uncertainty of the exact timing of these splitting events. The uncertainty arises from the fact that the branches that terminate on each species are not all the exact same distance from the root of the entire tree.

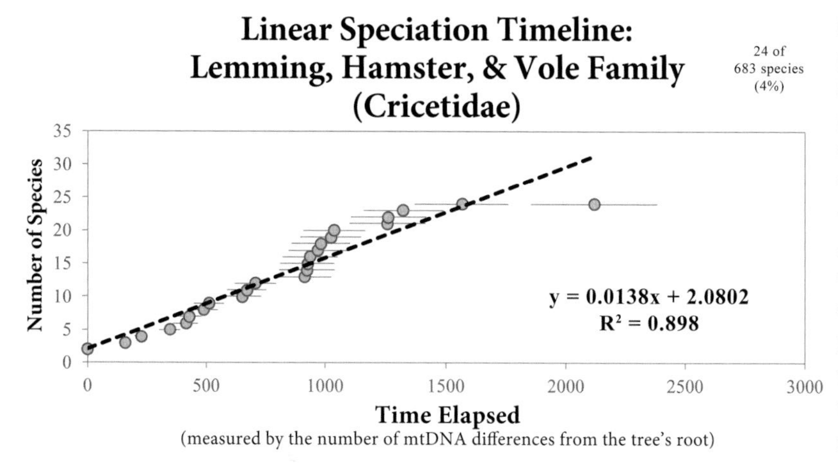

Figure 10.12. The pattern of speciation in the family Cricetidae matches a linear function. Splitting events on the Cricetidae mtDNA tree were scored as speciation events, and the timing of these events was obtained from the branch length values of the Cricetidae mtDNA tree. Each circle represents a single speciation (splitting) event. The total number of species represented in this graph (i.e., those with publically available mtDNA genome sequences) is shown in the top right as a percentage of the total number of species within the family. The black dashed line represents the linear regression function (shown in the equation on the center-right) for these events. The light horizontal gray lines on either side of the circles approximate the statistical uncertainty of the exact timing of these splitting events. The uncertainty arises from the fact that the branches that terminate on each species are not all the exact same distance from the root of the entire tree.

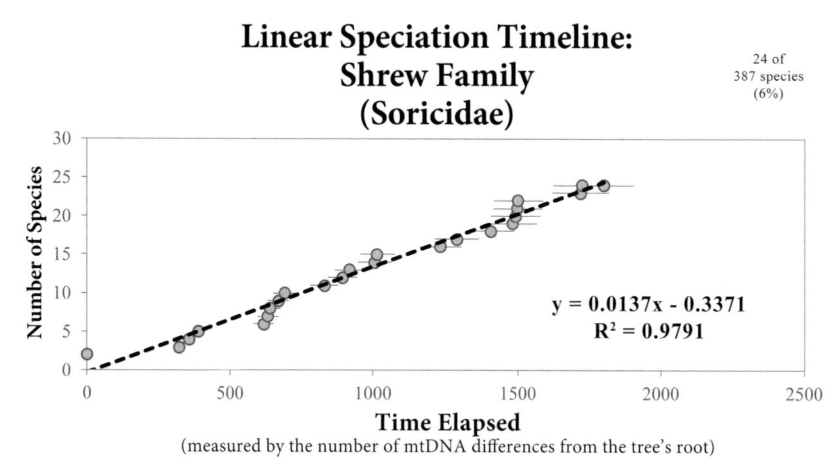

Figure 10.13. The pattern of speciation in the family Soricidae matches a linear function. Splitting events on the Soricidae mtDNA tree were scored as speciation events, and the timing of these events was obtained from the branch length values of the Soricidae mtDNA tree. Each circle represents a single speciation (splitting) event. The total number of species represented in this graph (i.e., those with publically available mtDNA genome sequences) is shown in the top right as a percentage of the total number of species within the family. The black dashed line represents the linear regression function (shown in the equation on the center-right) for these events. The light horizontal gray lines on either side of the circles approximate the statistical uncertainty of the exact timing of these splitting events. The uncertainty arises from the fact that the branches that terminate on each species are not all the exact same distance from the root of the entire tree.

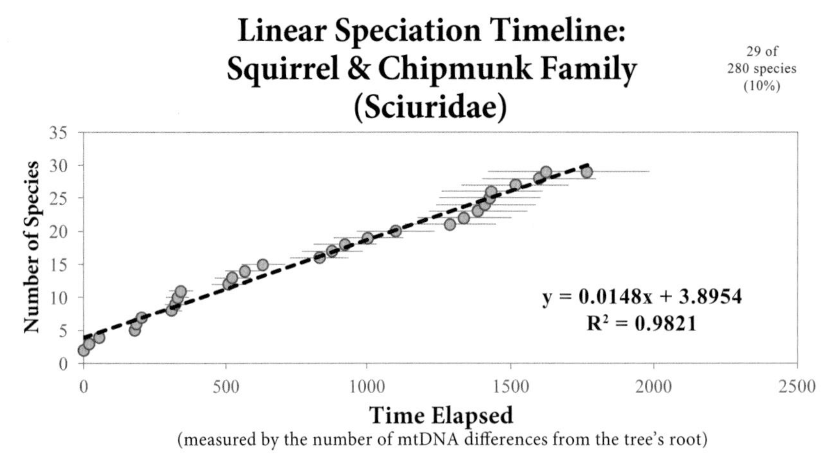

Figure 10.14. The pattern of speciation in the family Sciuridae matches a linear function. Splitting events on the Sciuridae mtDNA tree were scored as speciation events, and the timing of these events was obtained from the branch length values of the Sciuridae mtDNA tree. Each circle represents a single speciation (splitting) event. The total number of species represented in this graph (i.e., those with publically available mtDNA genome sequences) is shown in the top right as a percentage of the total number of species within the family. The black dashed line represents the linear regression function (shown in the equation on the center-right) for these events. The light horizontal gray lines on either side of the circles approximate the statistical uncertainty of the exact timing of these splitting events. The uncertainty arises from the fact that the branches that terminate on each species are not all the exact same distance from the root of the entire tree.

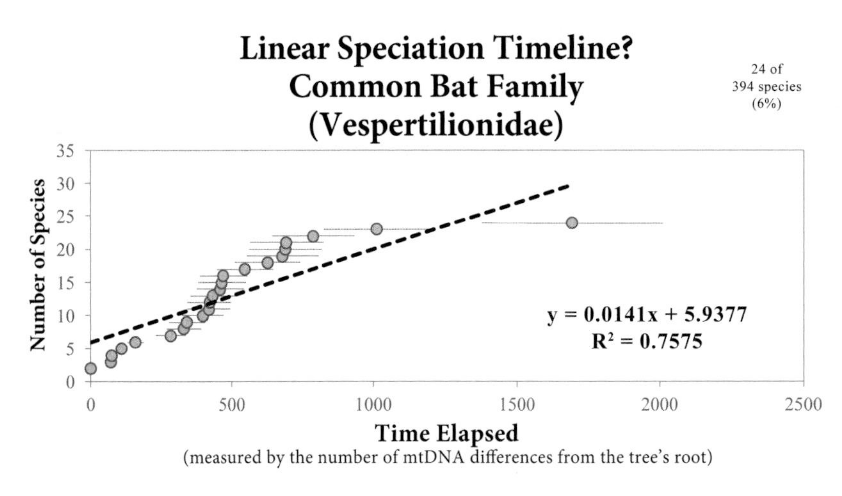

Figure 10.15. The pattern of speciation in the family Vespertilionidae shows resemblance to a linear function. Splitting events on the Vespertilionidae mtDNA tree were scored as speciation events, and the timing of these events was obtained from the branch length values of the Vespertilionidae mtDNA tree. Each circle represents a single speciation (splitting) event. The total number of species represented in this graph (i.e., those with publically available mtDNA genome sequences) is shown in the top right as a percentage of the total number of species within the family. The black dashed line represents the linear regression function (shown in the equation on the center-right) for these events. The light horizontal gray lines on either side of the circles approximate the statistical uncertainty of the exact timing of these splitting events. The uncertainty arises from the fact that the branches that terminate on each species are not all the exact same distance from the root of the entire tree.

carnivores, terrestrial species, aquatic species. Furthermore, the linear pattern was seen in families with high mtDNA species representation (e.g., Bovidae) and with lower mtDNA species representation (e.g., Canidae). These results suggested that the pattern would be generally true across mammals.

In light of these results, we can explore the pattern of speciation in families possessing many species — but with low species representation in the mtDNA dataset. For example, the five mammal families[24] with the most species (Table 10.1) had low mtDNA representation. Yet, despite this limitation, four (Figures 10.11–14) of the five families showed a linear pattern of speciation, at an R^2 value of 0.9 or higher.* Only the bat family (Figure 10.15) showed a lower R^2 value. This could be due to a different pattern of speciation within the bat family. Alternatively, it might simply be an artifact of poor mtDNA representation. If so, as more mtDNA sequences from species within the family are added to the analysis, I expect the R^2 value to increase.

Because nearly all of these patterns were linear, these results were generalizable into a predictive equation. The slope of the linear line was predictable from the number of species in the analyses divided by the total number of mtDNA differences between the start of speciation and the present. For example, in Bovidae, about 1,100 mtDNA differences separate the start of the speciation process from the present. Dividing the total number of species compared (i.e., 107)

* In the analysis of species within the family Cricetidae, rounding the R^2 value to the nearest hundredths place yields a value of 0.90.

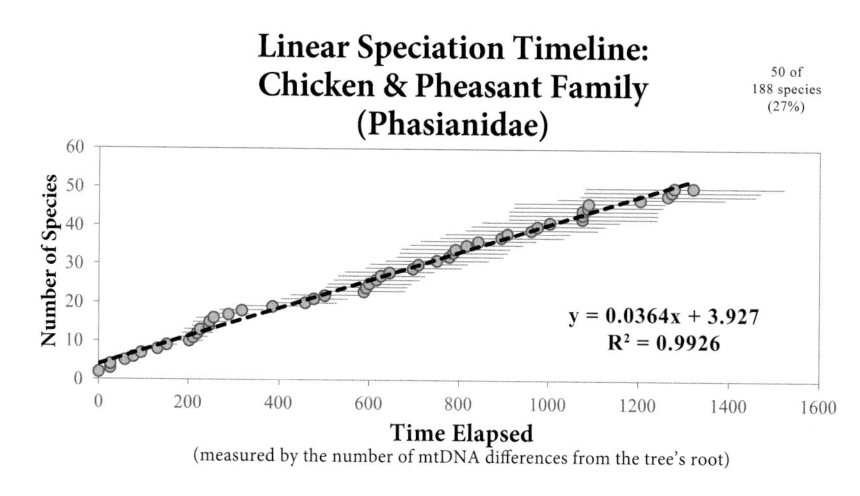

Figure 10.16. The pattern of speciation in the family Phasianidae matches a linear function. Splitting events on the Phasianidae mtDNA tree were scored as speciation events, and the timing of these events was obtained from the branch length values of the Phasianidae mtDNA tree. Each circle represents a single speciation (splitting) event. The total number of species represented in this graph (i.e., those with publically available mtDNA genome sequences) is shown in the top right as a percentage of the total number of species within the family. The black dashed line represents the linear regression function (shown in the equation on the center-right) for these events. The light horizontal gray lines on either side of the circles approximate the statistical uncertainty of the exact timing of these splitting events. The uncertainty arises from the fact that the branches that terminate on each species are not all the exact same distance from the root of the entire tree.

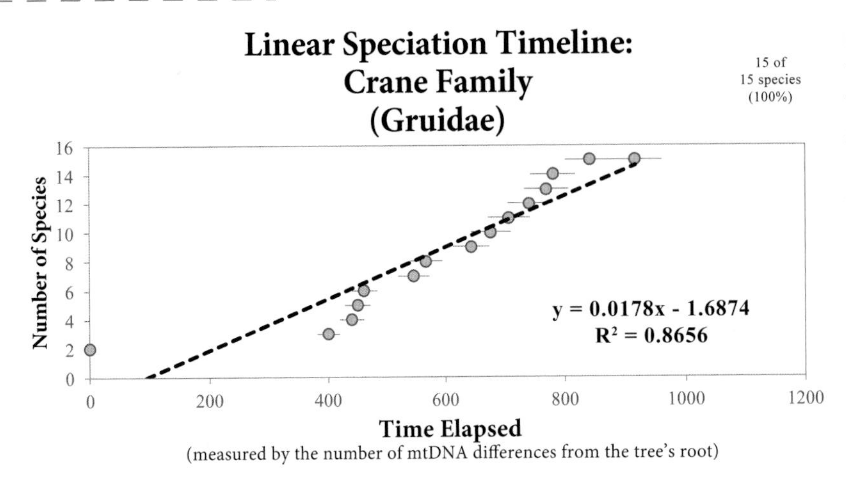

Figure 10.17. The pattern of speciation in the family Gruidae shows strong resemblance to a linear function. Splitting events on the Gruidae mtDNA tree were scored as speciation events, and the timing of these events was obtained from the branch length values of the Gruidae mtDNA tree. Each circle represents a single speciation (splitting) event. The total number of species represented in this graph (i.e., those with publically available mtDNA genome sequences) is shown in the top right as a percentage of the total number of species within the family. The black dashed line represents the linear regression function (shown in the equation on the center-right) for these events. The light horizontal gray lines on either side of the circles approximate the statistical uncertainty of the exact timing of these splitting events. The uncertainty arises from the fact that the branches that terminate on each species are not all the exact same distance from the root of the entire tree.

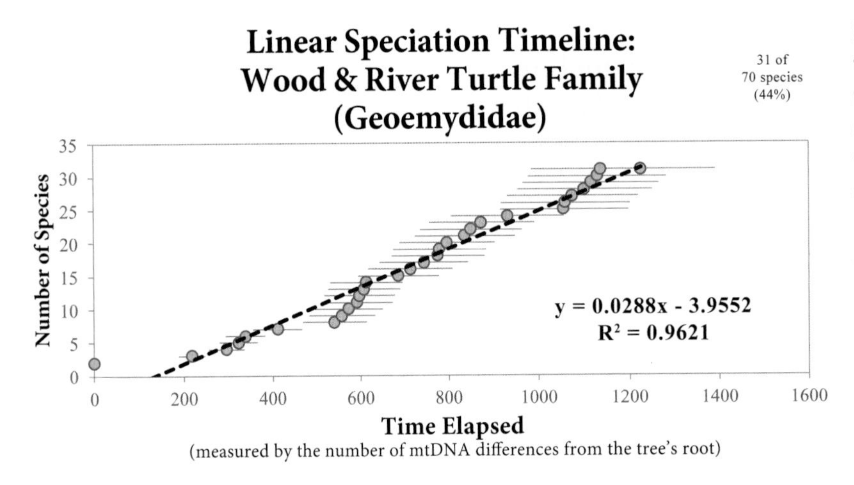

Figure 10.18. The pattern of speciation in the family Geoemydidae matches a linear function. Splitting events on the Geoemydidae mtDNA tree were scored as speciation events, and the timing of these events was obtained from the branch length values of the Geoemydidae mtDNA tree. Each circle represents a single speciation (splitting) event. The total number of species represented in this graph (i.e., those with publically available mtDNA genome sequences) is shown in the top right as a percentage of the total number of species within the family. The black dashed line represents the linear regression function (shown in the equation on the center-right) for these events. The light horizontal gray lines on either side of the circles approximate the statistical uncertainty of the exact timing of these splitting events. The uncertainty arises from the fact that the branches that terminate on each species are not all the exact same distance from the root of the entire tree.

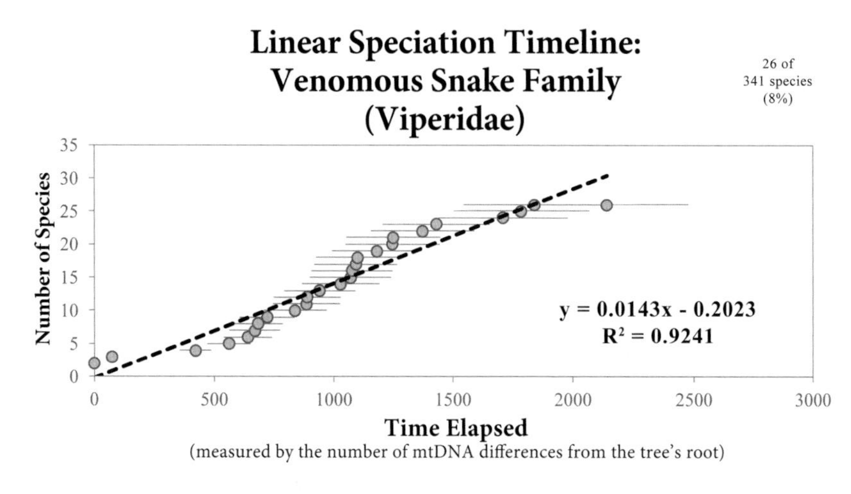

Figure 10.19. The pattern of speciation in the family Viperidae matches a linear function. Splitting events on the Viperidae mtDNA tree were scored as speciation events, and the timing of these events was obtained from the branch length values of the Viperidae mtDNA tree. Each circle represents a single speciation (splitting) event. The total number of species represented in this graph (i.e., those with publically available mtDNA genome sequences) is shown in the top right as a percentage of the total number of species within the family. The black dashed line represents the linear regression function (shown in the equation on the center-right) for these events. The light horizontal gray lines on either side of the circles approximate the statistical uncertainty of the exact timing of these splitting events. The uncertainty arises from the fact that the branches that terminate on each species are not all the exact same distance from the root of the entire tree.

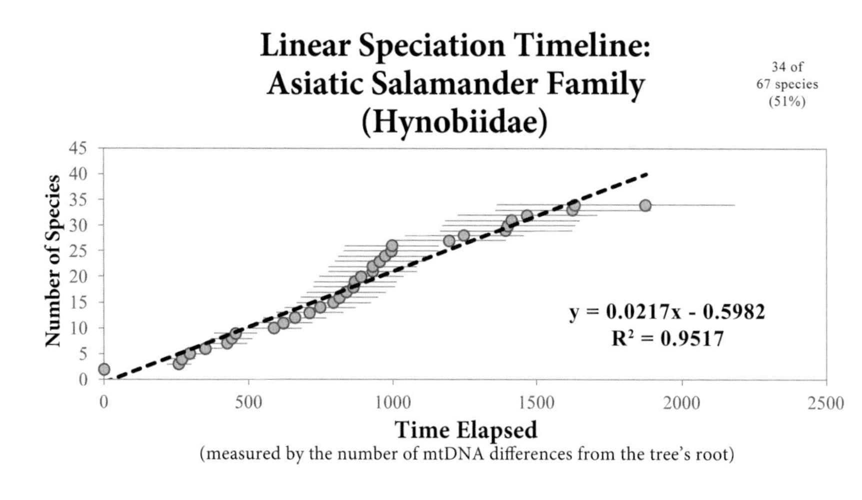

Figure 10.20. The pattern of speciation in the family Hynobiidae matches a linear function. Splitting events on the Hynobiidae mtDNA tree were scored as speciation events, and the timing of these events was obtained from the branch length values of the Hynobiidae mtDNA tree. Each circle represents a single speciation (splitting) event. The total number of species represented in this graph (i.e., those with publically available mtDNA genome sequences) is shown in the top right as a percentage of the total number of species within the family. The black dashed line represents the linear regression function (shown in the equation on the center-right) for these events. The light horizontal gray lines on either side of the circles approximate the statistical uncertainty of the exact timing of these splitting events. The uncertainty arises from the fact that the branches that terminate on each species are not all the exact same distance from the root of the entire tree.

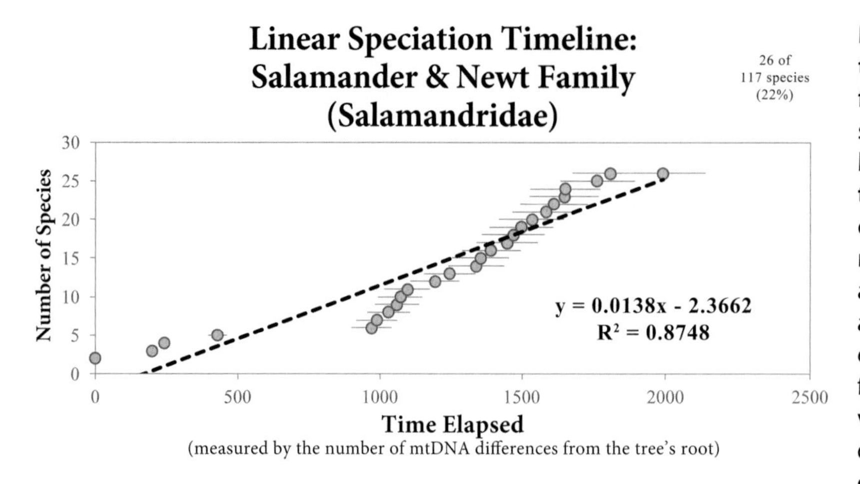

Figure 10.21. The pattern of speciation in the family Salamandridae shows strong resemblance to a linear function. Splitting events on the Salamandridae mtDNA tree were scored as speciation events, and the timing of these events was obtained from the branch length values of the Salamandridae mtDNA tree. Each circle represents a single speciation (splitting) event. The total number of species represented in this graph (i.e., those with publically available mtDNA genome sequences) is shown in the top right as a percentage of the total number of species within the family. The black dashed line represents the linear regression function (shown in the equation on the center-right) for these events. The light horizontal gray lines on either side of the circles approximate the statistical uncertainty of the exact timing of these splitting events. The uncertainty arises from the fact that the branches that terminate on each species are not all the exact same distance from the root of the entire tree.

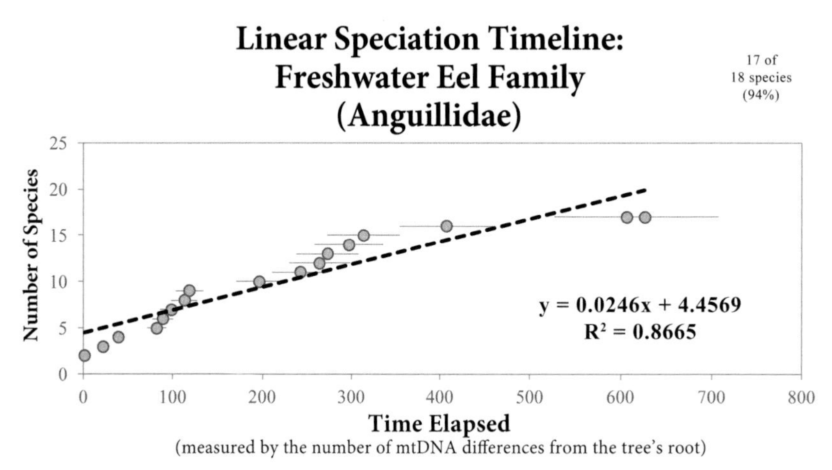

Figure 10.22. The pattern of speciation in the family Anguillidae shows strong resemblance to a linear function. Splitting events on the Anguillidae mtDNA tree were scored as speciation events, and the timing of these events was obtained from the branch length values of the Anguillidae mtDNA tree. Each circle represents a single speciation (splitting) event. The total number of species represented in this graph (i.e., those with publically available mtDNA genome sequences) is shown in the top right as a percentage of the total number of species within the family. The black dashed line represents the linear regression function (shown in the equation on the center-right) for these events. The light horizontal gray lines on either side of the circles approximate the statistical uncertainty of the exact timing of these splitting events. The uncertainty arises from the fact that the branches that terminate on each species are not all the exact same distance from the root of the entire tree.

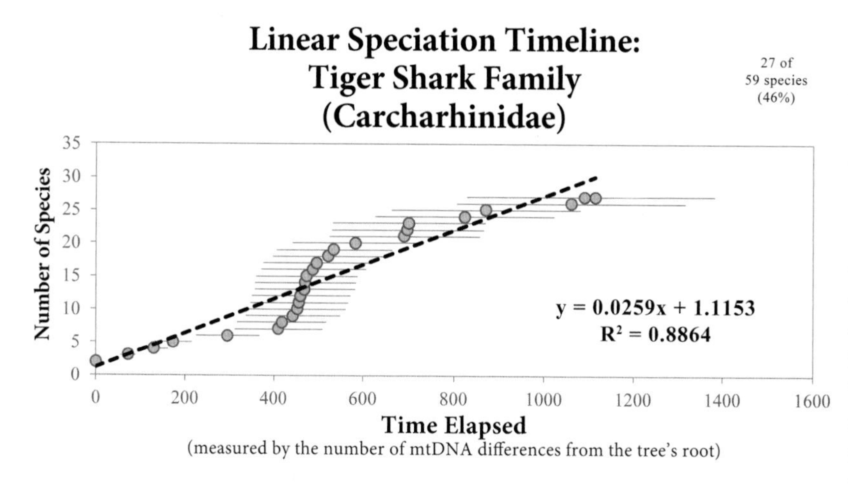

Figure 10.23. The pattern of speciation in the family Carcharhinidae matches a linear function. Splitting events on the Carcharhinidae mtDNA tree were scored as speciation events, and the timing of these events was obtained from the branch length values of the Carcharhinidae mtDNA tree. Each circle represents a single speciation (splitting) event. The total number of species represented in this graph (i.e., those with publically available mtDNA genome sequences) is shown in the top right as a percentage of the total number of species within the family. The black dashed line represents the linear regression function (shown in the equation on the center-right) for these events. The light horizontal gray lines on either side of the circles approximate the statistical uncertainty of the exact timing of these splitting events. The uncertainty arises from the fact that the branches that terminate on each species are not all the exact same distance from the root of the entire tree.

by 1,100 base pairs yields a result of 0.097 — which can be rounded to 0.1, which matches the slope (0.1022) of the linear function in the graph (Figure 10.3). A similar result held for the Old World monkey family. Dividing the total number of species compared (i.e., 62) by roughly 1,700 base pairs yields a result of 0.036 — which can be rounded to 0.04, which matches the slope (0.0436) of the linear function in the graph (Figure 10.5).

To clarify, these results will need to be updated as the mtDNA genomes of more species are sequenced. However, I do not anticipate the conclusion of linear patterns to change. Rather, I expect the slope of the line to increase, without actually changing the best-fit function from linear to some other mathematical equation (e.g., exponential, power, etc.).

In other words, the addition of mtDNA samples to the analysis of, say, Bovidae will probably result in a higher predicted rate of species forming per unit time. Instead of 0.10 new species per mtDNA mutation, it might rise to 0.12 new species per mtDNA mutation. But I don't expect it to change the function from linear to exponential.

The consistency among these results suggested that this pattern might extend beyond mammals. To test this, I sampled two families each from the bird,[25] reptile,[26] amphibian,[27] and fish (one bony fish family,[28] one cartilaginous fish family[29]) classes. I selected families which had (1) more than 12 total species (to make the statistics more robust); (2) at least the same number of species represented by mtDNA sequences (again, for statistical robustness); and (3) little branch

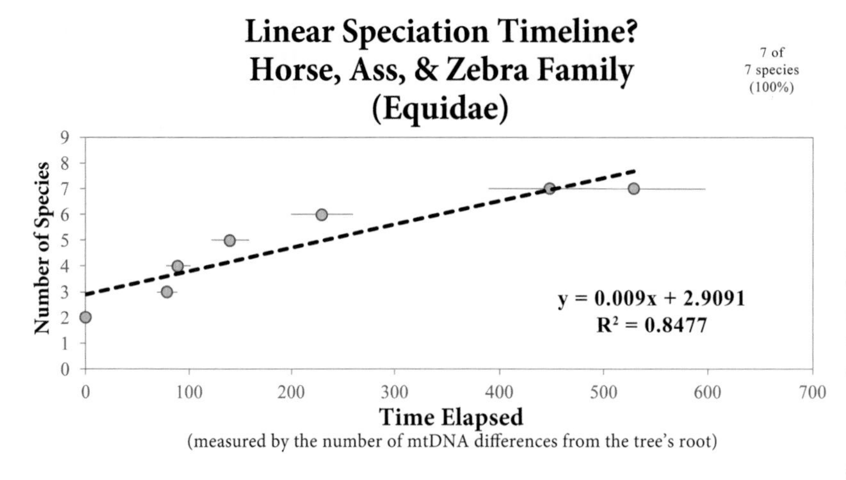

Figure 10.24. The pattern of speciation in the family Equidae shows resemblance to a linear function. Splitting events on the Equidae mtDNA tree were scored as speciation events, and the timing of these events was obtained from the branch length values of the Equidae mtDNA tree. Each circle represents a single speciation (splitting) event. The total number of species represented in this graph (i.e., those with publically available mtDNA genome sequences) is shown in the top right as a percentage of the total number of species within the family. The black dashed line represents the linear regression function (shown in the equation on the center-right) for these events. The light horizontal gray lines on either side of the circles approximate the statistical uncertainty of the exact timing of these splitting events. The uncertainty arises from the fact that the branches that terminate on each species are not all the exact same distance from the root of the entire tree.

length variety (to minimize statistical uncertainty). For example, some fish families had mtDNA trees with wildly different branch lengths; I excluded these from the analysis.

Regardless of which of these four vertebrate classes I examined, all families showed linear patterns of speciation. In each case, the R^2 value was at least 0.87 (Figures 10.16–10.23).

We can now return to the question of the pattern of speciation in the family Equidae. Like the other families that we explored, the pattern in equid species appeared to be linear[30] (Figure 10.24). Though the correlation coefficient (R^2 value = 0.85) was on the lower end of the spectrum, as compared to the other families that we examined, the equid family also possessed fewer data points. Consequently, each slight deviation from the overall mean carried more weight. Thus, despite the lower correlation coefficient, the precedence set in the more species-rich families suggested that a linear pattern of speciation was also true in equids.

In summary, among species on which creationists and evolutionist share the same view of common ancestry and of the means by which mtDNA differences among these species originated (i.e., via mutation), a consistent pattern of speciation was seen.

What might this pattern mean?

Consider what the results in the preceding section do *not* depict. Given the unpredictable nature of environments, we could easily have expected the speciation timelines to have jumped all over the graphs. For example, we might have predicted a few species to have formed shortly after speciation started, then to have done nothing for a long period of time, and then to have suddenly produced several more species. Graphing this pattern on a timeline would have revealed a line that jumped up, then was straight for a long time, and then jumped up again. In fact, we might have predicted the speciation process to be entirely *un*predictable.

Instead, in nearly every family that we examined, the pattern of speciation was a straight line.

This fact had implications for the mechanism of speciation. To see how, let's consider an analogy. For example, let's say that you and your neighbor decide to play the lottery. In addition, let's say that the two of you are the *only* people playing the lottery. For sake of simplicity, let's also say that the lottery awards one million dollars to a winner every year. Each year, since only two of you are playing, your chances of winning are 1 in 2 — or 50%. If you play the lottery two years in a row, statistics say that you should have a very good chance of winning at least once. If you play several years in a row, you are virtually guaranteed to win at least one million dollars.

Naturally, if you play the lottery for ten years in a row, you'll be winning cash hand over fist. If your other neighbors discover your bonanza in bucks, you and your fellow lottery player are probably not going to be the only ones playing for long. After ten years of winning, you might have an additional 98 participants in the lottery. In this scenario, your chances of winning drop precipitously. Instead of 1 in 2, your probability of winning is now only 1 in 100 — or 1%.

In reality, if you were winning the lottery frequently, a large mass of people would probably sign up to play. The more participants in the lottery, the lower your odds of winning.

The linear pattern of speciation implies a similar conclusion — but in the realm of speciation potential. Specifically, because the pattern of speciation is linear in each family that we examined, the rate of speciation per family is constant. This is the mathematical nature of a linear function. In other words, at any point in time, the rate of speciation, say, among bovid species, is about 0.1 new species per mtDNA mutation (or 1 new species per 10 mutations). If you examine speciation in bovids at early time points or at later time points, the rate is 0.1 new species per mtDNA mutation. The same principle holds true in other families that we examined.

However, while the rate of speciation within a family is constant with time, the number of species within a family is not. As each of these speciation timelines shows (Figures 10.3–10.24), the number of species within each family grows with time. For example, at the earliest time points, just a handful of bovid species existed (Figure 10.3). As time passed, the number of bovid species increased to, say, around 20 or 30. At later time points, the number approached 100 species.

Now consider again the lottery analogy, and apply it to species. For example, in Bovidae, when speciation first got going, only a few species were present. Yet the linear function demands that 0.1 new species form per mtDNA mutation. In other words, these few species would have formed new species after a certain period of time. With few initial species present, each one was virtually guaranteed to form another species. However, as time progressed, the number of species increased, and the chance that any one particular species formed another got lower and lower. Just like the lottery, the more that participated in the speciation process, the less the chance that any one species would "win" the speciation lottery and form another species.

Mathematically, we can reach the same conclusion. In the previous paragraph, I've made an argument from a fraction. The numerator is the rate of speciation *per family*. Since we derived this from the linear speciation function, it's constant with time. In contrast, the denominator is the number of species, which we derive from the timeline itself. Since the number of species increases with time, the denominator does as well. It is *not* constant with time. A constant numerator divided by an increasing denominator yields a quotient that is decreasing (with time). Speciation rates *per species* are declining with time.

In other words, whatever caused species to form initially seems to have declined in its abilities with time. The potential for speciation seems to have been diluted. In short, the ancestors of modern species look like they had much more potential for speciation than any of their descendants do today.

This conclusion is independent of an absolute timescale. It doesn't matter whether you believe species took millions of years or thousands of years to form. The only assumptions on which this conclusion rests are that mtDNA mutations occur in a clock-like manner and that we've correctly rooted each tree. Everything else follows from these assumptions.

Even though we've reached these conclusions by examining the mtDNA compartment, our results have implications far beyond mtDNA differences.

Recall what we've discussed about the mechanism by which species arise. If the ancestors to modern species possessed millions of heterozygous *nuclear* DNA sites, new species arise naturally (see chapters 8–9). Via the processes of recombination and gene conversion, new nuclear chromosome combinations arise each generation. When these processes involve millions of heterozygous sites, changes in visible traits are almost guaranteed to appear each generation.

To produce a new species, these traits and nuclear DNA combinations must be isolated. Small population sizes, inbreeding, and migration can lead to the founding of new, more homozygous populations. As these more homozygous groups grow to larger and larger sizes, they might eventually be recognized as new species.

Eventually, this process limits the ability of these new species to form additional species. Since, under this model, new species tend to arise via shifts from more heterozygous nuclear DNA states to more homozygous nuclear DNA states, these new species effectively lose a number of heterozygous DNA variants (for a simulation, see Color Plate 81). If speciation potential resides in the ability to spawn genetically and visually diverse offspring, then high heterozygosity (i.e., low homozygosity) represents high potential. Low heterozygosity (i.e., high homozygosity) represents low potential. The process of speciation that we've derived naturally dilutes the speciation potential in descendants.

Consistent with this model, most of the nuclear DNA differences among equid species represent homozygous differences.* For example, 26 million nuclear DNA differences exist between the imperial zebra and the domestic horse.[31] Just 2 million of these are heterozygous in the imperial zebra.[32] The mountain zebra also differs from the horse by about 26 million base pairs, and the plains zebra is almost 28 million base pairs away from the horse.[33] Yet only around 2 million and 4 million, respectively, of these sites are heterozygous in these species.[34] Once we factor in the number of heterozygous sites in the horse (just a few million base pairs[35]), we quickly recognize that the vast majority of the nuclear differences between zebras and horses are homozygous, not heterozygous.

Figure 10.25. Example of the semi-wild gayal species (*Bos frontalis*).

Other species display a similar pattern. The nuclear genome sequences for gayals (Figure 10.25) and cattle have been obtained. Of the nearly 24 million DNA differences between these species, over half are homozygous in gayals.[36] Tigers and housecats also possess known nuclear DNA sequences. Over 43 million DNA base pairs separate these two species. Yet only 1.5 million of these are heterozygous in the tiger,[37] and just a few million are heterozygous in cat breeds.[38] Fin whales differ from Minke whales at 34 million DNA sites. Between the two species, a combined total of less than 5.5 million of these are heterozygous.[39] Almost 80 million sites separate emperor penguins from Adélie penguins, yet, in each species, only 2–3 million base pairs are heterozygous.[40]

In short, the *patterns* of mtDNA and nuclear DNA differences exactly fit what we've concluded about the mechanism of speciation.

Even though we deliberately limited our analyses of speciation patterns to living species, these patterns indirectly suggested a potential explanation for extinct species.

The link between the two datasets is difficult to escape. For example, we know from the fossil record that living species represent a subset of all species that have ever existed. Whether comparisons are made at the species level of classification or higher, the conclusion is the same. For example, in mammals, around 16,000 species are present in the fossil record. Only around

* Just to clarify, nuclear DNA differences are scored regardless of whether the difference is homozygous or heterozygous. This concept is complex and best explained in Color Plate 81.

Figure 10.26. Tasmanian wolf. Now extinct.

Figure 10.27. Dodo bird. Now extinct.

5,400 exist today. At the classification level of family, over 550 mammal families exist in the fossil record. Only around 150 persist to the present.[41]

Even in recent recorded history, we've observed the fact of extinction. In the last few hundred years, the thylacine — the Tasmanian wolf (Figure 10.26) — disappeared from earth, as did the dodo bird (Figure 10.27). Many others have as well.[42]

Thus, speciation is just one stream in a two-way river. Speciation increases the number of species; extinction decreases it.

Extinction might explain a pattern that we hinted at in an earlier section. In mammals, we observed that most families are species-poor. Graphically, we can reach the same conclusion[43] (Figure 10.28). Most mammal families have few species. Few mammal families have many species.

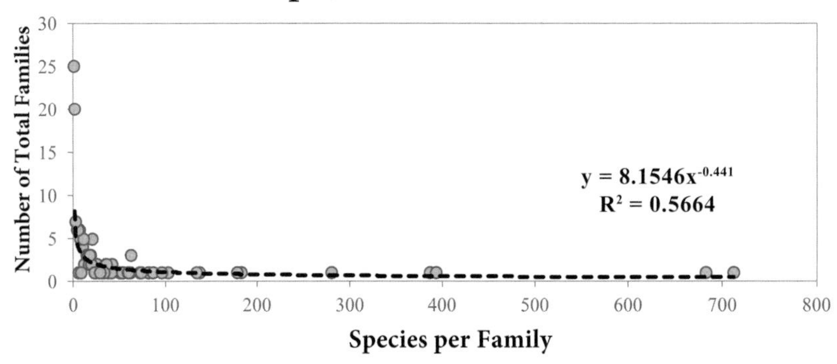

Figure 10.28. The asymmetric distribution of species within mammal families. The number of species within each living mammal family were obtained from publicly available sources and plotted. Each circle represents the data from one mammal family. A power function regression analyses was performed on these data—depicted by black dashed line and equation at the right. Few families had many species; many families had few species.

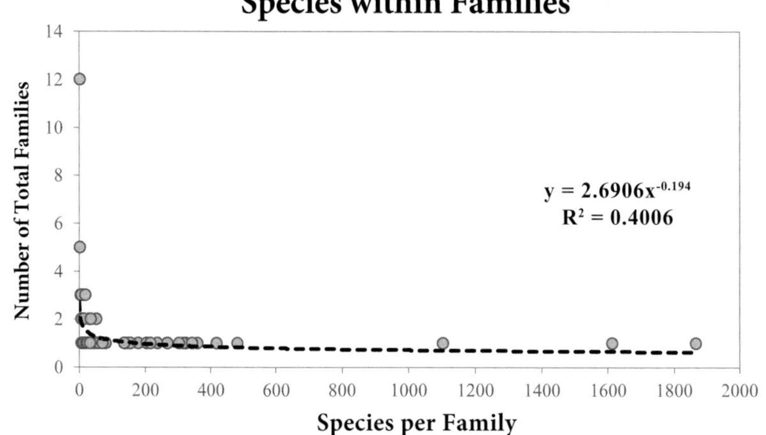

Figure 10.29. The asymmetric distribution of species within reptile families. The number of species within each living reptile family were obtained from publicly available sources and plotted. Each circle represents the data from one reptile family. A power function regression analyses was performed on these data — depicted by black dashed line and equation at the right. Few families had many species; many families had few species.

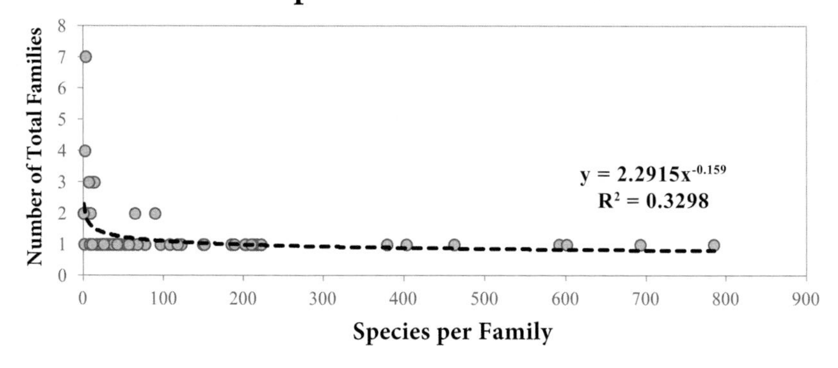

Figure 10.30. The asymmetric distribution of species within amphibian families. The number of species within each living amphibian family were obtained from publicly available sources and plotted. Each circle represents the data from one amphibian family. A power function regression analyses was performed on these data — depicted by black dashed line and equation at the right. Few families had many species; many families had few species.

Similar patterns exist in the other vertebrate classes. Most reptile families[44] have few species; few families have many species (Figure 10.29). Most amphibian families[45] have few species; few families have many species (Figure 10.30). Birds[46] and fish[47] display the same trend (Figures 10.31–10.32).[48]

Why?

Consider again the linear pattern of speciation that we've observed across all of these classes. Being a linear function, the rate of speciation is unchanging with time. While the addition of

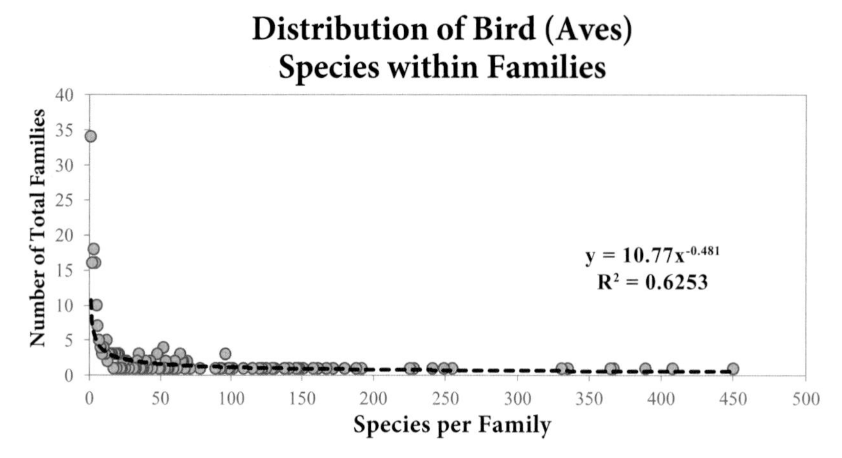

Figure 10.31. The asymmetric distribution of species within bird families. The number of species within each living bird family were obtained from publicly available sources and plotted. Each circle represents the data from one bird family. A power function regression analyses was performed on these data—depicted by black dashed line and equation at the right. Few families had many species; many families had few species.

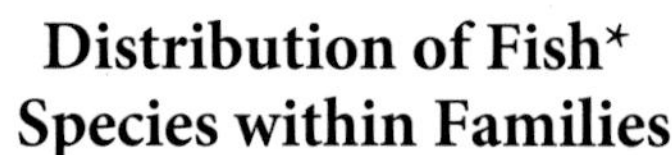

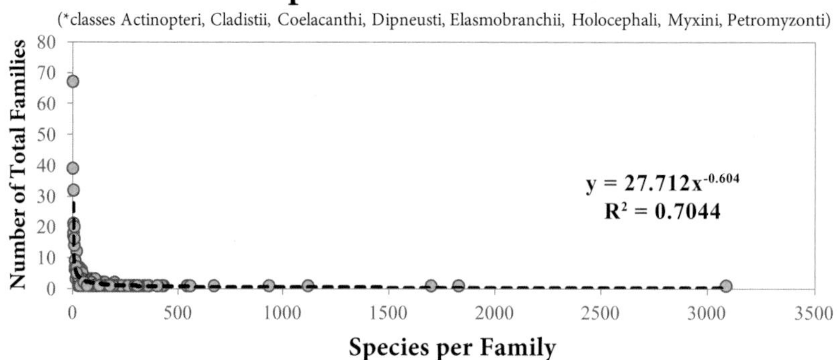

Figure 10.32. The asymmetric distribution of species within fish families. The number of species within each living fish family were obtained from publicly available sources and plotted. Each circle represents the data from one fish family. A power function regression analyses was performed on these data—depicted by black dashed line and equation at the right. Few families had many species; many families had few species.

mtDNA sequences from more species within these families will probably increase the value of the rate, our results thus far imply that these new mtDNA sequences will not affect the constancy of the rate with time. In other words, we observed that the rate at early time points was the same at late time points; the addition of more mtDNA sequences to our datasets will probably not change this fact. In short, the rate of speciation at early time points *predicts* the rate of speciation at late time points.

Since, by virtue of this mathematical relationship, the rate at early time points is the key determinant of the final number of living species in a family, high speciation rates at early time points will lead to species-rich families at late time points (i.e., in the present). Conversely, low speciation rates at early time points will produce species-poor families today. Early extinction events are critical determinants of final species numbers — perhaps more so than later extinction events. In other words, families that begin slow end slow. Families that begin speciating fast end with high species numbers. If a family loses many species early via extinction, this family will likely end up with few species today. Families that escape early extinction have the highest probability of ending with hundreds of species today. The pattern points to a very specific window of time where extinction would matter most.

All of these considerations fit the mechanism of speciation that we've discussed. We've argued that shifts from heterozygosity to homozygosity are the major genetic mechanism of speciation. In other words, heterozygosity levels are a surrogate measure of speciation potential. Today, species have low levels of heterozygosity. Extinction of a modern species would do little to decrease the overall pool of speciation potential among all the remaining species within a family. If the ancestors of modern species had the highest levels of heterozygosity, and if heterozygosity levels have declined with time, then extinction of some of the early ancestors would be much more impactful than extinction of recently formed species.

Early extinction events might be the reason some families are species-poor and others are species-rich.

Summed together, the observations of this chapter and of preceding chapters reveal the natural history of each species — from the origin of their genetic differences to the origin of their defining traits; from the origin of new species to the extinction of ancient ones. As a representative example, consider the species in the equid family. One of the defining traits in these species is the presence or absence of stripes. In a previous chapter, we observed that the hybridization of zebras with horses or asses produces striped offspring (Color Plates 8–9). Though the level of striping varies in the offspring, striping in some form appears to be dominant over the complete absence of stripes.

If the equid ancestor was highly heterozygous, and if one set of instructions specified stripes while the other did not, then the ancestor likely had some level of striping. However, it probably did not look like modern zebra species. Instead, it probably bore more resemblance to one of the zebra-horse or zebra-ass hybrids.

Early in the equid family history, many species appear to have formed. For example, in the fossil record, the Quaternary layers alone possess at least 5 genera and over 80 species of equids.[49]

Yet few of these species survive to this day. Only 7 species of equids exist in the present.

When did this extinction occur? Since early extinction events seem to be predictive of final species numbers, we can reason backward in time from current species numbers. Since 7 species is a comparatively low number, an early extinction event (rather than a late extinction event) would seem to be required to produce this low number. In other words, the extinction event(s) would seem to have occurred shortly after the burst of speciation events recorded in the Quaternary.

At a minimum, these events appear to have specifically affected the more heterozygous members of the equid family. Effectively, these deaths would have depleted the overall pool of heterozygous sites within the equid family, thereby reducing the potential for speciation in the remaining equid representatives. As a potential reflection of this fact, high levels of morphological similarity exist among the seven living descendants.

From the survivors of these early extinction events, the seven living equid species formed. From the mtDNA tree (Figure 10.1A), we can infer the relative timing of these events. The first speciation event split the horse lineage from the zebra/ass lineage.

Again, nuclear DNA comparisons are consistent with this hierarchy. Horses are the most distant species of all the equid species, as if the zebra and ass lineages were initially one lineage and then later split from one another.

The formation of the horse species likely involved a shift from heterozygosity to homozygosity at the genetic locus (or loci) controlling stripe formation. In the horse lineage, the information for stripes appears to have been almost completely lost.[50] Since the total absence of stripes appears to act like a recessive trait, the horse lineage appears to have become homozygous for the absence of striping at the relevant genetic loci.

After the formation of the horse lineage, the split among the ass and zebra lineages happened in fairly short succession. On the mtDNA tree (Figure 10.1A), the splits among the ass and zebra branches are fairly close together. Pinpointing exactly how close is complicated by the statistical noise of the analysis. This can be readily seen in the horizontal jaggedness of the branches — they don't all terminate in exactly the same horizontal distance from the root of the tree. Thus, precise resolution of the relative timing of these specific species splits is not possible at present.

Nevertheless, the zebra-ass lineage split around one-fourth of the way along the tree (Figure 10.1A). On a timescale of 4,500 years,* this split would have occurred around 3,400 years ago (i.e., 25% of 4,500 years) — or around 1,400 B.C. We know from historical records that the ancient Romans were aware of the existence of zebras.[51] Since the Roman Empire ruled only about 2,000 years ago (i.e., not 3,400 years ago), this speciation timescale is consistent with recorded history. Again, shifts from heterozygosity to homozygosity at the genetic locus (or loci) controlling stripe formation were probably involved in the formation of zebra species.

Since the split in the zebra and ass lineages, more minor speciation events have occurred. Eventually, the three zebra species split from one another. The mountain zebra appears to have split off first, followed by a divergence between the plains zebra and imperial zebra species (Figure 10.1A). Again, shifts from heterozygosity to homozygosity were likely involved. However, the locus for stripes was probably homozygous in the ancestor of all three species. Thus, in the for-

* Under the YEC model, modern equid species would have resulted from speciation events following Noah's Flood, which YEC scientists place around 4,500 years ago.

mation of three striped species with differing striped patterns, the shifts may have been at genetic loci that regulated the timing of laying down stripes, rather than at loci that controlled whether stripes appeared at all.[52]

Among asses, the African and Asian lineages split quickly. Much later, the two Asian species diverged from one another (Figure 10.1A). Like the other speciation events, shifts from heterozygosity to homozygosity probably controlled this process.

In theory, since equid species still possess millions of heterozygous DNA positions, this speciation process could continue to this day — and into the future. Consistent with this potential, the linear speciation results argue for a similar conclusion. If zebra species and their equid cousins formed at a constant rate, and if this pattern is still true today, then speciation within the equid family is ongoing. Just revisit the graph we've explored (Figure 10.24). Extend the linear line to the right — into the future. The trajectory is upward, implying that more species will form in the future.

If we put this process on the YEC timescale, this could suggest an absolute rate of speciation: 4,500 years / 7 species = about 640 years per speciation event. Given our discussion from chapter 6, this rate is entirely plausible at present.

At this point, though, it's worth revisiting some of the assumptions that we've made in our analyses, as well as some of the unanswered questions from previous chapters. In our investigation of the pattern of speciation, we implicitly assumed that mtDNA mutation rates were constant with time. If they were faster or slower in the past, this would change the relationship between mtDNA mutation rates and time, and it might lead to different conclusions on the pattern of speciation.

However, with respect to the evolutionary model, this consideration actually seems unlikely to change what we've concluded. Consider the evolutionary predictions for mtDNA differences in vertebrates from chapter 7: we observed that current rates of mutation predict too many mtDNA differences over the evolutionary timescale. Consequently, evolutionists will likely invoke natural selection to explain this discrepancy, which would lower the effective mutation rate. At first pass, this might suggest a non-linear relationship between mtDNA mutations and time.

Deeper reflection alters this conclusion. Under the evolutionary model, our analysis of the pattern of speciation in previous sections is an analysis, in essence, of the *results* of this mutations-plus-natural-selection process. In other words, the method we used in our analyses implicitly masked the difference between the measured mutation rate (i.e., measured in the present) and the historical, effective mutation rate (i.e., the rate that actually gave rise to mtDNA differences, after accounting for the process of natural selection). Thus, our analyses of the patterns of speciation are indifferent to evolutionary claims about slower mutation rates in the past.

But what about the YEC timescale? From chapter 7, we observed that, at current rates of mutation, the YEC model *under*predicted the number of mtDNA differences among vertebrate species. Consequently, you might expect the YEC model to postulate that mtDNA mutation rates were higher in the past.

What might have accelerated this rate? The linear speciation results from the previous sections suggested a potential mechanism. Since, within a family, mtDNA mutations track very carefully with the number of species, something about the speciation process might accelerate mtDNA mutations. From chapters 8–9, we have explored data that argued for a specific mechanism for this process — shifts from high levels of preexisting heterozygosity toward a more homozygous state. Since mtDNA mutations are linked to species numbers via the graphs in this chapter, and since species numbers are linked to levels of heterozygosity via the speciation process, you might speculate that mtDNA mutations are linked to levels of nuclear heterozygosity as well.

In chapter 8, we observed that *nuclear DNA* mutation rates are linked to levels of *nuclear* heterozygosity in plants. We also observed that this relationship in plants predicted the exact relationship between these two parameters, both in humans and chimpanzees, suggesting that a similar relationship held true in other species. Might levels of *nuclear DNA* heterozygosity be linked to *mtDNA* mutation rates?

This hypothesis can be tested.

If the results of this test reveal the relationship that I'm suggesting, we would need to revise the rate of speciation in equids. Rather than be best approximated by a rate of 1 new species every 640 years, the rate would be less — and would be calculable from the revised graph of the relationship between species numbers and time.

For example, in graphical terms, we can visualize how this discovery might change the timelines we observed in this chapter. As an illustration, consider one of the more species-rich mammal families, Bovidae. An accelerated mtDNA mutation rate in the past would not change the relationship between mtDNA mutations and the number of species. This relationship has

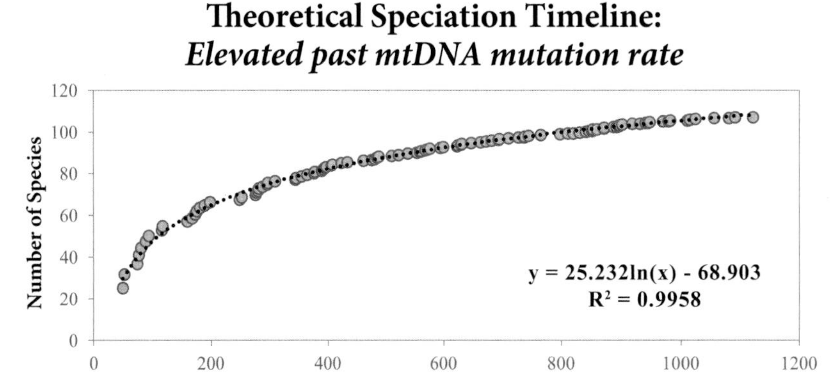

Figure 10.33. Theoretical depiction of the timeline of speciation under the hypothesis of non-uniform mutation rates. If mitochondrial mutation rates were higher in the past, then each mitochondrial mutation would not have the same absolute time value. Early mutations would represent smaller units of time; later mutations would correspond to larger blocks of time. Thus, given the mitochondrial DNA trees that we observe today, a higher past mutation rate would change current speciation timelines by bending the left side of the timeline.

already been established in the graphs in preceding sections. Rather, an accelerated mtDNA mutation rate would change the relationship between mtDNA mutations and *time*. Effectively, this would compress a large number of mutations — and their linked speciation events — into a small window of time on the left side of the graph (Figure 10.33).

Nevertheless, other conclusions that we reached would remain unchanged. For example, the fact that speciation rates are declining with time would be unaltered. This follows from mathematical considerations. Mathematically, when the relationship between speciation and time is linear, the fact of declining speciation rates flows from two considerations: (1) the unchanging rate of speciation within a family and (2) the increasing number of species within a family. Dividing the first by the second yields a rate of speciation *per species* that is declining with time. If the relationship were more logarithmic, the math would change as follows: (1) the rate of speciation within a family would be declining, while (2) the number of species within a family would still be increasing. Dividing the first by the second yields a rate of speciation *per species* that is declining with time — even more so than under the first scenario. In other words, if mtDNA mutation rates were a function of nuclear heterozygosity, the pattern of speciation would still underscore the conclusions we reached in previous chapters about preexisting nuclear DNA differences. In a sense, it would be a confirmed prediction of the model of preexisting nuclear DNA differences.

Regardless of what the actual rate of speciation in equids is at present, it appears that speciation is ongoing.

What might this look like?

Recall from chapter 6 that at least eight subspecies of zebras exist among the three living species (see Color Plates 18, 72–74 for some examples of these subspecies). Now recall the elements of the speciation process from chapter 9. The first element is the formation of visibly distinct offspring. The fact that the scientific community recognizes these eight groups of zebras as subspecies suggests that this step has already occurred. Furthermore, as subspecies, these groups of zebras represent *populations* of distinct individuals — which means that the third element of the speciation process (population regrowth) has also occurred. To form new species, it seems that the only step left to take is the isolation of one or more of these subspecies from other populations. When this occurs, a new species will have formed.

If it happens in our lifetime, we will have witnessed the answer to the "mystery of mysteries."[53] And we will have done it in an unexpected way. We won't have moved land and sea to uncover a long-lost clue to the origin of species. We will have discovered an answer simply by watching species form in real time.

Though I've used equids as a specific example, the natural history that I've described should be generally applicable to diverse species. To derive these histories for other vertebrate families, all that is needed is (1) a mtDNA tree from species within the family, (2) an analysis on the relative timing of speciation from the mtDNA tree to test for a linear pattern of speciation, (3) the current levels of nuclear DNA heterozygosity among species within the family and of overall nuclear DNA differences between species within the family, (4) the fossil record for species within the family, (5) some data on the dominance or recessiveness of various traits of interest, and (6) a catalog of the current number of subspecies within the family.

From previous sections, it already appears that linear speciation is the rule across animal taxa. Furthermore, in these previous sections we've also seen significant levels of heterozygosity in living species across diverse vertebrate classes. Finally, in mammals, the total number of subspecies (6,348) exceeds the total number of living species (5,415).[54]

The natural history of equids seems as if it might be paradigmatic for mammals, if not for vertebrates and animals in general.

Thus, the last category of corner piece — the linear patterns of speciation within families — rounds out the puzzle of the origin of species. As expected, this independent observation naturally connects mtDNA clocks, nuclear DNA clocks, and the genetics of populations.

To be sure, this discovery does not complete the puzzle of the origin of species. Massive holes in the puzzle still exist. The genetics of millions of species have yet to be determined, and the mutations rates in each of these species must still be measured. Millions of edge pieces have yet to be discovered. The majority of existing species are probably still undocumented. Furthermore, in light of what we observed in this chapter, more species will likely form.

Nevertheless, among the puzzle pieces that we possess, a consistent pattern has emerged. The genetic results we uncovered in the last four chapters are highly suggestive of future discoveries. If nothing else, they make testable predictions. The biggest prediction is the sheer number of corner pieces to be found.

If a large number of corner pieces are indeed found, then the shape of the puzzle of the origin of species will begin to emerge. Actually, I should use the plural — the *shapes* of the many *puzzles*. In other words, the discovery of numerous corner pieces would suggest that many puzzles exist. Species within a family (or within a genus) might connect as part of a single puzzle. But species from different families (or genera) might reside in puzzles that are sharply disconnected from one another by hard edges.

Only time will tell.

Afterword

Throughout this book, I have attempted to narrate the last 130 years of biology by relying heavily on an analogy to jigsaw puzzles. I would be remiss if I didn't address an even bigger puzzle that, itself, frames the puzzle of the origin of species.

In this debate over the origin of species, any regular participant knows that it has a long history of deep polarization. New discoveries are not made calmly, peacefully, and without fanfare. Rather, both sides vigorously dispute each new scientific finding. The fact that the origins debate remains so polarized hints at something deeper — it hints at a larger controversy of which the origin of species is simply a small part.

This larger controversy has always been present in the discussion of the origin of species, though it has often lingered in the background, unacknowledged. To identify it, let's revisit my explanation for the origin of species and perform a thought experiment. As I retell the narrative, at each point of the explanation, see if you can measure your emotional reaction to what I'm saying:

> In the beginning, around 6,000 years ago, God created "kinds" of creatures — the original *min*. Representing creatures somewhere between the rank of subgenus and order, these *min* contained millions of heterozygous sites in their genomes. As they reproduced, shifts from heterozygosity to homozygosity led to diverse offspring.
>
> Less than 1,700 years after the creation of these *min*, their population sizes were dramatically reduced. At least for the land-dwelling, air-breathing *min*, their

population sizes were reduced to no more than 14 individuals.* In some cases, their populations declined to just 2. However, because this population bottleneck was so short, the heterozygosity of the Ark passengers would have been minimally affected. For sexually reproducing *min*, a male and female could have possessed a combined four copies of nuclear DNA. These copies could have been very different, preserving a massive amount of speciation potential.

After the Flood, the process of speciation resumed. With a dramatically re-sculpted earth, migration could begin anew. Due to the consequent small population sizes and inbreeding, migration would have had profound effects on the genetics of each *min*. Effectively, conditions post-Flood were ripe for shifts from heterozygosity to homozygosity, which could have resulted in many new species.

Each shift would have decreased the speciation potential of the resultant species. Fewer heterozygous sites would have meant fewer possible shifts to homozygosity. Over time, this decrease in heterozygosity within individuals would have led to a decline in the rate of speciation per species.

However, to date, heterozygosity has not been completely extinguished. The potential to form new species still exists. Therefore, speciation is ongoing today, though at rates that typically exceed the average human lifespan or at rates that exceed our current abilities to measure them.

Did any of the above statements give you pause? From a scientific perspective, they shouldn't. If you doubt the claims that I made about ancestry, we can test them by the predictions that they make about the function of individual DNA base pairs. In addition, my claims about ancestry make predictions about the mtDNA mutation rate. Furthermore, the purported heterozygosity levels in these ancestors can be tested against levels of function across the genomes of various species. Migration fits the current geographic distribution of species. The timescale makes very specific predictions for mtDNA mutation rates. In other words, nothing about this model is scientifically contrived.

Yet, you might object that my explanation begins with a miracle. How can miracles have a place in science? If you're an evolutionist, I would respond by pointing to Michael Behe's published work.[1] In short, based on Behe's work, which is simply a modern restatement of Darwin's own test for evolution, the evolutionary history of life also seems to require a number of miracles for it to proceed. Why should creationists be the only ones who are prohibited from invoking miracles?

Furthermore, though my explanation begins with a miracle, the scientific model that flows from this initial position makes very specific testable predictions by which its major claims can be evaluated.

* If you're familiar with the Noah's Ark narrative, you've probably heard the phrase "two-by-two," as if Noah brought animals on board the Ark only in groups of 2. For some animals, Noah brought at least 7 individuals on board (Genesis 7:1–3). However, some biblical scholars suspect that the "7" might refer to pairs (rather than to individuals), implying that Noah brought 14 individuals (7 * 2 = 14) of these types of animals.

Perhaps the biggest emotive word in my narrative is "God." But is this entirely outside the domain of science? Consider the details of embryonic development that we discussed in chapter 3. Since creatures are like biological construction projects, we made an analogy between this process and the process of constructing a skyscraper.

Despite the many parallels between the two, the operation of the cell actually exceeds our analogy.[2] In one sense, creatures aren't like construction sites. They don't have intelligent human or animal participants. The zygote builds itself. A building doesn't.

What would it take for a building to build itself? For one, all direct human participants would have to be removed from the construction site. No construction workers, no foremen, no engineers could be present. Instead, some sort of programmed plan would have to be dropped at the site, after which the self-assembly could commence.

In other words, the development of a creature is more like dropping a computer chip at a potential construction site, and then watching the chip assemble a power plant, factory, and tools, and then connect all the pieces in the correct temporal and spatial order to produce the final product.

Actually, the analogy goes beyond this. Cells have the ability to sense and fix errors. The DNA machinery of the cells detects mutations and fixes them. Though the process is slightly imperfect, it is normally extremely accurate. Our computer chip would also have to possess the same error detection and correction ability.

To make our analogy even more realistic, we would need to take it one step further. The development of a creature doesn't yield a static entity. Adult creatures reproduce. Therefore, our computer chip would not only need to spontaneously produce a skyscraper, it would then need to spit out a second computer chip at the end of the process, which could go on by itself and repeat the processes of construction and reproduction.

Obviously, what I've just described is science fiction. No human computer programmer has achieved this task. In fact, the collective programming intelligence on the planet has yet to reach this goal. Consequently, the fact of development demands an intelligence that is greater than the collective human intelligence on earth. You could say that the existence and operation of the process of embryonic development points to the existence of a super-intelligence. Is it too great a leap to call the intelligence by the name "God"?

At this stage, you might object that "God" is too generic. Which "God"? How can "God" be seriously entertained as an explanation when religion has caused so much grief and destruction?

If these are your questions, then we've uncovered the larger puzzle. The questions in the preceding paragraph are not scientific questions. They're philosophical ones. The reason that the origins debate provokes such emotional responses is because it inexorably converges on philosophical and religious themes. The science of origins reaches to our deepest viscera. It's inescapable.

The debate over these philosophical and religious questions is the larger puzzle which frames the smaller debate over the origin of species.

This larger framework has always surrounded the puzzle of the origin of species. For example, we began this book with William Paley's design argument. To him, the design in nature naturally implied the existence of a supreme Designer. God and species went hand in hand. Conversely, when Darwin scientifically rejected the view of species' fixity, many atheists saw the design of life replaced by evolution and, therefore, the Designer replaced by mindless natural processes. Atheism and evolution went hand in hand. Ever since Linnaeus, it has been impossible to escape philosophy and religion when discussing the origin of species.

Since we've reached the realm of philosophy and religion, let's briefly consider the philosophical and religious elements of both evolution and creation. If atheistic evolution is true, then everything is the product of time, chance, and physical forces. By extension, our brains are the result of the same. Consequently, our thoughts are simply the product of chemical reactions. Therefore, we can't possibly know if they're true. In fact, how do you know that you're actually reading words right now? You might respond by saying that your eyes are telling you so. But your eyes eventually transmit this information via neurons and chemicals. If these neurons and chemical reactions are simply the result of time, chance, and physical forces, how can you trust them to convey accurate information? In short, atheistic evolution leads to philosophical incoherence.

Theistic evolution doesn't fare much better. In the United States, professing Christians are among the most vocal proponents of this view.* Yet the contradictions between evolution and the Bible are stark. For example, between evolution and the Bible, the order of origins events differs; the time of origins events differs; the mechanism of certain origins events differs; etc.[3] It's not possible to harmonize a plain, straightforward reading of the Bible with evolution. Either the words of Scripture must be mangled, or the standard view of evolution must be twisted to harmonize the two.

In contrast, the position that I've traced in this book results in a natural harmony with the plain, straightforward reading of the Bible.

To clarify, I did not argue for my conclusions from Scripture, and then selectively corral scientific data to justify it. Rather, I justified all of my scientific conclusions about the origin of species with science, not theology. Where I did reference the Bible, it was to derive testable predictions, or to show agreement between these predictions and the observations that we made.

Since the plain, straightforward reading of the Bible is the most internally consistent way to understand the text of Scripture,[4] the conclusions in this book are based on a philosophically and religiously coherent framework.

* I don't mean that all professing Christians espouse theistic evolution. I mean that the most prominent advocates of theistic evolution happen to be professing Christians, rather than, say, Muslims or Jews.

If I ended the discussion here, I would again be remiss. Simply stating that an alternative framework exists is unsatisfying. For example, in the preceding chapters, we took this point for granted. I didn't spend ten chapters simply stating that an alternative scientific framework exists for the puzzle of the origin of species. Instead, I spilled much ink describing the details of this framework and the justification for it.

Space does not permit me to spill as much ink describing the details of the philosophical and religious framework in which the scientific puzzle of this book fits. Instead, I will attempt to sketch the seminal points and encourage readers to read the Bible for themselves.

One of the biggest elements of this framework is the nature of the Designer/Creator. For example, the plain reading of the first chapter of Genesis reveals the Designer — God — to be an alien Being. I don't mean that He is foreign to the origin of everything. Rather, His *nature* is foreign to ours. As an illustration, consider God's relationship to time and space. The first chapter of Genesis says that God created time and space. Therefore, He must be outside of time and space. In other words, He must be eternal and present everywhere. A Being without beginning or end, and One who is present everywhere, is alien to human experience.

An additional aspect of God's nature follows from these facts. In order to create time and space, He must know everything about it. Therefore, His knowledge must exceed the collective knowledge of all humans that have ever existed on earth — an observation that is consistent with what we just observed about the engineering that is involved in the cells of our bodies. Our experience knows no human parallel to this level of intelligence.

Finally, when Genesis 1 describes how God creates, it shows Him speaking. When He speaks, His command is immediately enforced. In other words, Genesis 1 reveals that God sovereignly rules over all. No human has ever possessed such power.

If these characteristics were all that Genesis revealed about God, this revelation might be disconcerting. For instance, consider the possible ways in which this omnipotent potentate, God, could have designed the universe. As the sovereign Ruler and Creator, He could have made it in any way that He pleased. For example, He could have created an evil system of physical laws that tortures every resident of the earth. He could have planted malicious surprises at every turn. He could have made the human mind reside in a continual state of near insanity. Instead, Genesis 1 repeatedly describes the results of God's creation as good.[5] In fact, when God created Adam and Eve, He *blessed* them;[6] He didn't punish them. This level of unprompted beneficence is also foreign to our human experience.

In light of these facts, it would be appropriate for the humans that God created to respond to the magnanimous and good Creator with thanks and praise. Instead, Genesis 3 describes how the first humans, Adam and Eve, rebelled against God's commands.

Before considering the long-term ramifications of this fact, let's consider the theology of it. Some critics of the Bible have objected to the sequence of events in the first three chapters of

Genesis. They think Adam and Eve's rebellion nullifies at least one aspect of His character. They ask, "Couldn't God have prevented this tragedy? Either He didn't know it was coming, or He was powerless to stop it." If you pick either one of these options, you implicitly accept that God must *not* be all-knowing, or God must *not* be all-powerful, respectively. Yet, based on the events recorded just two chapters prior (i.e., in Genesis 1), it would seem that God *is* all-powerful and all-knowing.

Is there a resolution to this theological dilemma?

The answer is found in other aspects of God's nature. It is true that God could have stopped Adam and Eve from rebelling. After all, God is all-powerful. Yet He chose not to stop their actions. Furthermore, God knew that they would sin — He knows all things, even the future. Yet God deliberately designed the universe in the way that He did. To be sure, God is not the author of sin. Yet He was content with the events that transpired. I don't mean that He was content in the sense of taking pleasure in sin — to be sure, He was angry with Adam and Eve's rebellion. Rather, I mean that He was content in that He did not destroy Adam, Eve, and the entire universe, and He did not start over and make a new universe.

Why?

The Bible describes God as rich in mercy.[7] He loves to forgive.[8] When humans rebel against Him, He delights to shower them with mercy and forgiveness. A world in which no wrongdoing existed would be a world in which mercy and forgiveness were never shown.[9]

This aspect of God's nature is especially alien to us. We struggle to forgive. I don't mean that we reject forgiveness as a concept. In fact, when someone performs a gracious act of forgiveness, we applaud and admire them. But when we are required to show forgiveness, we know from experience just how emotionally trying and challenging it can be. Consequently, no human deliberately looks for an opportunity to be wronged, just so that they can show forgiveness. God is very unlike us.

Does this aspect of His character seem repulsive to you? It shouldn't. For one, the Bible describes God as glorious. He is resplendent and majestic. His attributes are not repulsive but beautiful. His mercy should delight us.

In addition, many of us have witnessed the happiness that accompanies forgiveness. We know from experience that, in the context of especially traumatic situations, forgiveness leads to resolution. It produces relief and joy. More personally, when we are the recipient of forgiveness, there is no debate in our hearts about whether forgiveness is a desirable trait. We long to be forgiven when we've wronged another. Thus, the glory of God's unparalleled mercy and forgiveness should attract us, not provoke us to anger.

To be sure, God is also perfectly just. In fact, the Bible describes Him as the habitation of justice.[10] Judgment for wrong-doing is always executed, and evil is never left unpunished. For example, the Flood was sent as a judgment on mankind because humans did *not* cherish God's glorious attributes. God's judgment might be delayed — as it was for 120 years, from the time that God announced the Flood to His actual commencement of it (Genesis 6:3). But judgment always comes.

Judgment also awaits each one of us. Though God's mercy, power, knowledge, eternality, and all of his other attributes should arrest us, we know that they have not. This is cosmic crime. God's glory is beyond description; the refusal to worship Him is a crime beyond description. No wonder that the Bible describes the fires of hell as the fate for those who have sinned against God.

Nevertheless, God's mercy and justice have manifested themselves in a way out. God sent His Son, Jesus, to take the punishment that we deserved. As God Himself, Jesus was (and is) glorious like God the Father. He was and is perfect in all His ways. Crucifying Him represents a cosmic injustice. Only His death could pay the penalty for our cosmic crimes. For those who disavow their rebellion against God and accept His Substitute as punished in their place, God promises a restoration of the very good creation — and even more. Heaven awaits those who have turned their back on their rebellious ways and put their faith in Jesus. Hell awaits those who don't.

I don't say this to scare people. I say it because this is the reality to which Genesis leads.

In summary, the puzzle of the origin of species is but one small part of a much larger religious and philosophical puzzle that each person must solve. In this book, I've argued for a new solution to the puzzle of the origin of species. It happens to fit into a larger religious and philosophical framework which is based on the plain, straightforward reading of the Bible. This larger framework has sobering implications — the full description of which is beyond this book. But it exists in another book, the Bible. This latter Book, unlike my own, contains a compelling invitation to read and study it: Scripture promises that God rewards those who diligently seek Him (Hebrews 11:6).

Will your search stop here? Or will you try to put together the pieces of the larger puzzle?

Endnotes

Chapter 1

1. Linnaeus' *Systema Naturae* can be found online at http://www.biodiversitylibrary.org/item/10277.
2. Wilson & Reeder's *Mammal Species of the World*, 3rd Ed., was accessed electronically at https://www.departments.bucknell.edu/biology/resources/msw3/ on October 7, 2016. From the .csv file of the complete taxonomy, the data were filtered for "SPECIES" via the "TaxonLevel" column, and then filtered for "FALSE" via the "Extinct?" column. A timeline of species' naming was created from the data in the "Date" column. For those species with a range of date listings, the earliest date was chosen.
3. In general, notwithstanding seasonal changes in pelage.
4. Again, in general, notwithstanding pelage variation among individuals, such as among black bear and red fox individuals.

   D.L. Garshelis, B.K. Scheick, D.L. Doan-Crider, J.J. Beecham, and M.E. Obbard, *Ursus americanus*, 2016 (errata version published in 2017). "The IUCN Red List of Threatened Species," 2016, e.T41687A114251609, downloaded on May 23, 2017; M. Hoffmann and C. Sillero-Zubiri, *Vulpes vulpes*, "The IUCN Red List of Threatened Species," 2016, e.T23062A46190249, http://dx.doi.org/10.2305/IUCN.UK.2016-1.RLTS.T23062A46190249.en, downloaded on May 23, 2017.
5. Paley's famous work, *Natural Theology*, can found online at http://darwin-online.org.uk/supplementary_works.html.
6. Darwin's *On the Origin of Species* is freely available online http://darwin-online.org.uk/.
7. http://www.catalogueoflife.org/.
8. If you go to the Annual Checklist of the Catalogue of Life (www.catalogueoflife.org/annual-checklist/), browse the Taxonomic Classification, search for specific taxonomic categories, and export results (this works for only modestly sized datasets — e.g., ~50,000 species or less, as of December 2016), you can create a timeline for which each species was named in the category selected. If you do so, you will discover that at least the majority of modern animal species were not named in 1859.

   See also Figure 1F and Figure S1 of the following reference, which suggests that perhaps only 15% of today's species were known in Darwin's day: C. Mora et al., "How Many Species Are There on Earth and in the Ocean?" 2011, *PLoS Biol*, 9(8):1–8.
9. http://data.uis.unesco.org/Index.aspx?DataSetCode=SCN_DS&popupcustomise=true&lang=en.
10. Some estimate that the total number of species alive today is close to 10 million. In other words, the vast majority of species are *still* undocumented. See C. Mora et al., "How Many Species Are There on Earth and in the Ocean?" *PLoS Biol.*, 2011, 9(8):1–8.

Chapter 2

1. Helpful background reading on the state of genetics leading up to 1859 can be found in Part III of Ernst Mayr's book, *The Growth of Biological Thought* (Cambridge, MA: The Belknap Press, 1982).
2. The English translation ("Experiments in Plant Hybridisation") of Mendel's original paper ("Versuche über Pflanzen-Hybriden") can be found online at http://www.esp.org/books/bateson/mendel/facsimile/contents/bateson-mendel-3-peas.pdf.
3. For discussion of whether this is actually true, see the following: S. Blixt, "Why Didn't Gregor Mendel Find Linkage?" *Nature*, 1975, 256:206; P. Smýkal, "Pea (*Pisum sativum* L.) in Biology Prior and After Mendel's Discovery," *Czech J. Genet, Plant Breed*, 2014, 50:52–64.
4. W.S. Sutton, "The Chromosomes in Heredity," *Biological Bulletin*, 1903, 4:231–251, available online at http://dev.esp.org/foundations/genetics/classical/holdings/s/wss-03.pdf.
5. For reasons explained in later chapters, Sutton's calculations were technically tallies of permutations, not combinations. Later discoveries would suggest that a tally of combinations would have been more appropriate.
6. Again, this calculation is technically a permutation — for each chromosomes, it treats the parent-of-origin as significant.
7. Chromosome numbers were obtained from the NCBI Genome Database (https://www.ncbi.nlm.nih.gov/genome/browse/) on May 3, 2017; or from associated papers (i.e., J. Alföldi et al, "The Genome of the Green Anole Lizard and a Comparative Analysis with Birds and Mammals," *Nature*, 2011, 477(7366):587–591; International Chicken Genome Sequencing Consortium, "Sequence and Comparative Analysis of the Chicken Genome Provide Unique Perspectives on Vertebrate Evolution," *Nature*, 2004, 432(7018):695–716.)

8. Again, these numbers follow Sutton's lead in treating these calculations as permutation tallies.
9. For further reading on the genetic developments in the early 1900s, see the following references: K.R. Benson, "T.H. Morgan's Resistance to the Chromosome Theory," *Nat Rev Genet,* 2001, 2(6):469–74; T. Boveri, "On Multipolar Mitosis as a Means of Analysis of the Cell Nucleus" [English translation], 1902, accessible online at http://10e.devbio.com/article.php?id=74; E.W. Crow and J.F. Crow, "100 Years Ago: Walter Sutton and the Chromosome Theory of Heredity," *Genetics,* 2002, 160(1):1–4; M. Hegreness and M. Meselson, "What Did Sutton See? Thirty Years of Confusion Over the Chromosomal Basis of Mendelism," *Genetics,* 2007, 176(4):1939–44, L.A.-C.P. Martins, "Did Sutton and Boveri Propose the So-Called Sutton-Boveri Chromosome Hypothesis?" *Genetics and Molecular Biology,* 1999, 22(2):261–271; R. Moore, "The 'Rediscovery' of Mendel's work," *Bioscene,* 2001, 27(2):13–24; T.H. Morgan, "Sex-limited Inheritance in *Drosophila,*" *Science,* 1910, 32:120–122; T.H. Morgan, "Complete Linkage in the Second Chromosome of the Male in *Drosophlia,*" *Science,* 1912, 36:719–720; T.H. Morgan et al, *The Mechanism of Mendelian Heredity*, chapter 1, 1915, available at http://www.esp.org/books/morgan/mechanism/facsimile/contents/morgan-mechanism-ch01-i.pdf; R.C. Olby, "Horticulture: the Font for the Baptism of Genetics," *Nat Rev Genet.,* 2000, 1(1):65–70; P. Smýkal, "Pea (*Pisum sativum* L.) in Biology Prior and After Mendel's Discovery," *Czech J. Genet. Plant Breed,* 2014, 50:52–64; A.H. Sturtevant, "The Linear Arrangement of Six Sex-linked Factors in *Drosophila,* as Shown by Their Mode of Association," *Journal of Experimental Zoology,* 14:43–59; W.S. Sutton, 1902, "On the Morphology of the Chromosome Group in *Brachystola magna,*" *Biological Bulletin,* 1913, 4:24–39. Available online at http://dev.esp.org/foundations/genetics/classical/wss-02.pdf; W.S. Sutton, "The Chromosomes in Heredity," *Biological Bulletin,* 1903, 4:231–251, available online at http://dev.esp.org/foundations/genetics/classical/holdings/s/wss-03.pdf; E.B. Wilson, "Mendel's Principles of Heredity and the Maturation of Germ Cells," *Science,* 1902, 16(416):991–3, available online at http://www.esp.org/foundations/genetics/classical/holdings/w/ebw-02.pdf.
10. For much of the rest of this chapter, many details can be found in most genetics textbooks. However, I am indebted to the following work for the depth in which this history has been told: H.F. Judson, *The Eighth Day of Creation: Makers of the Revolution in Biology* (New York: Simon and Schuster, 1979).
11. For example, see I. Hargittai, "The Tetranucleotide Hypothesis: A Centennial. *Struct,*" *Chem.,* 2009, 20:753–756.
12. F. Griffith, "The Significance of Pneumococcal Types," *J Hyg (Lond),* 1928, 27(2):113–159.
13. O.T. Avery, C.M. MacLeod, and M. McCarty, "Studies on the Chemical Nature of the Substance Inducing Transformation of Pneumococcal Types," *J Exp Med.,* 1944, 79(2):137–158.
14. E. Chargaff et al. "The Composition of the Deoxyribonucleic Acid of Salmon Sperm," *J Biol Chem.,* 1951, 192(1):223–30.
15. A.D. Hershey and M. Chase, "Independent Functions of Viral Protein and Nucleic Acid in Growth of Bacteriophage," *J Gen Physiol.,* 1952, 36(1):39–56.
16. Most genetics textbooks contain the history elaborated in this chapter. For example, see W.S. Klug and M.R. Cummings, *Concepts of Genetics,* 6th Ed. (Upper Saddle River, NJ: Prentice Hall, 2000).
17. J.D. Watson and F.H. Crick, "Molecular Structure of Nucleic Acids; a Structure for Deoxyribose Nucleic Acid," *Nature,* 1953, 171(4356):737–738.
18. For the critical experiment establishing how DNA is replicated, see M. Meselson and F.W. Stahl, "The Replication of DNA in *Escherichia coli,*" *Proc Natl Acad Sci USA,* 1958, 44(7):671–682.

Chapter 3

1. "If a process occurs in a *closed* system, the entropy of the system increases for irreversible processes and remains constant for reversible processes. It never decreases." D. Halliday, R. Resnick, and J. Walker, *Fundamentals of Physics* (New York: John Wiley & Sons, Inc., 2001).
2. W.P. Jencks and R.V. Wolfenden, "Fritz Albert Lipmann, 12 June 1899–24 July 1986," *Biographical Memoirs of the Royal Society,* 2000, 46:335–34.
3. And other enzymes.
4. For further reading on cellular energetics, see any standard biochemistry textbook.
5. Again, for much of the historical content of this section, see any standard genetics textbook. However, I am deeply indebted to the following work for the lucid detail in which this history has been conveyed: H.F. Judson, *The Eighth Day of Creation: Makers of the Revolution in Biology* (New York: Simon and Schuster, 1979).
6. A.E. Garrod, "The Incidence of Alkaptonuria: A Study in Chemical Individuality," *Lancet,* 1902, 2:1616–1620.

7. G.W. Beadle and E.L. Tatum, "Genetic Control of Biochemical Reactions in Neurospora," *Proc Natl Acad Sci USA,* 1941, 27(11):499–506.
8. B. Foltmann, "Protein Sequencing: Past and Present," *Biochemical Education,* 1981, 9(1):2–7; S. de Chadarevian, "Protein Sequencing and the Making of Molecular Genetics," *Trends Biochem Sci.,* 1999, 24(5):203–6; A.O.W. Stretton, "The First Sequence. Fred Sanger and Insulin," *Genetics,* 2002, 162(2):527–532.
9. V.M. Ingram, "A Specific Chemical Difference Between the Globins of Normal Human and Sickle-cell Anaemia Haemoglobin," *Nature,* 1956, 178(4537):792–4.
10. Table 3.1 adapted from Figure 7.1 of B. Lewin, *Genes VII* (New York: Oxford University Press, 2000).
11. Exceptions to the central dogma exist. For example, the human immunodeficiency virus (HIV) is an RNA virus that makes a DNA copy of itself and inserts itself into DNA.
12. For further detail, consult any standard genetics textbook.
13. A.E. Mirsky and H. Ris, "The Desoxyribonucleic Acid Content of Animal Cells and Its Evolutionary Significance," *J Gen Physiol.,* 1951, 34(4):451–462.
14. For the history of DNA sequencing, I have drawn heavily on the following references: E.N. Trifonov, "Earliest Pages of Bioinformatics," *Bioinformatics,* 2000, 16(1):5–9; D.R. Smith, "Goodbye Genome Paper, Hello Genome Report: The Increasing Popularity of 'Genome Announcements' and Their Impact on Science," *Brief Funct Genomics,* June 23, 2016. pii: elw026; L. Ernster and G. Schatz, 1981. "Mitochondria: A Historical Review," *J Cell Biol.,* 1981, 91(3 Pt 2):227s–255s; D.R. Smith, "The Past, Present, and Future of Mitochondrial Genomics: Have We Sequenced Enough mtDNAs?" *Briefings in Functional Genomics,* 2015, 1–8; J.L. Boore, 1999. "Animal Mitochondrial Genomes," *Nucleic Acids Res.,* 1999, 27(8):1767–80; J.M. Heather and B. Chain, "The Sequence of Sequencers: The History of Sequencing DNA," *Genomics,* 2016, 107(1):1–8; C.A. Hutchison III, "DNA Sequencing: Bench to Bedside and Beyond," *Nucleic Acids Res.,* 2007, 35(18):6227–37; M.S. Guyer and F.S. Collins, "How Is the Human Genome Project Doing, and What Have We Learned So Far?" *Proc Natl Acad Sci USA,* 1995, 92(24):10841–10848; E.D. Green, J.D. Watson, and F.S. Collins, "Human Genome Project: Twenty-five Years of Big Biology" *Nature,* 2015, 526(7571):29–31; E.D. Green, "Strategies for the Systematic Sequencing of Complex Genomes," *Nature Reviews Genetics,* 2001, 2:573–583; F. Sanger, "The Early Days of DNA Sequences," *Nat Med.,* 2001, 7(3):267–8.
15. S. Anderson et al., "Sequence and Organization of the Human Mitochondrial Genome," *Nature,* 1981, 290(5806):457–65.
16. A. Goffeau et al., "Life with 6000 Genes," *Science,* 1996, 274:546–567.
17. F.R. Blattner et al., "The Complete Genome Sequence of *Escherichia coli* K-12," *Science,* 1997, 277(5331):1453–1462. Technically, the first bacterial genome sequences were from a different species: R.C. Fleischmann et al., "Whole-genome Random Sequencing and Assembly of *Haemophilus influenzae* Rd," *Science,* 1995, 269(5223):496–512.
18. The C. elegans Genome Sequencing Consortium, "Genome Sequence of the Nematode *C. elegans*: A Platform for Investigating Biology," *Science,* 1998, 282:2012–2018.
19. M.D. Adams et al., "The Genome Sequence of *Drosophila melanogaster*," *Science,* 2000, 287(5461):2185–95.
20. E.S. Lander et al., "Initial Sequencing and Analysis of the Human Genome," *Nature,* 2001, 409(6822):860–921; J.C. Venter et al., "The Sequence of the Human Genome," *Science,* 2001, 291(5507):1304–1351.
21. Mouse Genome Sequencing Consortium et al., "Initial Sequencing and Comparative Analysis of the Mouse Genome," *Nature,* 2002, 420(6915):520–62.
22. M.A. Nóbrega et al., "Megabase Deletions of Gene Deserts Result in Viable Mice," *Nature,* 2004, 431:988–993.
23. For example see the following papers: W. Bender et al., "Molecular Genetics of the Bithorax Complex in *Drosophila melanogaster*," *Science,* 1983, 221(4605):23–29; M.P. Scott et al., "The Molecular Organization of the *Antennapedia* Locus of Drosophila," *Cell,* 1983, 35(3 Pt 2):763–76; L.E. Frischer, F.S. Hagen, and R.L. Garber, "An Inversion That Disrupts the *Antennapedia* Gene Causes Abnormal Structure and Localization of RNAs," *Cell,* 1986, 47(6):1017–23.
24. See any standard genetics textbook for further detail.
25. Summarized in B.C. Carthon et al., "Genetic Replacement of Cyclin D1 Function in Mouse Development by Cyclin D2," *Mol Cell Biol.,* 2005, 25(3):1081–1088.
26. K. Kozar et al., "Mouse Development and Cell Proliferation in the Absence of D-cyclins," *Cell,* 2004, 118(4):477–91.
27. J.K. White et al., "Genome-wide Generation and Systematic Phenotyping of Knockout Mice Reveals New Roles for Many Genes," *Cell,* 2013, 154(2):452–64.
28. E.A. Winzeler et al., "Functional Characterization of the *S. cerevisiae* Genome by Gene Deletion and Parallel Analysis," *Science,* 1999, 285(5429):901–6.

29. R.S. Kamath et al., "Systematic Functional Analysis of the *Caenorhabditis elegans* Genome Using RNAi," *Nature,* 2003, 421(6920):231–7.
30. M.E. Hillenmeyer et al., "The Chemical Genomic Portrait of Yeast: Uncovering a Phenotype for all Genes," *Science,* 2008, 320(5874):362–5.
31. A.K. Ramani et al., "The Majority of Animal Genes Are Required for Wild-type Fitness," *Cell,* 2012, 148:792–802.
32. modENCODE Consortium et al., "Identification of Functional Elements and Regulatory Circuits by *Drosophila* modENCODE," *Science,* 2010, 330:1787–1797.
33. M.B. Gerstein et al., "Integrative Analysis of the *Caenorhabditis elegans* Genome by the modENCODE Project," *Science,* 2010, 330:1775–1787.
34. F. Yue et al., "A Comparative Encyclopedia of DNA Elements in the Mouse Genome," *Nature,* 2014, 515:355–364.
35. ENCODE Project Consortium, "An Integrated Encyclopedia of DNA Elements in the Human Genome," *Nature,* 2012, 489:57–74.
36. ENCODE Project Consortium, "Identification and Analysis of Functional Elements in 1% of the Human Genome by the ENCODE Pilot Project," *Nature,* 2007, 447:799–816.
37. ENCODE Project Consortium, "An Integrated Encyclopedia of DNA Elements in the Human Genome," *Nature,* 2012, 489:57–74.
38. The researchers for the fruit fly biochemical analyses also published a summary, and claimed preliminary evidence for function for 89% of the fruit fly genome — a number remarkably similar to the human number. See page 1792 of the following: modENCODE Consortium et al., "Identification of Functional Elements and Regulatory Circuits by Drosophila modENCODE," *Science,* 2010, 330:1787–1797.
39. http://blogs.discovermagazine.com/notrocketscience/2012/09/05/encode-the-rough-guide-to-the-human-genome/#.WFGFQX3is-w.
40. G.Liu, J.S. Mattick, and R.J. Taft, "A Meta-analysis of the Genomic and Transcriptomic Composition of Complex Life," *Cell Cycle,* 2013, 12(13):2061–2072.
41. For historical overview, see the following: J.B. Gurdon, "From Nuclear Transfer to Nuclear Reprogramming: The Reversal of Cell Differentiation," *Annu Rev Cell Dev Biol.,* 2006, 22:1–22; J.B. Gurdon, "The Egg and the Nucleus: a Battle for Supremacy," *Development,* 2013, 140:2449–2456.
42. At least, in those cells of the body that he tested.
43. I. Wilmut et al., "Viable Offspring Derived from Fetal and Adult Mammalian Cells," *Nature,* 1997, 385(6619):810–3; K. Hochedlinger and R. Jaenisch, "Monoclonal Mice Generated by Nuclear Transfer from Mature B and T Donor Cells," *Nature,* 2002, 415(6875):1035–1038.
44. M. Melé et al., "The Human Transcriptome across Tissues and Individuals," *Science,* 2015, 348(6235):660–5.
45. See any standard genetics textbook for details on this process.
46. While initially controversial, the role of non-coding RNA in gene regulation is one of the hottest areas of current research in molecular biology. The peer-reviewed literature is now so vast that it would be irresponsible to try to list every relevant paper in this section. Instead, the reader is advised to consult one of the many current review papers.
47. Even more surprising is the role that each of the three bases plays in coding for protein. For example, since the genetic code is redundant (i.e., a total of sixty-four, 3-letter combinations in DNA code for only 20 amino acids — see Table 3.1), the third position in the 3-letter code for each amino acid was thought to be non-functional. Some evolutionists have even pointed to third-position sequences that are shared between humans and chimpanzees as evidence of common ancestry. For example, the evolutionary organization BioLogos (http://biologos.org/) claims as evidence for human evolution, "In a certain context, the words 'round' and 'circular' mean the same thing to an English speaker — they are synonyms. So too, there are 'synonyms' in the genetic code — different sequences of DNA bases that mean the same thing to cells (that is, they cause the production of the same proteins). Mutations in the genetic code are often harmful, resulting in an organism not being able to successfully reproduce. But if the mutation results in a 'synonym,' the organism would function the same and continue passing on its genes. Because of this we would expect the synonymous changes to be passed on much more effectively than non-synonymous changes. That is exactly what we find among the DNA of humans and chimpanzees: there are many more synonymous differences between the two species than non-synonymous ones. This is exactly what we would expect if the two species had a common ancestor, and so it provides further evidence that humans and chimpanzees were created through common descent from a single ancestral species." (http://biologos.org/common-questions/human-origins/what-scientific-evidence-do-we-have-about-the-first-humans, accessed May 4, 2017.)

However, like the overall function of the genome, the function of the third position is equally under-investigated. Nevertheless, early experimental results suggest that the third position is not, in fact, functionless. Instead, it might play a role in controlling the speed with which a messenger RNA molecule is translated and the speed with which a messenger RNA molecule is degraded. For example, see the following papers: E.M. Novoa et al., "A Role for tRNA Modifications in Genome Structure and Codon Usage," *Cell,* 2012, 149(1):202–13; M. Zhou et al., "Non-optimal Codon Usage Affects Expression, Structure and Function of Clock Protein FRQ," *Nature,* 2013, 495(7439):111–5; Y. Xu et al., "Non-optimal Codon Usage Is a Mechanism to Achieve Circadian Clock Conditionality," *Nature,* 2013, 495(7439):116–20; A.B. Stergachis et al., "Exonic Transcription Factor Binding Directs Codon Choice and Affects Protein Evolution," *Science,* 2013, 342(6164):1367–72; H. Gingold et al., "A Dual Program for Translation Regulation in Cellular Proliferation and Differentiation," *Cell,* 2014, 158(6):1281–92; V. Presnyak et al., "Codon Optimality Is a Major Determinant of mRNA Stability," *Cell,* 2015, 160(6):1111–24; D.D. Nedialkova and S.A. Leidel, "Optimization of Codon Translation Rates via tRNA Modifications Maintains Proteome Integrity," *Cell,* 2015, 161(7):1606–18; G. Boël et al., "Codon Influence on Protein Expression in *E. coli* Correlates with mRNA Levels," *Nature,* 2016, 529(7586):358–63; C.E. Gamble et al., "Adjacent Codons Act in Concert to Modulate Translation Efficiency in Yeast," *Cell,* 2016, 166(3):679–90; A. Radhakrishnan et al., "The DEAD-Box Protein Dhh1p Couples mRNA Decay and Translation by Monitoring Codon Optimality," *Cell,* 2016, 167(1):122–132.e9.

48. Again, the diversity and scope of RNA-based regulation is being documented at an unprecedented pace. See any of a number of current reviews.
49. All of these discoveries have shed light on the relationship between genome size and organismal complexity. Currently, organismal complexity appears to have much more to do with a creature's non-protein-coding DNA content than its protein-coding DNA content. However, these observations don't fully explain some of the genomic paradoxes that we've uncovered. Why do amphibians and lungfish have so much more DNA than humans? What are the former species doing with so much DNA? Is there a functional explanation for this excessive DNA? Or is the result of accidental DNA duplication? The explanations are not yet clear.
50. For an intriguing discussion of stripe formation, see J.B.L. Bard, "A Unity Underlying the Different Zebra Striping Patterns," *J. Zool., Lond.,* 1977, 183:527–539.
51. N.L. Badlangana, J.W. Adams, and P.R. Manger, "The Giraffe (*Giraffa camelopardalis*) Cervical Vertebral Column: a Heuristic Example in Understanding Evolutionary Processes?" *Zool. J. Linn. Soc.,* 2009, 155 (3):736–757.
52. As will be discussed further in later chapters, different categories of DNA differences exist (e.g., "SNPs", "indels", "large insertions and deletions", etc.). In one category of DNA difference, the percentage is 0.1%. In another category of DNA difference, the percentage is 0.6%. For primary documentation on the various types of DNA differences among humans, see the following papers: 1000 Genomes Project Consortium et al., "An Integrated Map of Genetic Variation from 1,092 Human Genomes," *Nature,* 2012, 491(7422):56–65; 1000 Genomes Project Consortium, "A Global Reference for Human Genetic Variation," *Nature,* 2015, 526:68–74; P.H. Sudmant et al., "An Integrated Map of Structural Variation in 2,504 Human Genomes," *Nature,* 2015, 526(7571):75–81.
53. L. Orlando et al., 2013. "Recalibrating Equus Evolution Using the Genome Sequence of an Early Middle Pleistocene Horse," *Nature,* 2013, 499(7456):74–8; J. Huang et al., "Analysis of Horse Genomes Provides Insight into the Diversification and Adaptive Evolution of Karyotype," *Sci Rep.,* 2014, 4:4958.
54. J. Huang et al., "Donkey Genome and Insight Into the Imprinting of Fast Karyotype Evolution," *Sci Rep.,* 2015, 5:14106.
55. K. Lindblad-Toh et al., "Genome Sequence, Comparative Analysis and Haplotype Structure of the Domestic Dog," *Nature,* 2005, 438(7069):803–19.

Chapter 4

1. O.T. Avery, C.M. MacLeod, and M. McCarty, "Studies on the Chemical Nature of the Substance Inducing Transformation of Pneumococcal Types," *J Exp Med.,* 1944, 79(2):137–158.
2. A.D. Hershey and M. Chase, "Independent Functions of Viral Protein and Nucleic Acid in Growth of Bacteriophage," *J Gen Physiol.,* 1952, 36(1):39–56.
3. In other words, while this chapter relies heavily on structure and logic of the arguments in *On the Origin of Species,* I'm using primarily modern examples as illustrations, not necessarily the ones that Darwin himself used.

4. Unless otherwise noted, the source for species' geography in this chapter is the IUCN Red List database http://www.iucnredlist.org/.
5. B. Larison et al., "How the Zebra Got Its Stripes: a Problem with Too Many Solutions," *R Soc Open Sci.,* 2015, 2(1):140452; A.D. Melin et al., "Zebra Stripes through the Eyes of Their Predators, Zebras, and Humans," *PLoS One,* 2016, 11(1):e0145679.
6. I. Singleton, S.A. Wich , M. Nowak, and G. Usher, "*Pongo abelii* (errata version published in 2016), The IUCN Red List of Threatened Species," 2016, e.T39780A102329901, downloaded on January 6, 2017; M. Ancrenaz, M. Gumal, A.J. Marshall, E. Meijaard, S.A. Wich , and S. Husson, "*Pongo pygmaeus,* The IUCN Red List of Threatened Species," 2016, e.T17975A17966347, http://dx.doi.org/10.2305/IUCN.UK.2016-1.RLTS.T17975A17966347.en, downloaded on January 6, 2017; W. Brockelman, S. Molur, and T. Geissmann, "*Hoolock hoolock,* The IUCN Red List of Threatened Species," 2008, e.T39876A10278553, http://dx.doi.org/10.2305/IUCN.UK.2008.RLTS.T39876A10278553.en, downloaded on January 6, 2017; T. Geissmann and V. Nijman, "*Hylobates agilis,* The IUCN Red List of Threatened Species," 2008, e.T10543A3198943, http://dx.doi.org/10.2305/IUCN.UK.2008.RLTS.T10543A3198943.en, downloaded on January 6, 2017; V. Nijman, M. Richardson, and T. Geissmann, "*Hylobates albibarbis,* The IUCN Red List of Threatened Species," 2008, e.T39879A10279127, http://dx.doi.org/10.2305/IUCN.UK.2008.RLTS.T39879A10279127.en, downloaded on January 6, 2017; D. Whittaker, T. Geissmann, "*Hylobates klossii,* The IUCN Red List of Threatened Species," 2008, e.T10547A3199263. http://dx.doi.org/10.2305/IUCN.UK.2008.RLTS.T10547A3199263.en, downloaded on January 6, 2017; N. Andayani, W. Brockelman, T. Geissmann, V. Nijman, and J. Supriatna, "*Hylobates moloch,* The IUCN Red List of Threatened Species," 2008, e.T10550A3199941, http://dx.doi.org/10.2305/IUCN.UK.2008.RLTS.T10550A3199941.en, downloaded on January 6, 2017; T. Geissmann and V. Nijman, "*Hylobates muelleri,* The IUCN Red List of Threatened Species," 2008, e.T10551A3200262, http://dx.doi.org/10.2305/IUCN.UK.2008.RLTS.T10551A3200262.en, downloaded on January 6, 2017; B. Bleisch and T. Geissmann, "*Nomascus nasutus,* The IUCN Red List of Threatened Species," 2008, e.T41642A10526189, http://dx.doi.org/10.2305/IUCN.UK.2008.RLTS.T41642A10526189.en, downloaded on January 6, 2017; N. Manh Ha, B. Rawson, T. Geissmann, and R.J. Timmins, "*Nomascus siki,* The IUCN Red List of Threatened Species," 2008, e.T39896A10272362, http://dx.doi.org/10.2305/IUCN.UK.2008.RLTS.T39896A10272362.en, downloaded on January 6, 2017; V. Nijman and T. Geissman, "*Symphalangus syndactylus,* The IUCN Red List of Threatened Species," 2008, e.T39779A10266335, http://dx.doi.org/10.2305/IUCN.UK.2008.RLTS.T39779A10266335.en, downloaded on January 6, 2017.
7. H. Bauer, C. Packer, P.F. Funston, P. Henschel, and K. Nowell, "*Panthera leo,* The IUCN Red List of Threatened Species," 2016, e.T15951A107265605, downloaded on December 15, 2016.
8. N. Myers et al., "Biodiversity Hotspots for Conservation Priorities," *Nature,* 2000, 403(6772):853-858.
9. http://dbedt.hawaii.gov/economic/databook/2014-individual/_05/.
10. http://www.ri.gov/facts/history.php.
11. A.R. Wallace, *The Geographical Distribution of Animals* (New York: Harper & Brothers, 1876).
12. http://www.wildmadagascar.org/overview/geography.html.
13. http://www.britannica.com/place/Galapagos-Islands.
14. http://islands.unep.ch/CBT.htm; http://islands.unep.ch/CCJ.htm.
15. My Australian friends insist on this.
16. http://www.newworldencyclopedia.org/entry/Bering_Strait.
17. http://files.hawaii.gov/dbedt/economic/databook/2014-individual/05/050114.pdf.
18. http://www.britannica.com/place/Galapagos-Islands.
19. The fossil record and other historical data add layers of complexity to the discussion of migration — according to both modern creationists and evolutionists. For examples from modern creationists, see the following: K.P. Wise and M. Croxton, "Rafting: a Post-Flood Biogeographic Dispersal Mechanism," in R.L. Ivey, ed., *Proceedings of the Fifth International Conference on Creationism* (Pittsburgh, PA: Creation Science Fellowship, 2003), p. 465–477; D. Statham, "Phytogeography and Zoogeography — Rafting vs Continental Drift," *Journal of Creation,* 2015, 29(1):80–87.
20. At least, he would have argued this point — two of these five species were not documented in his day (see the BirdLife International checklist, http://www.birdlife.org/datazone/userfiles/file/Species/Taxonomy/BirdLife_Checklist_Version_80.zip).

Chapter 5

1. Again, while this chapter relies heavily on the structure and logic of the arguments in *On the Origin of Species*, I'm using primarily modern examples as illustrations, not necessarily the ones that Darwin himself used.
2. Technically, for creatures that are the result of asexual reproduction (including one form of asexual reproduction termed *parthenogenesis*), the offspring is identical to the parent. But for purposes of this chapter, which focuses primarily on sexually reproducing creatures, the original statement still applies.
3. I'm following the more traditional classification system in this chapter and ignoring other schemes, such as those that invoke separate *domains*.
4. As per the Paleobiology Database (http://fossilworks.org/).
5. Ibid.
6. N. Shubin, *Your Inner Fish* (New York: Pantheon Books, 2009).
7. See the following reference for further discussion: N. Irie and S. Kuratani, "The Developmental Hourglass Model: a Predictor of the Basic Body Plan?" *Development,* 2014, 141(24):4649–55.
8. E.B. Daeschler, N.H. Shubin, and F.A. Jenkins Jr., "A Devonian Tetrapod-like Fish and the Evolution of the Tetrapod Body Plan," *Nature,* 2006, 440(7085):757–63.
9. Some have objected to the parallels between the products of human design and the Linnaean classification system. For example, some might say that another product of human design is the pencil. If you study pencils for a long time, no obvious hierarchy in pencil classification will appear.

   We can extend this objection to the example I used in this chapter. Some might object that sedans don't fall into an obvious hierarchical classification categories — that sedans themselves don't form an obvious hierarchy among themselves. For example, individual Honda Civics don't seem to form a nested hierarchy among themselves.

   In response, I would probably agree. Sets of individual sedans don't seem to form obvious groups-within-groups. But then I would point out that the same scenario exists in biology. For example, the individual members of each species don't form an obvious groups-within-groups pattern. The individual members of the giant panda species don't form an obvious groups-within-groups pattern. Why not? Because they're part of the same level of classification. If they did indeed form a groups-within-groups pattern, they would be elevated or demoted to different levels of classification, almost by definition.

   Take another potential objection: "But individual members of a species *do* form slight hierarchies — look at their *DNA* similarities and differences." In response, I would point out that a strong parallel exists in the design realm. Sedans can be classified based on their make and year. We can create hierarchies and levels of relatedness based on the history of their design.

   At this stage, one final objection might arise: "But animals reproduce, and sedans don't." This is a valid observation. But the point of my argument is not that sedans and animals are *identical.* My point is that they are parallel — specifically, parallel in the pattern in which they can be classified. Pointing out basic differences effectively changes the subject — it doesn't address the fact that parallels exist between the two, specifically in the patterns in which they can be classified.

   In other words, the objections to the parallels I've drawn stem largely from a myopic look at human designs. They don't represent a logically persuasive argument.
10. I am indebted to the Intelligent Design community for first bringing to my attention many of these parallels.
11. In some embryological cases, the coccyx fails to develop. As a workaround, the body finds a different site of attachment for the pelvic floor muscles. Consequently, some critics see this as evidence of non-function for the coccyx. However, if my computer broke down as I typed this sentence, I could resume my writing with a legal pad and a pen. The fact that I can find a workaround does nothing to undermine the extreme design present in my laptop. Similar logic holds true in the human body.
12. J.P. Dines et al., "Sexual Selection Targets Cetacean Pelvic Bones," *Evolution,* 2014, 68(11):3296–306.
13. Journal articles: http://creation.com/is-our-inverted-retina-really-bad-design; http://creation.com/is-the-human-pharynx-poorly-designed; http://creation.com/images/pdfs/tj/j26_3/j26_3_60-67.pdf; http://creation.com/recurrent-laryngeal-nerve-design

    Lay-level, but book form: https://answersingenesis.org/human-body/vestigial-organs/vestigial-organs-evidence-for-evolution/.
14. J. Coyne, *Why Evolution is True* (New York: Viking, 2009).

15. H.F. Smith et al., "Multiple Independent Appearances of the Cecal Appendix in Mammalian Evolution and an Investigation of Related Ecological and Anatomical Factors," *Comptes Rendus Palevol,* 2013, 12(6):339–354.
16. As further evidence of poor design, Coyne also points out the medical problems that the appendix can cause. For example, after acknowledging the function of the appendix in the immune system, Coyne says, "But these minor benefits are surely outweighed by the severe problems that come with the human appendix" (p.61).

    This type of argument is common among evolutionists. Neil Shubin makes a similar claim in *Your Inner Fish*: "In a perfectly designed world . . . we would not have to suffer everything from hemorrhoids to cancer" (p.185).

    Coyne and Shubin both seem to think that we must choose between either perfectly flawless design or evolution. But is this a logical dichotomy? If a 747 crashes because it tries to fly through a thunderstorm, does this mean the 747 is the result of evolution and not of design? If a computer catches fire, should we conclude that it must have been the product of time, chance, and natural selection? Clearly, flawed design is still design. Coyne and Shubin have created a false dichotomy.

    Nevertheless, flawed design still makes many uncomfortable — particularly if the Designer of life is an all-knowing, all-powerful God. However, this aspect of this argument is a theological debate, not a scientific one. For theological answers, I recommend reading this book through to the end, and then consulting some of the following references: http://www.icr.org/article/did-lions-roam-garden-eden; https://answersingenesis.org/evidence-for-creation/design-in-nature/design-in-the-curse/ https://answersingenesis.org/evidence-for-creation/design-in-nature/how-did-defense-attack-structures-come-about/; https://answersingenesis.org/evidence-for-creation/design-in-nature/designed-to-kill-in-a-fallen-world/; https://answersingenesis.org/evidence-for-creation/design-in-nature/good-designs-gone-bad/; https://answersingenesis.org/suffering/.
17. Reference for breed numbers: FAO. 2009. *Status and trends of animal genetic resources–2008*, Intergovernmental Technical Working Group on Animal Genetic Resources for Food and Agriculture, Fifth Session, Rome, January 28–30, 2009, CGRFA/WG-AnGR-5/09/Inf. 7 (available at ftp://ftp.fao.org/docrep/fao/meeting/016/ak220e.pdf). For explicit derivation of the numbers from the FAO document, see Supplemental Table 4 of the following reference: N.T. Jeanson and J. Lisle, "On the Origin of Eukaryotic Species' Genotypic and Phenotypic Diversity: Genetic Clocks, Population Growth Curves, and Comparative Nuclear Genome Analyses Suggest Created Heterozygosity in Combination with Natural Processes as a Major Mechanism," *Answers Research Journal,* 2016, 9:81–122. Available online: https://answersingenesis.org/natural-selection/speciation/on-the-origin-of-eukaryotic-species-genotypic-and-phenotypic-diversity/.
18. 18.C. Darwin, *On the Origin of Species*, p.16–17, http://darwin-online.org.uk/Variorum/1859/1859-16-c-1860.html.
19. As per the citation in R. Higuchi et al., "DNA Sequences from the Quagga, an Extinct Member of the Horse Family," *Nature,* 1984, 312(5991):282–4.
20. Reference for breed numbers: FAO. 2009. *Status and trends of animal genetic resources–2008*, Intergovernmental Technical Working Group on Animal Genetic Resources for Food and Agriculture, Fifth Session, Rome, January 28–30, 2009, CGRFA/WG-AnGR-5/09/Inf. 7 (available at ftp://ftp.fao.org/docrep/fao/meeting/016/ak220e.pdf). For explicit derivation of the numbers from the FAO document, see Supplemental Table 4 of the following reference: N.T. Jeanson and J. Lisle, "On the Origin of Eukaryotic Species' Genotypic and Phenotypic Diversity: Genetic Clocks, Population Growth Curves, and Comparative Nuclear Genome Analyses Suggest Created Heterozygosity in Combination with Natural Processes as a Major Mechanism," *Answers Research Journal,* 2016, 9:81–122. Available online: https://answersingenesis.org/natural-selection/speciation/on-the-origin-of-eukaryotic-species-genotypic-and-phenotypic-diversity/.

    See also the following: Dog breeds: http://www.fci.be/en/Nomenclature/, accessed March 14, 2017; Cat breeds: http://www.tica.org/cat-breeds, accessed March 14, 2017.
21. The mammal species classification dataset was downloaded from the IUCN Red List (http://www.iucnredlist.org/) on February 24, 2017. Only extant species were considered.
22. See previous endnote for documentation.
23. The number of avian species per family was downloaded from the BirdLife Taxonomic Checklist (version 9; http://datazone.birdlife.org/species/taxonomy) on February 23, 2017.
24. This test naturally raises the question of where reproductively incompatible creatures fit within the modern creationist paradigm. In short, the answer is still being sought. Within the modern creationist model, hybridization represents a good positive test for common ancestry; it doesn't represent a good negative test — i.e., it fails to

identify which species do *not* share a common ancestor. Since genetics is the only direct scientific record of a species' ancestry, modern creationist research is exploring this field for answers (see chapters 7–10).

25. Strictly speaking, the statement that "*min* appear to be best approximated by the classification level of family or order" applies only to those groups of creatures in which hybridization studies have been performed. Also, since the Bible never speaks of humans in terms of *min*, modern creationists do not apply the family/order rule to humans. (Also, humans cannot successfully breed with any other creature.) Nevertheless, since the results of these studies appear to be consistently arriving at the classification rank of family or order, and since this is true across several vertebrate classes, I have generalized the results. See the following papers for more discussion: T.C. Wood, 2006. "The Current Status of Baraminology," *Creation Research Society Quarterly*, 2006, 43:149–158, available online: http://www.creationresearch.org/crsq/articles/43/43_3/baraminology.htm; G. Fankhauser and K.B. Cummings, 2008. "Snake Hybridization: A Case for Intrabaraminic Diversity," *Proceedings of the Sixth International Conference on Creationism*, Creation Science Fellowship, Pittsburgh, PA, and Institute for Creation Research, Dallas, TX, 2008, p. 117–132, available online: http://www.icr.org/article/snake-hybridization-intrabaraminic-diversity/; J.K. Lightner, "Identification of a Large Sparrow-finch Monobaramin in Perching Birds (Aves: Passeriformes), *Journal of Creation*, 2010, 24(3):117–121, available online: https://creation.com/images/pdfs/tj/j24_3/j24_3_117-121.pdf; B. Pendragon and N. Winkler, "The Family of Cats — Delineation of the Feline Basic Type," *Journal of Creation*, 2011, 25(2):118–124, available online: http://creation.com/images/pdfs/tj/j25_2/j25_2_118-124.pdf.
26. This is one of the reasons Noah could have easily fit the animals on board the ark. The claim that the ark was too small to fit all the animals on board has been a long-standing, poorly reasoned objection to Genesis 6–9. See the following for more answers to other popular (but erroneous) stereotypes of this account, such as the ability of Noah and his family to build the ark, the ability of Noah and his family to care for the animals, the survival of freshwater fish during the Flood, etc.: https://answersingenesis.org/noahs-ark/reimagining-ark-animals/; https://answersingenesis.org/noahs-ark/what-did-noahs-ark-look-like/; https://answersingenesis.org/noahs-ark/caring-for-the-animals-on-the-ark/; https://answersingenesis.org/the-flood/how-could-fish-survive-the-genesis-flood/.
27. For invertebrates, plants, fungi, and microbes, the best taxonomic approximation for *min* is still uncertain, but probably above the level of species. At a minimum, modern creationists would have little problem endorsing the formation of new species within invertebrate, plant, fungal, and microbial genera or subgenera.
28. One of the major differences between modern creationist views and Darwin's view is on the limits of speciation. Modern creationists do not think that one *min* can naturally change into another *min*. For a more comprehensive derivation of this conclusion within a creationist framework, see the following: https://answersingenesis.org/creation-science/baraminology/which-animals-were-on-the-ark-with-noah/. In other words, modern creationists would tend to generally reject the origin of one family or order from another, but would accept the origin of species within families or orders.

    Conversely, Darwin's breed-species argument naturally has limits. For example, while all the diversity among horse breeds and donkey breeds might capture diversity among modern equid species, it fails to bridge the gap between living equids and living rhinos. Thus, in this sense, modern creationists would agree with Darwin's breed-species argument.

Chapter 6

1. C. Darwin, *On the Origin of Species,* p.108–110, http://darwin-online.org.uk/Variorum/1860/1860-108-c-1859.html.
2. G. Larson et al., "Current Perspectives and the Future of Domestication Studies," *Proc Natl Acad Sci USA,* 2014, 111(17):6139–6146.
3. R. Dawkins, *The Greatest Show on Earth* (New York: Free Press, 2009), p.28.
4. G. Larson et al., "Current Perspectives and the Future of Domestication Studies," *Proc Natl Acad Sci USA,* 2014, 111(17):6139–6146.
5. The mammal species classification dataset was downloaded from the IUCN Red List (http://www.iucnredlist.org/) on February 24, 2017. Only extant species were considered.
6. Ibid.
7. See previous endnote for source of mammal species numbers and mammal classification. For the other vertebrate classes, the number of reptile species per family was downloaded from Reptile Database (http://www.reptile-database.org/data/) on February 23, 2017. At that time, the website listed the most updated version as "24 Dec 2016." The number of amphibian species per family was downloaded from the Amphibian Species of the World database

(http://research.amnh.org/vz/herpetology/amphibia/) on February 23, 2017. The number of avian species per family was downloaded from the BirdLife Taxonomic Checklist (version 9; http://datazone.birdlife.org/species/taxonomy) on February 23, 2017. The number of fish species per family was downloaded from the Catalog of Fishes (http://researcharchive.calacademy.org/research/ichthyology/catalog/SpeciesByFamily.asp) on February 23, 2017.

8. Since the answer to the timescale question has typically been derived from the fields of geology and paleontology, the role of fossils in this discussion naturally arises. However, the origin of extinct species is a separate question from the origin of living species, which is what chapter 6 addresses. Later chapters cover fossils in more detail; hence, they are omitted here.
9. See earlier endnote for justification of species numbers.
10. http://www.australia.gov.au/about-australia/australian-story/european-discovery-and-colonisation.
11. I focus on Europe since the start of the modern origin of species debate traces to Carolus Linnaeus, who was a European (a Swede).
12. https://blog.oup.com/2012/08/facts-silk-road-peak-trader-camel-travel/.
13. http://www.newworldencyclopedia.org/entry/Scramble_for_Africa.
14. Wilson & Reeder's *Mammal Species of the World*, 3rd Ed., was accessed electronically at https://www.departments.bucknell.edu/biology/resources/msw3/ on October 7, 2016. From the .csv file of the complete taxonomy, the data were filtered for "SPECIES" via the "TaxonLevel" column, and then filtered for "FALSE" via the "Extinct?" column. A timeline of species' naming was created from the data in the "Date" column. For those species with a range of date listings, the earliest date was chosen. See the following reference, especially the endnotes, for a similar analysis: https://answersingenesis.org/reviews/articles/does-biologos-strive-for-dialogue/.
15. As per Wilson & Reeder's *Mammal Species of the World*, 3rd Ed.
16. See previous endnotes for details on derivation of the species dataset.
17. http://www.australia.gov.au/about-australia/australian-story/european-discovery-and-colonisation.
18. http://www.newworldencyclopedia.org/entry/David_Livingstone.
19. Again, the derivation of these numbers are from the Wilson & Reeder dataset cited in a previous endnote.
20. Again, timeline comes from Wilson & Reeder dataset using the derivation in a previous endnote.
21. Wilson & Reeder, *Mammal Species of the World*, 2nd Ed. (Washington: Smithsonian Institution Press, 1993).
22. Derived from 3rd Ed. of Wilson & Reeder, extant species only.
23. Similar conclusions can be reached with other animal datasets, such as reptiles (see the Reptile Database, http://www.reptile-database.org/db-info/news.html) and birds (see the BirdLife International checklist, http://www.birdlife.org/datazone/userfiles/file/Species/Taxonomy/BirdLife_Checklist_Version_80.zip). Furthermore, since at least half of all living animal species were unknown in Darwin's day (see the endnote from the Introduction), these conclusions appear to be applicable across the animal kingdom.
24. D.J. Futuyma, *Evolution* (Sunderland, MA: Sinauer Associates, Inc., 2013), chapter 17.
25. B. Glanzmann et al., "The Complete Genome Sequence of the African Buffalo (*Syncerus caffer*)," *BMC Genomics,* 2016, 17(1):1001.
26. D.M. Brown et al., "Extensive Population Genetic Structure in the Giraffe," *BMC Biology,* 2007, 5(1):57.
27. J. Fennessy et al., "Multi-locus Analyses Reveal Four Giraffe Species Instead of One," *Curr Biol.* 26(18):2543–9.
28. P. Novellie, "*Equus zebra.* The IUCN Red List of Threatened Species 2008," e.T7960A12876787. http://dx.doi.org/10.2305/IUCN.UK.2008.RLTS.T7960A12876787.en. Downloaded on 15 December 2016.
29. S.R.B. King and P.D. Moehlman, "*Equus quagga.* The IUCN Red List of Threatened Species 2016," e.T41013A45172424, http://dx.doi.org/10.2305/IUCN.UK.2016-2.RLTS.T41013A45172424.en, downloaded on December 15, 2016.
30. IUCN number; see earlier endnote for documentation.
31. IUCN number; see earlier endnote for documentation.
32. IUCN number; see earlier endnote for documentation.
33. IUCN data; see documentation in an earlier endnote.

Chapter 7

1. Preliminary versions of the findings in this chapter can be found in the following references: N.T. Jeanson, 2013. "Recent, Functionally Diverse Origin for Mitochondrial Genes from ~2700 Metazoan Species," *Answers Research*

Journal, 2013, 6:467–501, available online at https://answersingenesis.org/genetics/mitochondrial-dna/recent-functionally-diverse-origin-for-mitochondrial-genes-from--2700-metazoan-species/; N.T. Jeanson, "Mitochondrial DNA Clocks Imply Linear Speciation Rates within 'Kinds,' " *Answers Research Journal,* 2015, 8:273–304, available online at https://answersingenesis.org/natural-selection/speciation/clocks-imply-linear-speciation-rates-with-in-kinds/; N.T. Jeanson, "A Young-earth Creation Human Mitochondrial DNA 'Clock': Whole Mitochondrial Genome Mutation Rate Confirms D-loop Results," *Answers Research Journal,* 2015, 8:375–378, available online at https://answersingenesis.org/genetics/mitochondrial-genome-mutation-rate/; N.T. Jeanson and J. Lisle, "On the Origin of Eukaryotic Species' Genotypic and Phenotypic Diversity: Genetic Clocks, Population Growth Curves, and Comparative Nuclear Genome Analyses Suggest Created Heterozygosity in Combination with Natural Processes as a Major Mechanism," *Answers Research Journal,* 2016, 9:81–122, available online at https://answersingenesis.org/natural-selection/speciation/on-the-origin-of-eukaryotic-species-genotypic-and-phenotypic-diversity/; N.T. Jeanson, "On the Origin of Human Mitochondrial DNA Differences, New Generation Time Data Both Suggest a Unified Young-Earth Creation Model and Challenge the Evolutionary Out-of-Africa Model," *Answers Research Journal,* 2016, 9:123–130, available online at https://answersingenesis.org/genetics/mitochondrial-dna/origin-human-mitochondrial-dna-differences-new-generation-time-data-both-suggest-unified-young-earth/.

2. For example, see the following: D.F. Conrad et al., "Variation in Genome-wide Mutation Rates within and between Human Families," *Nature Genetics,* 2011, 43(7):712–714.
3. As examples, see the following papers: D.R. Denver et al., "High Mutation Rate and Predominance of Insertions in the Caenorhabditis Elegans Nuclear Genome," *Nature,* 2004, 430(7000):679–682; C. Haag-Liautard et al., "Direct Estimation of Per Nucleotide and Genomic Deleterious Mutation Rates in *Drosophila,*" *Nature,* 2007, 445(7123):82–85; P.D. Keightley et al., "Estimation of the Spontaneous Mutation Rate in *Heliconius melpomene,*" *Mol. Biol. Evol.,* 2015, 32(1):239–243; A. Uchimura et al., "Germline Mutation Rates and the Long-term Phenotypic Effects of Mutation Accumulation in Wild-type Laboratory Mice and Mutator Mice," *Genome Research,* 2015, 25(8):1125–34.
4. As an example, see the following paper: S. Ossowski et al., "The Rate and Molecular Spectrum of Spontaneous Mutations in *Arabidopsis thaliana,*" *Science,* 2010, 327(5961):92–94.
5. As an example, see the following paper: A. Farlow et al., "The Spontaneous Mutation Rate in the Fission Yeast *Schizosaccharomyces pombe,*" *Genetics,* 2015, 201(2):737–44.
6. One of the most prominent examples of bacterial mutation can be found in the following papers: J.E. Barrick et al., "Genome Evolution and Adaptation in a Long-term Experiment with *Escherichia coli,*" *Nature,* 2009, 461(7268):1243–7; Z.D. Blount et al., "Genomic Analysis of a Key Innovation in an Experimental *Escherichia coli* Population," *Nature,* 2012, 489(7417):513–8.
7. S. Anderson et al., "Sequence and Organization of the Human Mitochondrial Genome," *Nature,* 1981, 290(5806):457–65.
8. E.S. Lander et al., "Initial Sequencing and Analysis of the Human Genome," *Nature,* 2001, 409(6822):860–921; J.C. Venter et al., "The Sequence of the Human Genome," *Science,* 2001, 291(5507):1304–1351.
9. M.J. Bibb et al., "Sequence and Gene Organization of Mouse Mitochondrial DNA," *Cell,* 1981, 26(2 Pt 2):167–80.
10. S. Anderson et al., "Complete Sequence of Bovine Mitochondrial DNA. Conserved Features of the Mammalian Mitochondrial Genome," *J Mol Biol.,* 1982, 156(4):683–717.
11. J.L. Boore, "Animal Mitochondrial Genomes," *Nucleic Acids Res.,* 1999, 27(8):1767–80.
12. The RefSeq database at https://www.ncbi.nlm.nih.gov/nuccore/; data as of March 9, 2017.
13. As of March 9, 2017.
14. That is, when you search for mitochondrial DNA genomes at NCBI (https://www.ncbi.nlm.nih.gov/nuccore/) without using the RefSeq filter, and excluding sequences below 10,000 base pairs and above 50,000 base pairs in length. Data as of March 9, 2017.
15. For example, see the data in the following paper: N.T. Jeanson, "Recent, Functionally Diverse Origin for Mitochondrial Genes from ~2700 Metazoan Species," *Answers Research Journal,* 2013, 6:467–501, available online at https://answersingenesis.org/genetics/mitochondrial-dna/recent-functionally-diverse-origin-for-mitochondrial-genes-from--2700-metazoan-species/.
16. Again, see the following paper for more extensive documentation: N.T. Jeanson, "Recent, Functionally Diverse Origin for Mitochondrial Genes from ~2700 Metazoan Species," *Answers Research Journal,* 2013, 6:467–501, available online at https://answersingenesis.org/genetics/mitochondrial-dna/recent-functionally-diverse-origin-for-mitochondrial-genes-from--2700-metazoan-species/.

17. In Table 7.1, only some of the species within the rhino and tapir families are shown because mtDNA sequences have not yet been obtained from all species within these families. Comparisons were done among all species with mtDNA sequences available at the time of analysis.
18. These results were obtained through my own personal analysis. Sequences were downloaded from NCBI database (NCBI accession numbers for equid sequences = KM881681.1, NC_016061.1, NC_018780.1, NC_018781.1, NC_020433.1, NC_024030.1, NC_020432.2; NCBI accession numbers for rhino, tapir, and sheep sequences = NC_012682.1, NC_001808.1, NC_012684.1, NC_012683.1, NC_001779.1, NC_023838.1, NC_015889.1). Where appropriate, sequences were manually adjusted to share the same position #1. Sequences were aligned in-house with CLUSTALX2 (http://www.clustal.org/clustal2/) or CLUSTALW-MTV (http://www4a.biotec.or.th/GI/tools/clustalw-mtv). In BioEdit (http://www.mbio.ncsu.edu/BioEdit/bioedit.html) software, all non-standard nucleotides (e.g., N, M, R, Y, B, W, S, V, H, D) were replaced with gaps. Then all columns containing gaps were stripped, after which a sequence difference count matrix was created.
19. Again, for much more extensive documentation, see the following paper: N.T. Jeanson, "Recent, Functionally Diverse Origin for Mitochondrial Gens from ~2700 Metazoan Species," *Answers Research Journal,* 2013, 6:467–501, available online: https://answersingenesis.org/genetics/mitochondrial-dna/recent-functionally-diverse-origin-for-mitochondrial-genes-from-~2700-metazoan-species/.
20. N.T. Jeanson, "Recent, Functionally Diverse Origin for Mitochondrial Genes from ~2700 Metazoan Species," *Answers Research Journal,* 2013, 6:467–501, available online at https://answersingenesis.org/genetics/mitochondrial-dna/recent-functionally-diverse-origin-for-mitochondrial-genes-from-~2700-metazoan-species/.
21. For further discussion, see any standard biochemistry textbook.
22. See the following reviews for more details on moonlighting, and for primary references: C.J. Jeffery, "Moonlighting Proteins: Old Proteins Learning New Tricks," *Trends Genet.,* 2003, 19:415–417.
    D.H.E.W. Huberts and I.J. van der Klei, "Moonlighting Proteins: an Intriguing Mode of Multitasking," *Biochim. Biophys. Acta,* 2010, 1803:520–525.
23. At the whole mtDNA genome level, one of the first comparisons in humans can be found in the following paper: M. Ingman et al., "Mitochondrial Genome Variation and the Origin of Modern Humans," *Nature,* 2000, 408:708–713. Since then, far too many papers have been published on human mtDNA diversity to list them all. The following websites are helpful databases of human mtDNA genome diversity: http://www.mtdb.igp.uu.se/, http://www.phylotree.org/.
24. See Tables 1–2 of the following paper: N.T. Jeanson, "Recent, Functionally Diverse Origin for Mitochondrial Genes from ~2700 Metazoan Species," *Answers Research Journal,* 2013, 6:467–501, available online at https://answersingenesis.org/genetics/mitochondrial-dna/recent-functionally-diverse-origin-for-mitochondrial-genes-from-~2700-metazoan-species/. See also Table 1 of the following paper: N.T. Jeanson, "A Young-earth Creation Human Mitochondrial DNA 'Clock': Whole Mitochondrial Genome Mutation Rate Confirms D-loop Results," *Answers Research Journal,* 2015, 8:375–378, available online at https://answersingenesis.org/genetics/mitochondrial-genome-mutation-rate/.
25. Again, see the references in the previous endnote for the back-and-forth among investigators.
26. Again, see the references in the previous endnote for justification.
27. D.J. Futuyma, *Evolution* (Sunderland, MA: Sinauer Associates, Inc., 2013).
28. Ibid.
29. D.P. Locke et al., "Comparative and Demographic Analysis of Orang-utan Genomes," *Nature,* 2011, 469(7331):529–533.
30. C. Stringer, "The Origin and Evolution of *Homo sapiens,*" *Philos Trans R Soc Lond B Biol Sci.,* 2016, 371(1698). pii: 20150237.
31. The discussion over salt flow has actually been a significant flashpoint in the debate over the age of the earth. See the following references: S.A. Austin and D.R. Humphreys, "The Sea's Missing Salt: A Dilemma for Evolutionists," in *Proceedings of the Second International Conference on Creationism,* R.E. Walsh and C.L. Brooks, eds., volume 2 (Pittsburgh, PA: Creation Science Fellowship, 1990), p. 17–33, available online at http://www.icr.org/article/sea-missing-salt/. The above reference contains discussion of critics at the end. See also Addendum in http://creation.com/salty-seas-evidence-for-a-young-earth#albite.
32. Again, this is another flashpoint in the debate over the age of the earth. See the following references: L. Vardiman, "The Sands of Time: A Biblical Model of Deep Sea-Floor Sedimentation," *Creation Research Society Quarterly,* 1996,

33(3):191–198, available online at http://www.icr.org/article/deep-sea-floor-sedimentation/; A.A. Snelling, *Earth's Catastrophic Past: Geology, Creation and the Flood* (Dallas, TX: Institute for Creation Research, 2009), p. 881–884.

33. But see the following for how carbon-14 is revolutionizing our understanding of the age of the earth: J.R. Baumgardner, "Measurable 14C in Fossilized Organic Materials: Confirming the Young Earth Creation-Flood Model," in *Proceedings of the Fifth International Conference on Creationism*, R.L. Ivey Jr., ed. (Pittsburgh, PA: Creation Science Fellowship, 2003), p. 127–142; J. Baumgardner, "Carbon-14 Evidence for a Recent Global Flood and a Young Earth," in L. Vardiman, A.A. Snelling, and E.F. Chaffin, eds. *Radioisotopes and the Age of the Earth: Results of a Young-Earth Research Initiative*, Vol. 2 (El Cajon, CA: Institute for Creation Research and Chino Valley, AZ: Creation Research Society, 2005). Response to critics: https://answersingenesis.org/geology/radiometric-dating/are-the-rate-results-caused-by-contamination/; see also https://answersingenesis.org/geology/carbon-14/a-creationist-puzzle/ and https://answersingenesis.org/geology/carbon-14/radiocarbon-dating/.
34. D.J. Futuyma, *Evolution* (Sunderland, MA: Sinauer Associates, Inc., 2013).
35. For example, see the following and the references therein: http://biologos.org/resources/infographic/how-do-we-know-the-earth-is-old.
36. Hypotheses: L. Vardiman, A.A. Snelling, and E.F. Chaffin, eds., *Radioisotopes and the Age of the Earth: A Young-Earth Research Initiative*. Vol. 1 (El Cajon, CA: Institute for Creation Research and St. Joseph, MO: Creation Research Society, 2000).

    Experimental Results: L. Vardiman, A.A. Snelling, and E.F. Chaffin, eds., *Radioisotopes and the Age of the Earth: Results of a Young-Earth Research Initiative*. Vol. 2 (El Cajon, CA: Institute for Creation Research and Chino Valley, AZ: Creation Research Society, 2005).

    Responses to critics: https://answersingenesis.org/geology/radiometric-dating/are-the-rate-results-caused-by-contamination/, http://creation.com/argon-diffusion-age, http://creation.com/russ-humphreys-refutes-joe-meerts-false-claims-about-helium-diffusion, http://creation.com/helium-evidence-for-a-young-world-continues-to-confound-critics, https://creation.com/images/pdfs/tj/j24_1/j24_1_14-16.pdf, https://creation.com/images/pdfs/tj/j24_3/j24_3_34-39.pdf, http://creation.com/images/pdfs/tj/j26_2/j26_2_45-49.pdf, https://www.trueorigin.org/helium01.php, https://www.trueorigin.org/helium02.php.
37. Prediction: J.R. Baumgardner, "Numerical Simulation of the Large-scale Tectonic Changes Accompanying the Flood," in R.E. Walsh, C.L. Brooks, and R.S. Crowell, eds., *Proceedings of the First International Conference on Creationism*, Vol. 2 (Pittsburgh, PA: Creation Science Fellowship, 1986), p. 17–30, available online at http://www.icr.org/article/large-scale-tectonic-change-flood/.

    Fulfillment: S.P. Grand, "Mantle Shear Structure Beneath the Americas and Surrounding Oceans," *Journal of Geophysical Research,* 1994, 99:11591–11621; J.E. Vidale, "A Snapshot of Whole Mantle Flow," *Nature,* 1994, 370:16–17; P. Voosen, 2016. "Graveyard of Cold Slabs Mapped in Earth's Mantle," *Science,* 2016, 354(6315):954–955.

    For a broader overview, see: A.A. Snelling, "Can Catastrophic Plate Tectonics Explain Flood Geology?" in K. Ham, ed., *New Answers Book* (Green Forest, AR: Master Books, 2006), p. 186–197, available online at https://answersingenesis.org/geology/plate-tectonics/can-catastrophic-plate-tectonics-explain-flood-geology/.
38. S.A. Austin, 1986. *Proceedings of the First International Conference on Creationism,* R.S. Crowell, editor (Pittsburgh, PA: Creation Science Fellowship, 1986), p. 3–9, available online at http://static.icr.org/i/pdf/technical/Mount-St-Helens-and-Catastrophism.pdf; S.A. Austin, "Rapid Erosion at Mount St. Helens," *Origins,* 1984, 11(2):90–98, available online at http://static.icr.org/i/pdf/technical/Rapid-Erosion-at-Mount-St-Helens.pdf; S.A. Austin, "Excess Argon within Mineral Concentrates from the New Dacite Lava Dome at Mount St Helens Volcano," *Journal of Creation,* 1996, 10(3):335–343, available online at http://creation.com/excess-argon-within-mineral-concentrates.

    For a broader overview, see the following: S.A. Austin, "Why Is Mount St. Helens Important to the Origins Controversy?" in K. Ham, ed., *New Answers Book 3* (Green Forest, AR: Master Books, 2009), p. 253–262, available online at https://answersingenesis.org/geology/mount-st-helens/why-is-mount-st-helens-important-to-the-origins-controversy/; J. Morris and S. Austin, *Footprints in the Ash: The Explosive Story of Mount St. Helens* (Green Forest, AR: Master Books, 2003).

    See also the following references: J. Schieber, J. Southard, and K. Thaisen, "Accretion of Mudstone Beds from Migrating Floccule Ripples," *Science,* 2007, 318(5857):1760–1763; J. Schieber and Z. Yawar, "A New Twist on Mud Deposition — Mud Ripples in Experiment and Rock Record," *The Sedimentary Record,* 2009, 7(2):4–8.
39. D.R. Humphreys, "The Creation of Planetary Magnetic Fields," *CRSQ,* 1984, 21(3):140–149, available online at http://www.creationresearch.org/crsq/articles/21/21_3/21_3.html; D.R. Humphreys, "The Creation of Cosmic

Magnetic Fields," *Proceedings of the Sixth International Conference on Creationism* (Pittsburgh, PA: Creation Science Fellowship and Dallas, TX: Institute for Creation Research, 2008), p. 213–230, available online at http://www.icr.org/article/cosmic-magnetic-fields-creation/; D.R. Humphreys, "Planetary Magnetic Dynamo Theories: a Century of Failure," *Proceedings of the Seventh International Conference on Creationism*, Pittsburgh, PA, August 4–8, 2013, information and proceedings available at: creationicc.org.

See also the following: D.R. Humphreys, 2002. "The Earth's Magnetic Field is Still Losing Energy," *CRSQ*, 2002, 39(1):1–11, available online at http://www.creationresearch.org/crsq/articles/39/39_1/GeoMag.htm; D.R. Humphreys, "Creationist Cosmologies Explain the Anomalous Acceleration of Pioneer Spacecraft," *Journal of Creation,* 2007, 21(2):61–70, available online at http://creation.com/creationist-cosmologies-explain-the-anomalous-acceleration-of-pioneer-spacecraft; D.R. Humphreys, "Earth's Magnetic Field Is Decaying Steadily — with a Little Rhythm," *CRSQ,* 2011, 47(3):193–201. Available online at http://www.creationresearch.org/crsq/articles/47/47_3/CRSQ%20Winter%202011%20Humphreys.pdf; D.R. Humphreys, "More Secular Confusion about Moon's Former Magnetic Field," *Journal of Creation,* 2013, 27(2):12–13, available online at http://creation.com/confusion-about-moon-magnetic-field.

See also: http://creation.com/search-continues-for-a-non-creationist-solution-to-the-pioneer-anomaly; http://creation.com/mercurys-magnetic-field-is-young; https://www.trueorigin.org/ca_rh_01.php.

40. Those experiments that seem to measure the speed of light in only one direction are, in reality, only measuring the round-trip speed since they have tacitly assumed the one-way speed in the way that the clocks have been synchronized. See the following papers: J. Lisle, "Anisotropic Synchrony Convention — A Solution to the Distant Starlight Problem," *Answers Research Journal,* 2010, 3:191–207, available online at https://answersingenesis.org/astronomy/starlight/anisotropic-synchrony-convention-distant-starlight-problem/; J.G. Hartnett, "A Biblical Creationist Cosmogony," *Answers Research Journal,* 2015, 8:13–20, available online at https://answersingenesis.org/astronomy/starlight/a-biblical-creationist-cosmogony/.
41. D.J. Futuyma, *Evolution* (Sunderland, MA: Sinauer Associates, Inc., 2013).
42. For example, evolutionists used a divergence calculation to date the human-chimpanzee ancestral split. See O. Venn et al., "Strong Male Bias Drives Germline Mutation in Chimpanzees," *Science,* 2014, 344:1272–1275. Though the Venn et al. study deals with nuclear DNA rather than mtDNA, it illustrates the type of equation that is necessary for inter-species comparisons.
43. D.P. Locke et al., "Comparative and Demographic Analysis of Orang-utan Genomes," *Nature,* 2011, 469(7331):529–533; K.E. Langergraber et al., "Generation Times in Wild Chimpanzees and Gorillas Suggest Earlier Divergence Times in Great Ape and Human Evolution," *Proc Natl Acad Sci USA,* 2012, 109(39):15716–15721; O. Venn et al., "Strong Male Bias Drives Germline Mutation in Chimpanzees," *Science,* 2014, 344:1272–1275.
44. K.E. Langergraber et al., "Generation Times in Wild Chimpanzees and Gorillas Suggest Earlier Divergence Times in Great Ape and Human Evolution," *Proc Natl Acad Sci USA,* 2012, 109(39):15716–15721.
45. These numbers were calculated using more precise estimates of the human mtDNA mutation rate. See the following reference: N.T. Jeanson, "A Young-earth Creation Human Mitochondrial DNA 'Clock': Whole Mitochondrial Genome Mutation Rate Confirms D-loop Results," *Answers Research Journal,* 2015, 8:375–378, available online at https://answersingenesis.org/genetics/mitochondrial-genome-mutation-rate/.
46. These results were obtained through my own personal analysis. Sequences were downloaded from NCBI RefSeq database (NCBI accession numbers = NC_012920.1, NC_001643.1). Sequences were aligned in-house with CLUSTALX2 (http://www.clustal.org/clustal2/) or CLUSTALW-MTV (http://www4a.biotec.or.th/GI/tools/clustalw-mtv). BioEdit (http://www.mbio.ncsu.edu/BioEdit/bioedit.html) was used to score sequence differences.
47. N.T. Jeanson, "A Young-earth Creation Human Mitochondrial DNA 'Clock': Whole Mitochondrial Genome Mutation Rate Confirms D-loop Results," *Answers Research Journal,* 2015, 8:375–378, available online at https://answersingenesis.org/genetics/mitochondrial-genome-mutation-rate/.
48. C. Stringer, 2016. "The Origin and Evolution of *Homo sapiens,*" *Philos Trans R Soc Lond B Biol Sci.,* 2016, 371(1698). pii: 20150237.
49. We could treat them as separate species, but this just makes the discrepancy between the evolutionary timescale and actual differences even worse. See the paragraphs that follow in the text.
50. These results were obtained through my own personal analysis. Sequences were downloaded from NCBI RefSeq database (NCBI accession numbers = NC_012920.1, NC_011137.1). Sequences were aligned in-house with CLUSTALX2 (http://www.clustal.org/clustal2/) or CLUSTALW-MTV (http://www4a.biotec.or.th/GI/tools/clustalw-mtv). BioEdit (http://www.mbio.ncsu.edu/BioEdit/bioedit.html) was used to score sequence differences.

51. N.T. Jeanson, 2015. "A Young-earth Creation Human Mitochondrial DNA 'Clock': Whole Mitochondrial Genome Mutation Rate Confirms D-loop Results," *Answers Research Journal,* 2015, 8:375–378, available online at https://answersingenesis.org/genetics/mitochondrial-genome-mutation-rate/.
52. M. Ingman et al., "Mitochondrial Genome Variation and the Origin of Modern Humans," *Nature,* 2000, 408:708–713.
53. D.J. Futuyma, 2013. *Evolution* (Sunderland, MA: Sinauer Associates, Inc., 2013), p.634–635.
54. Ibid., p.634.
55. See the following paper for a much more technically detailed discussion: N.T. Jeanson, "On the Origin of Human Mitochondrial DNA Differences, New Generation Time Data Both Suggest a Unified Young-Earth Creation Model and Challenge the Evolutionary Out-of-Africa Model," *Answers Research Journal,* 2016, 9:123–130, available online at https://answersingenesis.org/genetics/mitochondrial-dna/origin-human-mitochondrial-dna-differences-new-generation-time-data-both-suggest-unified-young-earth/.
56. United Nations, Department of Economic and Social Affairs, Population Division. 2013. *World Marriage Data 2012* (POP/DB/Marr/Rev2012). These data are analyzed at length in the following paper: N.T. Jeanson, "On the Origin of Human Mitochondrial DNA Differences, New Generation Time Data Both Suggest a Unified Young-Earth Creation Model and Challenge the Evolutionary Out-of-Africa Model," *Answers Research Journal,* 2016, 9:123–130, available online at https://answersingenesis.org/genetics/mitochondrial-dna/origin-human-mitochondrial-dna-differences-new-generation-time-data-both-suggest-unified-young-earth/.
57. A.G. Hinch et al., "The Landscape of Recombination in African Americans," *Nature,* 2011, 476:170–175.
58. Adapted with permission from Figure 1 of N.T. Jeanson, "On the Origin of Human Mitochondrial DNA Differences, New Generation Time Data Both Suggest a Unified Young-Earth Creation Model and Challenge the Evolutionary Out-of-Africa Model," *Answers Research Journal,* 2016, 9:123–130. See the following link for a diagram with much higher resolution: https://assets.answersingenesis.org/doc/articles/arj/v9/out-of-africa/Figure-1.pdf.
59. You will likely have noticed that subgroups are also identifiable in the human mtDNA tree. The evolutionary community has recognized the fact of subnodes with additional haplogroup labels. For example, in the human mtDNA literature, you will find discussions of haplogroups H, R, X, etc. Despite the existence of many subnodes, evolutionists still agree that these other haplogroups coalesce into the three major nodes — L, M, and N.
60. P.A. Underhill and T. Kivisild, "Use of Y Chromosome and Mitochondrial DNA Population Structure in Tracing Human Migrations," *Annu Rev Genet,* 2007, 41:539–64. See also any recent human mtDNA paper.
61. Alfred Russel Wallace knew this to be true in the 1800s (http://people.wku.edu/charles.smith/wallace/S453.htm).
62. Derived from my own analysis of the fauna in the Paleobiology Database (http://fossilworks.org/).
63. Again, from my own analysis of the fauna in the Paleobiology Database (http://fossilworks.org/).
W. Kiessling, A.J.W. Hendy, M.E. Clapham, A.I. Miller, M. Aberhan, F.T. Fursich, J. Alroy, P.J. Wagner, M.E. Patzkowsky, M. Foote, J.A. Sessa, M.D. Uhen, D.J. Bottjer, L.C. Ivany, M.T. Carrano, L. Villier, S.M. Holland, Björ Kröger, A.J. McGowan, and T.D. Olszewski, Taxonomic occurrences of Bivalvia recorded in Fossilworks, the Evolution of Terrestrial Ecosystems database, the Paleobiology Database, and the Paleogeographic Atlas Project database, Fossilworks, 2017, http://fossilworks.org.
64. For much more in-depth treatment of the geologic observations discussed in this section, see the following: A.A. Snelling, *Earth's Catastrophic Past: Geology, Creation and the Flood* (Dallas, TX: Institute for Creation Research, 2009).
65. S.A. Austin, "Nautiloid Mass Kill and Burial Event, Redwall Limestone (Lower Mississippian), Grand Canyon Region, Arizona and Nevada," in *Proceedings of the Fifth International Conference on Creationism*, R.. Ivey, Jr., ed. (Pittsburgh, PA: Creation Science Fellowship, 2003), p. 55–100.
66. J.H. Whitmore et al., "The Petrology of the Coconino Sandstone (Permian), Arizona, USA," *Answers Research Journal,* 2014, 7:499–532, available online at https://answersingenesis.org/geology/rock-layers/petrology-of-the-coconino-sandstone/; J.H. Whitmore, G. Forsythe, and P.A. Garner, 2015. "Intraformational Parabolic Recumbent Folds in the Coconino Sandstone (Permian) and Two Other Formations in Sedona, Arizona (USA)," *Answers Research Journal,* 2015, 8:21–40, available online at https://answersingenesis.org/geology/rock-layers/intraformational-parabolic-recumbent-folds/.
67. S.A. Austin, 1986. *Proceedings of the First International Conference on Creationism*, R.S. Crowell, editor (Pittsburgh, PA: Creation Science Fellowship, 1986), p. 3–9, available online at http://static.icr.org/i/pdf/technical/Mount-St-Helens-and-Catastrophism.pdf; S.A. Austin, "Rapid Erosion at Mount St. Helens," *Origins,* 1984, 11(2):90–98, available online at http://static.icr.org/i/pdf/technical/Rapid-Erosion-at-Mount-St-Helens.pdf; S.A. Austin, "Excess Argon within Mineral Concentrates from the New Dacite Lava Dome at Mount St Helens Volcano,"

Journal of Creation, 1996, 10(3):335–343, available online at http://creation.com/excess-argon-within-mineral-concentrates.

For a broader overview, see the following: S.A. Austin, "Why Is Mount St. Helens Important to the Origins Controversy?" in K. Ham, ed., *New Answers Book 3* (Green Forest, AR: Master Books, 2009), p. 253–262, available online at https://answersingenesis.org/geology/mount-st-helens/why-is-mount-st-helens-important-to-the-origins-controversy/; S. Austin and J. Morris, *Footprints in the Ash: The Explosive Story of Mount St. Helens.* (Green Forest AR: Master Books, 2003).

68. M.H. Schweitzer et al., "Soft-Tissue Vessels and Cellular Preservation in *Tyrannosaurus rex*," *Science*, 2005, 307(5717):1952–1955. See also B. Thomas, "A Review of Original Tissue Fossils and Their Age Implications," *Proceedings of the Seventh International Conference on Creationism* (Pittsburgh, PA: Creation Science Fellowship, 2013).
69. J.R. Baumgardner, "Numerical Simulation of the Large-scale Tectonic Changes Accompanying the Flood," in R.E. Walsh, C.L. Brooks, and R.S. Crowell, eds., *Proceedings of the First International Conference on Creationism,* Vol. 2 (Pittsburgh, PA: Creation Science Fellowship, 1986), p. 17–30.; J.R. Baumgardner, "3-D Finite Element Simulation of the Global Tectonic Changes Accompanying Noah's Flood," in R.E. Walsh, C.L. Brooks, and R.S. Crowell, eds., *Proceedings of the Second International Conference on Creationism,* Vol. 2 (Pittsburgh, PA: Creation Science Fellowship, 1990), p.35–45; S.A. Austin et al., "Catastrophic Plate Tectonics: a Global Flood Model of Earth History," in R.E. Walsh, ed., *Proceedings of the Third International Conference on Creationism* (Pittsburgh, PA: Creation Science Fellowship, 1994), p.609–621; J.R. Baumgardner, 1994. "Computer Modeling of the Large-scale Tectonics Associated with the Genesis Flood," in R.E. Walsh, ed., *Proceedings of the Third International Conference on Creationism* (Pittsburgh, PA: Creation Science Fellowship, 1994), p.49–62; J.R. Baumgardner, "Runaway Subduction as the Driving Mechanism for the Genesis Flood," in R.E. Walsh, ed., *Proceedings of the Third International Conference on Creationism* (Pittsburgh, PA: Creation Science Fellowship, 1994), p.63–75; J.R. Baumgardner, 2003. 'The physics Behind the Flood,' in R.L. Ivey Jr., ed., *Proceedings of the Fifth International Conference on Creationism* (Pittsburgh, PA: Creation Science Fellowship, 2003), p.113–126.
70. See previous endnotes.
71. "And God said to Noah, 'The end of all flesh has come before Me, for the earth is filled with violence through them; and behold, I will destroy them with the earth. . . . But I will establish My covenant with you; and you shall go into the ark — you, your sons, your wife, and your sons' wives with you' " (Genesis 6:13–18).

    "Then the LORD said to Noah, 'Come into the ark, you and all your household, because I have seen that you are righteous before Me in this generation' " (Genesis 7:1).

    "So Noah, with his sons, his wife, and his sons' wives, went into the ark because of the waters of the flood. . . . And it came to pass after seven days that the waters of the flood were on the earth. . . . On the very same day Noah and Noah's sons, Shem, Ham, and Japheth, and Noah's wife and the three wives of his sons with them, entered the ark" (Genesis 7:7–13).

    "Now the flood was on the earth forty days. The waters increased and lifted up the ark, and it rose high above the earth. . . . And the waters prevailed exceedingly on the earth, and all the high hills under the whole heaven were covered. The waters prevailed fifteen cubits upward, and the mountains were covered. And all flesh died that moved on the earth: birds and cattle and beasts and every creeping thing that creeps on the earth, and every man. All in whose nostrils was the breath of the spirit of life, all that was on the dry land, died. So He destroyed all living things which were on the face of the ground: both man and cattle, creeping thing and bird of the air. They were destroyed from the earth. Only Noah and those who were with him in the ark remained alive." (Genesis 7:17–23).

    "Now the sons of Noah who went out of the ark were Shem, Ham, and Japheth. . . . These three were the sons of Noah, and from these the whole earth was populated" (Genesis 9:18–19).
72. See Genesis 11:1–9.
73. See the following: https://answersingenesis.org/bible-timeline/when-did-animals-exit-ark/. C. Hardy, and R. Carter, "The Biblical Minimum and Maximum Age of the Earth, *Journal of Creation,* 2014, 28(2):89–96. The debate over what the biblical text actually says about the age of the universe has been vigorously debated within the Christian community for over two centuries. Since I am a biologist and since this book is focused biology, this is not the place to discuss these details. Instead, for a more in-depth, scholarly treatment of these biblical issues, see the following works: T. Mortenson and T.H. Ury, eds., 2008. *Coming to Grips with Genesis: Biblical Authority and the Age of the Earth* (Green Forest, AR: Master Books, 2008); T. Mortenson, ed., *Searching for Adam: Genesis & the Truth About Man's Origin* (Green Forest, AR: Master Books, 2016).

74. The fact that mutations separate the three nodes suggest that genealogical time separated these three women. In other words, they do not appear to have been sisters. For example, at current rates of mtDNA mutation, around 1 in every 6 individuals acquires a mutation. Thus, from a statistical perspective, three sisters would not be expected to have any mtDNA differences among them. Conversely, since these three nodes *are* separated by mtDNA differences, this fact suggests that the three women were *not* sisters.
75. N.T. Jeanson, "Mitochondrial DNA Clocks Imply Linear Speciation Rates within 'Kinds,' " *Answers Research Journal,* 2015, 8:273–304. Available online at https://answersingenesis.org/natural-selection/speciation/clocks-imply-linear-speciation-rates-within-kinds/.
76. Ibid.
77. E.A. Hardouin and D. Tautz, "Increased Mitochondrial Mutation Frequency after an Island Colonization: Positive Selection or Accumulation of Slightly Deleterious Mutations?" *Biol Lett.,* 2013, 9(2):20121123.
78. M. Alexander et al., "Mitogenomic Analysis of a 50-generation Chicken Pedigree Reveals a Rapid Rate of Mitochondrial Evolution and Evidence for Paternal mtDNA Inheritance," *Biol Lett.* 11(10):20150561.
79. C.D. Millar et al., "Mutation and Evolutionary Rates in Adélie Penguins from the Antarctic," *PLoS Genetics,* 2008, 4(10):e1000209.
80. Since different species were being compared, I made predictions using a divergence equation.
81. Mouse-rat evolutionary split was set to 11 million years (see M.J. Benton and P.C. Donoghue, "Paleontological Evidence to Date the Tree of Life," *Mol Biol Evol.,* 2007, 24(1):26–53).

    The first splits among living species within the family Phasianidae were set to 37.1 million years (see the lowest reported time in Table 1, Node C in X-Z Kan et al., "Estimation of Divergence Times for Major Lineages of Galliform Birds: Evidence from Complete Mitochondrial Genome Sequences," *African Journal of Biotechnology,* 2010, 9(21):3073–3078).

    The first splits among living species within the family Spheniscidae were set to 17 million years (see the lowest reported time in Table 1, Node 4 in S. Subramanian et al., "Evidence for a Recent Origin of Penguins," *Biol Lett.,* 2013, 9(6):20130748).
82. The mtDNA diversity among species within each of the three families was obtained as follows:

    1. Sequences were obtained from NCBI
    (NCBI accession numbers for Muridae: NC_001665.2, NC_010650.1, NC_011638.1, NC_012374.1, NC_012387.1, NC_012389.1, NC_012461.1, NC_014696.1, NC_014698.1, NC_014855.1, NC_014858.1, NC_014861.1, NC_014864.1, NC_014867.1, NC_014871.1, NC_016060.1, NC_016428.1, NC_016662.1, NC_017599.1, NC_019584.1, NC_019585.1, NC_019617.1, NC_020758.1, NC_023263.1, NC_023347.1, NC_023960.1, NC_025268.1, NC_025269.1, NC_025270.1, NC_025287.1, NC_025670.1, NC_025952.1, NC_027683.1, NC_027684.1, NC_027932.1, NC_028335.1, NC_028715.1, NC_029888.1, NC_030342.1, NC_032286.1)
    (NCBI accession numbers for Phasianidae: NC_003408.1, NC_004575.1, NC_007238.1, NC_007239.1, NC_007240.1, NC_010767.1, NC_010770.1, NC_010771.1, NC_010774.1, NC_010778.1, NC_010781.1, NC_011816.1, NC_011817.1, NC_012453.1, NC_012895.1, NC_012897.1, NC_012900.1, NC_013619.1, NC_013979.1, NC_014576.1, NC_015526.1, NC_015897.1, NC_016679.1, NC_018033.1, NC_018034.1, NC_020583.1, NC_020584.1, NC_020585.1, NC_020586.1, NC_020587.1, NC_020588.1, NC_020589.1, NC_020590.1, NC_020591.1, NC_020613.1, NC_022683.1, NC_022684.1, NC_023264.1, NC_023779.1, NC_023939.1, NC_023940.1, NC_024533.1, NC_024554.1, NC_024615.1, NC_024616.1, NC_024619.1, NC_025318.1, NC_026547.1, NC_026548.1, NC_027279.1)
    (NCBI accession numbers for Spheniscidae: NC_004538.1, NC_008138.1, NC_021137.1, NC_021474.1, NC_022817.1, NC_027938.1)

    2. Where appropriate, sequences were manually adjusted to share the same position #1.

    3. Sequences from species within a family were aligned in-house with CLUSTALX2 (http://www.clustal.org/clustal2/) or CLUSTALW-MTV (http://www4a.biotec.or.th/GI/tools/clustalw-mtv).

    4. In BioEdit (http://www.mbio.ncsu.edu/BioEdit/bioedit.html) software, all non-standard nucleotides (e.g., N, M, R, Y, B, W, S, V, H, D) were replaced with gaps. Then all columns containing gaps were stripped, after which a sequence difference count matrix was created.

    5. For each sequence difference count matrix for species within each family, Microsoft Excel was used to calculate the average sequence difference and to identify the maximum pairwise sequence difference.

6. These numbers were compared to the predictions for each family. Predictions were made by taking the full statistical range of the published mtDNA mutation rate for species within these families, and multiplying these rates by the appropriate timescale and by a factor of 2 (i.e., a divergence calculation). (Since the published rates were reported in units of *mutations per year*, there was no need to convert the units of these mutation rates with species-specific generation times.)

83. D.R. Denver et al., "High Direct Estimate of the Mutation Rate in the Mitochondrial Genome of *Caenorhabditis elegans*," *Science,* 2000, 289:2342–2344.
84. D.K. Howe, C.F. Baer, and D.R. Denver, "High Rate of Large Deletions in *Caenorhabditis briggsae* Mitochondrial Genome Mutation Processes," *Genome Biol Evol.,* 2009, 2:29–38.
85. R.I. Molnar et al., "Mutation Rates and Intraspecific Divergence of the Mitochondrial Genome of *Pristionchus pacificus*," *Mol Biol Evol.,* 2011, 28(8):2317–26.
86. C. Haag-Liautard et al., "Direct Estimation of the Mitochondrial DNA Mutation Rate in *Drosophila melanogaster*," *PLoS Biology,* 2008, 6:1706–1714.
87. S. Xu et al., "High Mutation Rates in the Mitochondrial Genomes of *Daphnia pulex*," *Mol. Biol. Evol.,* 2012, 29(2):763–769.
88. M. Lynch et al., "A Genome-wide View of the Spectrum of Spontaneous Mutations in Yeast," *Proc. Natl. Acad. Sci. USA,* 2008, 105(27):9272–9277.
89. As per my search of the NCBI RefSeq database on March 18, 2017.
90. NCBI accession numbers: NC_009885.1, NC_001328.1, NC_025756.1
91. NCBI accession numbers for all eleven species: NC_001322.1, NC_005779.1, NC_005780.1, NC_005781.1, NC_011596.1, NC_018348.1, NC_023825.1, NC_024511.2, NC_025936.1, NC_027937.1, NC_028518.1
92. NCBI accession numbers for all three species: NC_028152.1, NC_026914.1, NC_000844.1
93. NCBI accession numbers: NC_001224.1, NC_012145.1, NC_018044.1, NC_027264.1, NC_031184.1, NC_031185.1, NC_031511.1, NC_031512.1, NC_031513.1, NC_031514.1, NC_031515.1, NW_017264706.1
94. The algorithms that I use to compare DNA sequences (e.g., CLUSTALW) are designed to detect single base pair differences. However, other types of DNA mutations occur, like genome rearrangements. These rearrangements complicate the scoring of individual base pair differences. For example, let's say a species with a 50,000 base pair mitochondrial genome undergoes an inversion of a single 5,000 base pair segment of its genome that leaves the rest of the genome intact. If the mitochondrial genome of a closely related species possesses a sequence that is very similar to the original (i.e., pre-inversion) genome of the first species, genome alignment will be straightforward. After all, around 45,000 base pairs of the two genomes would easily align. Under the parameters of the CLUSTALW algorithm, the 5,000 base pair inversion will be scored as a bunch of individual mismatches — even though a single mutational event led to the mismatches in this region. Consequently, if we didn't know any better, we might erroneously conclude that many single nucleotide mutational events gave rise to the sequence differences between these species. Thus, for CLUSTALW-based analyses (and analyses based on algorithms like it), the DNA sequences to be aligned must be first inspected for large-scale genome rearrangements before single nucleotide differences can be scored.

    For our immediate purposes, among the available mitochondrial genome sequences for *Saccharomyces* species, many contain apparent genome rearrangements. Thus, my analysis was limited to a subset of the available sequences.
95. L. Frézal and M.A. Félix, "*C. elegans* outside the Petri dish," *Elife,* 2015, 4:e05849.
96. T.A. Markow, "The Secret Lives of *Drosophila* Flies," *Elife,* 2015, 4:e06793; V. Zonato et al., "Is Diapause an Ancient Adaptation in *Drosophila*? *J Insect Physiol.,* 2017, 98:267–274.
97. J. Pijanowska and G. Stolpe, "Summer Diapause in *Daphnia* as a Reaction to the Presence of Fish," *J Plankton Res,* 1996, 18(8):1407–1412.
98. A.M. Neiman, "Sporulation in the Budding Yeast *Saccharomyces cerevisiae*," *Genetics,* 2011, 189(3):737–65; G. Liti, "The Fascinating and Secret Wild Life of the Budding Yeast *S. cerevisiae*," *Elife,* 2015, 4:e05835.
99. Laboratory generation times obtained from the following sources:

    -*Caenorhabditis elegans*: D.R. Denver et al., "High Direct Estimate of the Mutation Rate in the Mitochondrial Genome of *Caenorhabditis elegans*," *Science,* 2000, 289:2342–2344; http://genome.wustl.edu/genomes/detail/caenorhabditis-elegans/, accessed May 5, 2017.

    -*Drosophila melanogaster*: http://flystocks.bio.indiana.edu/Fly_Work/culturing.htm, accessed May 5, 2017.

    -*Daphnia pulex*: S. Xu et al., "High Mutation Rates in the Mitochondrial Genomes of *Daphnia pulex*," *Mol. Biol. Evol.,* 2012, 29(2):763–769; http://genome.jgi.doe.gov/Dappu1/Dappu1.home.html, accessed May 5, 2017.

-*Saccharomyces cerevisiae*: I. Herskowitz, "Life Cycle of the Budding Yeast *Saccharomyces cerevisiae*," *Microbiological Reviews,* 1998, 52 (4): 536–553.

100. A.D. Cutter, "Divergence Times in *Caenorhabditis* and *Drosophila* Inferred from Direct Estimates of the Neutral Mutation Rate," *Mol. Biol. Evol.,* 2008, 25:778–786.
101. C.A.M. Russo et al., "Phylogenetic Analysis and a Time Tree for a Large Drosophilid Data Set (Diptera: Drosophilidae)," *Zool. J. Linn. Soc.,* 2013, 4:765–775.
102. C.R. Haag et al., "Nucleotide Polymorphism and Within-Gene Recombination in *Daphnia magna* and *D. pulex,* Two Cyclical Parthenogens," *Genetics,* 2009, 182:313–323.
103. The *Saccharomyces cerevisiae-Saccharomyces kudriavzevii* split was listed at 15 to 20 million years ago: C.R. Landry et al., "Ecological and Evolutionary Genomics of *Saccharomyces cerevisiae*," *Mol Ecol.,* 2006, 15(3):575–91.
104. To make predictions, the full statistical range of the published SNP mutation rates for *C. elegans* and *C. briggsae* were used. Using generation times measured in the lab (see previous endnotes) or estimated for the wild (i.e., lab generation time divided by 10), the units of the published mutation rates were converted from mutations per generation to mutations per year. This rate was multiplied by the length of the mtDNA alignment (see on), by a factor of 2, and by the evolutionary time of origin.

    The SNP mtDNA differences among the three *Caenorhabditis* species were calculated as follows:

    1. Where appropriate, sequences (NCBI accession numbers: NC_009885.1, NC_001328.1, NC_025756.1) were manually adjusted to share the same position #1. For example, after visual inspection, it was apparent that the *C. briggsae* sequence had a gene order that was the reverse complement of the *C. elegans* and *C. tropicalis* gene order. The *C. briggsae* sequence was manually adjusted to match the other two species using publicly available software (http://www.bioinformatics.org/sms/rev_comp.html).

    2. Sequences were aligned in-house with CLUSTALX2 (http://www.clustal.org/clustal2/) or CLUSTALW-MTV (http://www4a.biotec.or.th/GI/tools/clustalw-mtv).

    3. In BioEdit (http://www.mbio.ncsu.edu/BioEdit/bioedit.html) software, all non-standard nucleotides (e.g., N, M, R, Y, B, W, S, V, H, D) were replaced with gaps. Then all columns containing gaps were stripped, after which a sequence difference count matrix was created.

    4. Microsoft Excel was used to calculate the average sequence difference and standard deviation.
105. To make predictions, the full statistical range of the published SNP mutation rates for *D. melanogaster* were used. Using generation times measured in the lab (see previous endnotes) or estimated for the wild (i.e., lab generation time divided by 10), the units of the published mutation rates were converted from mutations per generation to mutations per year. This rate was multiplied by the length of the mtDNA alignment (see on), by a factor of 2, and by the evolutionary time of origin.

    The SNP mtDNA differences among the eleven *Drosophila* species were calculated as follows:

    1. Where appropriate, sequences (NCBI accession numbers: NC_001322.1, NC_005779.1, NC_005780.1, NC_005781.1, NC_011596.1, NC_018348.1, NC_023825.1, NC_024511.2, NC_025936.1, NC_027937.1, NC_028518.1) were manually adjusted to share the same position #1.

    2. Sequences were aligned in-house with CLUSTALX2 (http://www.clustal.org/clustal2/) or CLUSTALW-MTV (http://www4a.biotec.or.th/GI/tools/clustalw-mtv).

    3. In BioEdit (http://www.mbio.ncsu.edu/BioEdit/bioedit.html) software, all non-standard nucleotides (e.g., N, M, R, Y, B, W, S, V, H, D) were replaced with gaps. Then all columns containing gaps were stripped, after which a sequence difference count matrix was created.

    4. Microsoft Excel was used to calculate the average sequence difference and standard deviation.
106. To make predictions, the full statistical range of the published mutation rates (all mutational events, not just SNPs) for *D. pulex* were used. Unlike the other invertebrate and fungal species that I analyzed, nearly three-fourths of the measured mutations in *D. pulex* represent indels, rather than SNPs (S. Xu et al., "High Mutation Rates in the Mitochondrial Genomes of *Daphnia pulex*," *Mol. Biol. Evol.,* 2012, 29(2):763–769), which is why I made predictions for all mutational events instead of just SNPs.

    Using generation times measured in the lab (see previous endnotes) or estimated for the wild (i.e., lab generation time divided by 10), the units of the published mutation rates were converted from mutations per generation to mutations per year. This rate was multiplied by the length of the mtDNA alignment (see on), by a factor of 2, and by the evolutionary time of origin.

    The mtDNA differences among the three *Daphnia* species were calculated as follows:

1. Where appropriate, sequences (NCBI accession numbers: NC_028152.1, NC_026914.1, NC_000844.1) were manually adjusted to share the same position #1. Since visual inspection of the gene order in these species indicated discrepancies in the tRNA-Ile gene and the D-loop, the ends of these sequences were trimmed based on the designations in the NCBI Nucleotide entries for each species.

2. Sequences were aligned in-house with CLUSTALX2 (http://www.clustal.org/clustal2/) or CLUSTALW-MTV (http://www4a.biotec.or.th/GI/tools/clustalw-mtv).

3. In BioEdit (http://www.mbio.ncsu.edu/BioEdit/bioedit.html) software, all non-standard nucleotides (e.g., N, M, R, Y, B, W, S, V, H, D) were replaced with gaps. Then all columns containing gaps were stripped, after which a sequence difference count matrix was created.

4. Microsoft Excel was used to calculate the average sequence difference and standard deviation.

107. To make predictions, the full statistical range of the published SNP mutation rates for *S. cerevisiae* were used. Using generation times measured in the lab (see previous endnotes) or estimated for the wild (i.e., lab generation time divided by 10), the units of the published mutation rates were converted from mutations per generation to mutations per year. This rate was multiplied by a factor of 2, by the evolutionary time of origin, and by the average mtDNA genome size among the four *Saccharomyces* species: *S. kudriavzevii*, *S. mikatae*, *S. paradoxus*, and *S. cerevisiae*.

The mtDNA differences among the *Saccharomyces* species were calculated as follows:

1. Since visual inspection of the gene order in these species indicated multiple discrepancies, alignment was especially difficult and limited to a few pairwise alignments.

2. Before the *S. kudriavzevii* and *S. mikatae* sequences (NCBI accession numbers: NC_031184.1, NC_031185.1) were aligned, these sequences were trimmed were appropriate, based on the designations in the NCBI Nucleotide entries for each species to keep the gene order the same in both species.

3. Sequences were aligned in-house with CLUSTALX2 (http://www.clustal.org/clustal2/) or CLUSTALW-MTV (http://www4a.biotec.or.th/GI/tools/clustalw-mtv).

4. In BioEdit (http://www.mbio.ncsu.edu/BioEdit/bioedit.html) software, all non-standard nucleotides (e.g., N, M, R, Y, B, W, S, V, H, D) were replaced with gaps. Then all columns containing gaps were stripped, after which a sequence difference count matrix was created.

5. Microsoft Excel was used to calculate the percent sequence difference. This percentage was used to calculate a theoretical overall divergence between *S. kudriavzevii* and *S. mikatae*, assuming a mtDNA genome size of 73,963 base pairs.

6. The mtDNA sequence divergence between *S. paradoxus* and *S. cerevisiae* was calculated as in the following reference, but using a mtDNA genome size of 73,963 rather than 82,000: N.T. Jeanson and J. Lisle, "On the Origin of Eukaryotic Species' Genotypic and Phenotypic Diversity: Genetic Clocks, Population Growth Curves, and Comparative Nuclear Genome Analyses Suggest Created Heterozygosity in Combination with Natural Processes as a Major Mechanism," *Answers Research Journal,* 2016, 9:81–122, available online at https://answersingenesis.org/natural-selection/speciation/on-the-origin-of-eukaryotic-species-genotypic-and-phenotypic-diversity/.

108. Methods were identical to those used for making predictions on the evolutionary timescale, with one alteration: the 6,000-year timescale was used instead of the evolutionary timescale. Otherwise, the calculations were the same, and the determination of actual mtDNA differences was the same as well.

109. See Supplemental Table 2 of the following reference: N.T. Jeanson, "Mitochondrial DNA Clocks Imply Linear Speciation Rates within 'Kinds,' " *Answers Research Journal,* 2015, 8:273–304. Available online at https://answersingenesis.org/natural-selection/speciation/clocks-imply-linear-speciation-rates-within-kinds/.

110. The following papers formed the basis for the reclassification of these species into new families: C.M. Miller-Butterworth et al., "A Family Matter: Conclusive Resolution of the Taxonomic Position of the Long-fingered Bats," *Miniopterus. Mol Biol Evol.*, 2007, 24(7):1553–1561; J.B. Lack et al., "Molecular Phylogenetics of *Myotis* Indicate Familial-level Divergence for the Genus *Cistugo* (Chiroptera)," *Journal of Mammalogy*, 2010, 91(4):976–992.

111. Though chapter 6 invoked a 12,000-year timescale rather than a 6,000-year timescale, the former was derived from the evolutionary timescale, under evolutionary assumptions of constant rates of change in geology and astronomy. Conversely, the assumption of rate constancy in genetics created a severe contradiction between predictions and reality for the evolutionary timescale. In other words, the 12,000-year timescale for the origin of breeds is probably even less than what evolutionists have claimed. If so, this would likely bring the breed-species timescale into agreement with the 6,000-year timescale.

112. If you don't agree with this statement, then consider the main difference between the geologists and astronomers who accept the millions-of-years timescale, and the YEC geologists and astronomers who hold to a 6,000-year

timescale. The latter do not arrive at their conclusions by inventing laws of physics or by rejecting sound science. Rather, they simply allow the possibility that rates of geologic and astronomical change have not been constant.

113. Some might object to the logic of this paragraph on the grounds that the 6,000-year timescale has its own inconsistencies. After all, in the realm of geology, YEC scientists have invoked changing rates. Yet, in the realm of genetics, I have just shown data from calculations in which I assume constant rates of change. At first pass, it might seem that the YEC model has a logical problem as well.

    In fact, the apparent problems for each model are quite different. For example, when YEC scientists have invoked changing rates in geology, they have not insisted that changing rates must be the rule. Rather, they have pointed out that changing rates are an equally plausible hypothesis, one that happens to make testable, accurate predictions.

    Furthermore, the hypothesis of variable rates of geologic change was not an arbitrary one, invoked simply for sake of convenience. Instead, it flowed naturally from the straightforward reading of Genesis 1–9. These nine chapters describe universal or global events with potentially dramatic geologic and astronomical consequences. In addition, in Genesis 9, the text states that the global Flood will never happen again. Therefore, by definition, from a YEC perspective, the current rates of geologic processes have not always been the same.

    Applying this sort of thinking to biology does not automatically require variable rates of change. From a strictly textual perspective, little in the opening chapters of Genesis (aside from the initial creation act itself) implies that rates of genetic change in biology would be variable. Conversely, neither variable nor constant rates are inconsistent with the text, but neither are necessarily stipulated. Therefore, from the biblical text alone, the expectations for rates of change in biology are different from the expectations for rates of change in geology.

    Combined with the fact that YEC geologists have not arbitrarily insisted that variable rates must be the rule, the YEC model has no logical problem.

    In contrast, in the realm of geology, evolutionists have insisted that constant rates of change must be invoked — "the present is the key to the past" has long been the rule. Furthermore, when YEC geologists have pointed out the seemingly arbitrary nature of this rule, as well as the plausibility of the opposite hypothesis (variable rates), evolutionists have simply doubled down on their insistence. In light of this practice, it becomes more difficult for evolutionists to suddenly permit variable rates of change in genetics. If they now permit variable rates when data are not in their favor, then it would be logically necessary for them to permit YEC geologists to do the same. The latter would undermine the entire millions-of-years framework.

Chapter 8

1. J.H. Thomas, "Genome Evolution in *Caenorhabditis*," *Brief Funct Genomic Proteomic,* 2008, 7(3):211–216.
2. Cytochrome *c* comparisons are a classic example — e.g., see p.139–140 of the following reference: D.L. Nelson and M.M. Cox, *Lehninger Principles of Biochemistry*, 3rd ed. (New York: Worth Publishers, 2000).
3. However, it appears that the typical evolutionary explanation of protein differences is being modified. Rather than attributing changes primarily to a nuclear DNA "clock," evolutionists are beginning to invoke more of a role for natural selection. See the following paper as an example: J. Parker et al., "Genome-wide Signatures of Convergent Evolution in Echolocating Mammals," *Nature,* 2013, 502:228–231.
4. N.T. Jeanson, "Recent, Functionally Diverse Origin for Mitochondrial Genes from ~2700 Metazoan Species," *Answers Research Journal,* 2013, 6:467–501, available online at https://answersingenesis.org/genetics/mitochondrial-dna/recent-functionally-diverse-origin-for-mitochondrial-genes-from--2700-metazoan-species/.
5. D. Graur et al., "On the Immortality of Television Sets: 'Function' in the Human Genome According to the Evolution-free Gospel of ENCODE," *Genome Biology and Evolution,* 2013, 5(3):578–590.
6. See Table 2 of P. Sulem et al., "Identification of a Large Set of Rare Complete Human Knockouts," *Nat Genet.,* 2015, 47(5):448–52.
7. For example, see E.E. Max, "Plagiarized Errors and Molecular Genetics," http://www.talkorigins.org/faqs/molgen/.
8. See Figure 4A of the following paper: C. Sisu et al., "Comparative Analysis of Pseudogenes Across Three Phyla," *Proc Natl Acad Sci USA,* 2014, 111(37):13361–13366.
9. J. Bergman and J. Tomkins, "The Chromosome 2 Fusion Model of Human Evolution — Part 1: Re-evaluating the Evidence," *Journal of Creation,* 2011, 25(2):106–110, available online at http://creation.com/chromosome-2-fusion-1; J. Tomkins and J. Bergman, "The Chromosome 2 Fusion Model of Human Evolution — Part 2: Re-analysis

of the Genomic Data," *Journal of Creation,* 2011, 25(2):111–117, available online at http://creation.com/chromosome-2-fusion-2; J. Tomkins, "Alleged Human Chromosome 2 'Fusion Site' Encodes an Active DNA Binding Domain Inside a Complex and Highly Expressed Gene — Negating Fusion," *Answers Research Journal,* 2013, 6:367–375, available online at https://answersingenesis.org/genetics/dna-similarities/alleged-human-chromosome-2-fusion-site-encodes-an-active-dna-binding-domain-inside-a-complex-and-hig/; J.P. Tomkins, "Debunking the Debunkers: A Response to Criticism and Obfuscation Regarding Refutation of the Human Chromosome 2 Fusion," *Answers Research Journal,* 2017, 10:45–54, available online at https://answersingenesis.org/genetics/dna-similarities/debunking-the-debunkers/.

10. J. Tomkins, "The Human Beta-Globin Pseudogene is Non-Variable and Functional," *Answers Research Journal,* 2013, 6:293–301, available online at https://answersingenesis.org/genetics/human-genome/the-human-beta-globin-pseudogene-is-non-variable-and-functional/.
11. For example, see the following papers: D.F. Conrad et al., "Variation in Genome-wide Mutation Rates within and between Human Families," *Nature Genetics,* 2011, 43(7):712–714; A. Kong et al., "Rate of de novo Mutations and the Importance of Father's Age to Disease Risk," *Nature,* 2012, 488(7412):471–5; L.C. Francioli et al., "Genome-wide Patterns and Properties of de novo Mutations in Humans," *Nat Genet.,* 2015, 47(7):822–6; W.S. Wong, et al., "New Observations on Maternal Age Effect on Germline de novo Mutations," *Nat Commun.,* 2016, 7:10486. For a very helpful overview of the many types and rates of human mutations, see C.D. Campbell and E.E. Eichler, "Properties and Rates of Germline Mutations in Humans," *Trends Genet.,* 2013, 29(10):575–84.
12. Calculated assuming a mutation rate of $1.2x10^{-8}$ mutations per base pair per generation, and assuming a haploid genome size of $3.242x10^{9}$. Haploid genome size obtained from NCBI Genome database (https://www.ncbi.nlm.nih.gov/genome/browse/) on March 27, 2017.
13. O. Venn et al., "Strong Male Bias Drives Germline Mutation in Chimpanzees," *Science,* 2014, 344:1272–1275.
14. Calculated assuming a mutation rate of $1.2x10^{-8}$ mutations per base pair per generation, and assuming a haploid genome size of $3.231x10^{9}$. Haploid genome size obtained from NCBI Genome database (https://www.ncbi.nlm.nih.gov/genome/browse/) on March 27, 2017.
15. D.P. Locke et al., "Comparative and Demographic Analysis of Orang-utan Genomes," *Nature,* 2011, 469(7331):529–533; K.E. Langergraber et al., "Generation Times in Wild Chimpanzees and Gorillas Suggest Earlier Divergence Times in Great Ape and Human Evolution," *Proc Natl Acad Sci USA,* 2011, 109(39):15716–15721
16. O. Venn et al., "Strong Male Bias Drives Germline Mutation in Chimpanzees," *Science,* 2014, 344:1272–1275.
17. See the original paper on the chimpanzee genome sequence (The Chimpanzee Sequencing and Analysis Consortium, "Initial Sequence of the Chimpanzee Genome and Comparison with the Human Genome," *Nature,* 2005, 437:69–87), Tables S8 and S9 where 6-8% of the human genome is not covered by chimpanzee sequence. Added to the 4.23% difference that arises from SNP and indel calculations, the total difference between these species appears to be 10–12%. See also Table 1 of the following paper where only ~90% of the human sequence is covered by DNA alignments to chimpanzee or bonobo DNA: K. Prüfer et al., "The Bonobo Genome Compared with the Chimpanzee and Human Genomes," *Nature,* 2012, 486(7404):527–31. See also Table ST3.2 of the following paper, where only 89.5% of the human DNA sequence is included in the alignment of multiple great ape species and a macaque species: A. Scally et al., "Insights into Hominid Evolution from the Gorilla Genome Sequence," *Nature,* 2012, 483(7388):169–75. See also the alignment of human and chimpanzee genomes by Ensembl: http://useast.ensembl.org/info/genome/compara/mlss.html?mlss=688.

    See also J.P. Tomkins, "Documented Anomaly in Recent Versions of the BLASTN Algorithm and a Complete Reanalysis of Chimpanzee and Human Genome-Wide DNA Similarity Using Nucmer and LASTZ." *Answers Research Journal,* 2015, 8:379–390, available online at https://answersingenesis.org/genetics/dna-similarities/blastn-algorithm-anomaly/; see also J.P. Tomkins, "Analysis of 101 Chimpanzee Trace Read Data Sets: Assessment of Their Overall Similarity to Human and Possible Contamination With Human DNA," *Answers Research Journal,* 2016, 9:294–298, available online at https://answersingenesis.org/genetics/dna-similarities/analysis-101-chimpanzee-trace-read-data-sets-assessment-their-overall-similarity-human-and-possible-/.
18. The length of the aligned DNA is only 2.2 billion base pairs; mutational predictions were made for this same length of DNA.
19. O. Venn et al., "Strong Male Bias Drives Germline Mutation in Chimpanzees," *Science,* 2014, 344:1272–1275. I am indebted to Rob Carter of Creation Ministries International for his insights on this evolutionary discrepancy.
20. S.Y. Ho et al., "Time Dependency of Molecular Rate Estimates and Systematic Overestimation of Recent Divergence Times," *Mol Biol Evol.,* 2005, 22(7):1561–1568.

21. O. Venn et al., "Strong Male Bias Drives Germline Mutation in Chimpanzees," *Science,* 2014, 344:1272–1275.
22. Again, the length of the aligned DNA is only 2.2 billion base pairs; mutational predictions were made for this same length of DNA.
23. Methods were the same as those employed in chapter 7, except that the minimum timescale was raised from 4.5 million years to 11 million years.
24. I am indebted to Rob Carter of Creation Ministries International for his insights on the evolutionary conundrum articulated in this paragraph.
25. See Table ST3.3 of the following paper for the percent-difference among several primate species' genome sequences: A. Scally et al., "Insights into Hominid Evolution from the Gorilla Genome Sequence," *Nature,* 2012, 483(7388):169–175. See Table 1 of the following paper for chimpanzee-bonobo nucleotide substitutions: K. Prüfer et al., "The Bonobo Genome Compared with the Chimpanzee and Human Genomes," *Nature,* 2012, 486(7404):527–31.
26. To calculate each revised divergence time, the following steps were performed:

    1. Genome sizes for each species (*Gorilla gorilla, Pan paniscus, Pan troglodytes, Pongo abelii, Macaca mulatta*) were obtained from NCBI (https://www.ncbi.nlm.nih.gov/genome/browse/).

    2. These genome sizes were multiplied by the numbers in the column labeled "percentage included" from Table ST3.2 (A. Scally et al., "Insights into Hominid Evolution from the Gorilla Genome Sequence," *Nature,* 2012, 483(7388):169–175) to calculate the length of the alignment.

    3. The absolute nucleotide difference between a pair of species was calculated by multiplying the mismatch percentage (Table ST3.3 of A. Scally et al., "Insights into Hominid Evolution from the Gorilla Genome Sequence," *Nature,* 2012, 483(7388):169–175) by the shortest alignment length (which was calculated in step 2). Alternatively, for the chimpanzee-bonobo difference, the difference was obtained by adding together the "Lineage-specific substitutions" from K. Prüfer et al., "The Bonobo Genome Compared with the Chimpanzee and Human Genomes," *Nature,* 2012, 486(7404):527–31.

    4. The revised divergence time was calculated by dividing the absolute nucleotide difference (from step 3) by the average of the published human and chimpanzee mutation rates (4.25×10^{-10} mutations per base pair per year), by a factor of 2, and by the shortest alignment length (which was calculated in step 2). Alternatively, for the chimpanzee-bonobo divergence time, the absolute nucleotide difference (from step 3) was divided by the average of the published human and chimpanzee mutation rates (4.25×10^{-10} mutations per base pair per year), by a factor of 2, and by the chimpanzee genome length (3.26×10^{9} base pairs).
27. S. Tavaré et al., "Using the Fossil Record to Estimate the Age of the Last Common Ancestor of Extant Primates," *Nature,* 2002, 416(6882):726–729.
28. I made predictions for the diploid genome — in other words, for 6,483,900,000 base pairs of DNA per generation.
29. See Table 1 of the following paper: 1000 Genomes Project Consortium. "A Global Reference for Human Genetic Variation," *Nature,* 2015, 526:68–74.
30. Again, for a more complete discussion of the processes explored in the following paragraphs, consult any standard genetics textbook.
31. 1000 Genomes Project Consortium et al., "An Integrated Map of Genetic Variation from 1,092 Human Genomes," *Nature,* 2012, 491(7422):56–65; 1000 Genomes Project Consortium, "A Global Reference for Human Genetic Variation," *Nature,* 2015, 526:68–74.
32. As per the data in 1000 Genomes Project Consortium et al., "An Integrated Map of Genetic Variation from 1,092 Human Genomes," *Nature,* 2012, 491(7422):56–65; 1000 Genomes Project Consortium, "A Global Reference for Human Genetic Variation," *Nature,* 2015, 526:68–74.
33. The discussion in the following paragraphs is largely based on S. Yang et al., "Parent-progeny Sequencing Indicates Higher Mutation Rates in Heterozygotes," *Nature,* 2015, 523:463–467.
34. See the following paper for a much more in-depth discussion of these concepts, as well as for the references to the primary literature: N.T. Jeanson and J. Lisle, "On the Origin of Eukaryotic Species' Genotypic and Phenotypic Diversity: Genetic Clocks, Population Growth Curves, and Comparative Nuclear Genome Analyses Suggest Created Heterozygosity in Combination with Natural Processes as a Major Mechanism," *Answers Research Journal,* 2016, 9:81–122, available online at https://answersingenesis.org/natural-selection/speciation/on-the-origin-of-eukaryotic-species-genotypic-and-phenotypic-diversity/.
35. I made predictions for the diploid genome — in other words, for 6,483,900,000 base pairs of DNA per generation.

36. For example, see the following paper: H. Li and R. Durbin, "Inference of Human Population History from Individual Whole-genome Sequences," *Nature,* 2011, 475(7357):493–6.
37. I made predictions for the diploid genome — in other words, for 6,483,900,000 base pairs of DNA per generation.
38. See Table ST3.3 of the following paper for the percent-difference among several primate species' genome sequences: A. Scally et al., "Insights into Hominid Evolution from the Gorilla Genome Sequence," *Nature,* 2012, 483(7388):169–175.
39. The exact percentage might be less than 99%. If the relationship between nuclear DNA heterozygosity and nuclear DNA mutation rates is positive, then the highly heterozygous ancestors of modern Pongidae species would have mutated their nuclear DNA at faster rates that we observe today. Thus, perhaps slightly more than 0.02% of modern differences are mutational in origin.
40. Methods as follows:

    1. Predictions were made using the published mouse nuclear SNP mutation rate (A. Uchimura et al., "Germline Mutation Rates and the Long-term Phenotypic Effects of Mutation Accumulation in Wild-type Laboratory Mice and Mutator Mice," *Genome Research,* 2015, 25(8):1125–34), converted to mutations per year (assuming a generation time of 6 to 26 weeks). The converted mutation rate was multiplied by 2, by 4,500 years, and by the mouse nuclear genome size (2.8 billion base pairs).

    2. Predictions were compared to the SNP differences between the C57Bl/6 strain (derived from *Mus musculus*) and a strain (SPRET/EiJ) derived from *Mus spretus* (Table 1 of T.M. Keane et al., "Mouse Genomic Variation and Its Effect on Phenotypes and Gene Regulation," *Nature,* 2011, 477(7364):289–294).
41. Methods as follows:

    1. Predictions were made using the published collared flycatcher (*Ficedula albicollis*) nuclear SNP mutation rate (L. Smeds, A. Qvarnström, and H. Ellegren, "Direct Estimate of the Rate of Germline Mutation in a Bird," *Genome Res.,* 2016, 26(9):1211–1218), converted to mutations per year (assuming a generation time of 2 years, as per the Smeds 2016 *Genome Res* paper). The converted mutation rate was multiplied by 2, by 4,500 years, and by the collared flycatcher nuclear genome size (1.12 billion base pairs).

    2. Predictions were compared to the SNP differences between the collared flycatcher and the pied flycatcher (*Ficedula hypoleuca*), which was calculated by multiplying the pairwise nucleotide divergence rate (0.0046 plus or minus 0.0011; see H. Ellegren et al., "The Genomic Landscape of Species Divergence in *Ficedula* Flycatchers," *Nature,* 2012, 491(7426):756–760) by the *Ficedula albicollis* genome size.
42. Roundworms currently lack a published reference for genome-wide single base pair nuclear DNA differences among *Caenorhabditis* species. Water fleas currently possess only a single species (*Daphnia pulex*) with a sequenced nuclear DNA genome.
43. Methods as follows:

    1. Predictions were made using the published fruit fly (*Drosophila melanogaster*) nuclear SNP mutation rate (P.D. Keightley et al., "Estimation of the Spontaneous Mutation Rate Per Nucleotide Site in a *Drosophila melanogaster* Full-sib Family," *Genetics,* 2014, 196(1):313–320; C. Haag-Liautard et al., "Direct Estimation of Per Nucleotide and Genomic Deleterious Mutation Rates in *Drosophila,*" *Nature,* 2007, 445(7123):82–85), which was converted to units of mutations per year (assuming a generation time of 7 to 190 days). The converted mutation rate was multiplied by 2, by 6,000 years, and by the *D. melanogaster* nuclear genome size (144 million base pairs).

    2. Predictions were compared to the number of SNP differences between *D. melanogaster* and *D. yakuba* (Table S1 of D.J. Begun et al., "Population Genomics: Whole-genome Analysis of Polymorphism and Divergence in *Drosophila simulans,*" *PLoS Biology,* 2007, 5(11):e310). The number of SNP differences was calculated by multiplying the frequency of SNP differences between the two species by the *D. melanogaster* genome size.
44. Methods as follows:

    1. Predictions were made using the published *Saccharomyces* cerevisiae nuclear SNP mutation rate (Y.O. Zhu et al., "Precise Estimates of Mutation Rate and Spectrum in Yeast," *Proc. Natl. Acad. Sci. USA,* 2014, 111(22):E2310–2318; M. Lynch et al., "A Genome-wide View of the Spectrum of Spontaneous Mutations in Yeast," *Proc. Natl. Acad. Sci.* USA, 2008, 105(27):9272–9277), converted to mutations per year (assuming a generation time of 100 to 1,000 minutes). The converted mutation rate was multiplied by 2, by 6,000 years, and by the average genome size (11.1 million base pairs) of the four *Saccharomyces* species (*S. kudriavzevii, S. mikatae, S. paradoxus,* and *S. cerevisiae*).

    2. Predictions were compared to estimates of whole-genome divergence between *S. cerevisiae* and the remaining three species. For divergence between *S. cerevisiae* and *S. kudriavzevii,* see Table S1 of P. Cliften et al., "Finding

Functional Features in *Saccharomyces* Genomes by Phylogenetic Footprinting," *Science*, 2003, 301(5629):71–76. For divergence between *S. cerevisiae* and each of the remaining two species (*S. mikatae, S. paradoxus*), see Figure 3 of M. Kellis et al., "Sequencing and Comparison of Yeast Species to Identify Genes and Regulatory Elements," *Nature*, 2003, 423(6937):241–254.

45. Methods as follows:

    1. Predictions were made using the published *Saccharomyces cerevisiae* nuclear SNP mutation rate (Y.O. Zhu et al., "Precise Estimates of Mutation Rate and Spectrum in Yeast," *Proc. Natl. Acad. Sci. USA,* 2014, 111(22):E2310–2318; M. Lynch et al., "A Genome-wide View of the Spectrum of Spontaneous Mutations in Yeast," *Proc. Natl. Acad. Sci. USA,* 2008, 105(27):9272–9277), which was converted to units of mutations per year (assuming a generation time of 100 to 1,000 minutes). The converted mutation rate was multiplied by 2, by 15 million years, and by the average genome size (11.1 million base pairs) of the four *Saccharomyces* species (*S. kudriavzevii, S. mikatae, S. paradoxus,* and *S. cerevisiae*).

    2. Predictions were compared to estimates of whole-genome divergence between *S. cerevisiae* and the remaining three species. For divergence between *S. cerevisiae* and *S. kudriavzevii,* see Table S1 of P. Cliften et al., "Finding Functional Features in *Saccharomyces* Genomes by Phylogenetic Footprinting," *Science,* 2003, 301(5629):71–76. For divergence between *S. cerevisiae* and each of the remaining two species (*S. mikatae, S. paradoxus*), see Figure 3 of M. Kellis et al., "Sequencing and Comparison of Yeast Species to Identify Genes and Regulatory Elements," *Nature,* 2003, 423(6937):241–254.

46. Again, see any standard population genetics textbook for justification. See also the following papers: R. Carter and C. Hardy, "Modelling Biblical Human Population Growth," *Journal of Creation,* 2015, 29(1):72–79; R.W. Carter and M. Powell, "The Genetic Effects of the Population Bottleneck Associated with the Genesis Flood," *Journal of Creation,* 2016, 29(3):102–111.
47. 1000 Genomes Project Consortium et al., "An Integrated Map of Genetic Variation from 1,092 Human Genomes," *Nature,* 2012, 491(7422):56–65.
48. For example, see the following references: A. Coventry et al., "Deep Resequencing Reveals Excess Rare Recent Variants Consistent with Explosive Population Growth," *Nature Communications,* 2010, 1:131; A. Keinan and A.G. Clark, "Recent Explosive Human Population Growth Has Resulted in an Excess of Rare Genetic Variants," *Science,* 2012, 336(6082):740–743; M.R. Nelson et al., "An Abundance of Rare Functional Variants in 202 Drug Target Genes Sequenced in 14,002 People," *Science,* 2012, 337(6090):100–104; J.A. Tennessen et al., "Evolution and Functional Impact of Rare Coding Variation from Deep Sequencing of Human Exomes," *Science,* 2012, 337(6090):64–69; W. Fu et al., "Analysis of 6,515 Exomes Reveals the Recent Origin of Most Human Protein-coding Variants," *Nature,* 2013, 493(7431):216–20.
49. See also the following paper: N.T. Jeanson and J. Lisle, "On the Origin of Eukaryotic Species' Genotypic and Phenotypic Diversity: Genetic Clocks, Population Growth Curves, and Comparative Nuclear Genome Analyses Suggest Created Heterozygosity in Combination with Natural Processes as a Major Mechanism," *Answers Research Journal,* 2016, 9:81–122, available online at https://answersingenesis.org/natural-selection/speciation/on-the-origin-of-eukaryotic-species-genotypic-and-phenotypic-diversity/."
50. See Figure 4A of the following paper: C. Sisu et al., "Comparative Analysis of Pseudogenes Across Three Phyla," *Proc Natl Acad Sci USA,* 2014, 111(37):13361–13366.
51. Ibid., see Figure 4A.
52. modENCODE Consortium et al., "Identification of Functional Elements and Regulatory Circuits by *Drosophila* modENCODE," *Science,* 2010, 330(6012):1787–1797.

Chapter 9

1. See endnotes from chapter 6 for justification of these numbers.
2. A conceptually similar argument could be made from the numbers of invertebrate, plant, fungal, and bacterial species. However, since YEC scientists are still debating the limits of ancestry in these species, and since some still lean toward placing it as low as the subgenus level, it's more difficult to make a numerically robust argument. In other words, even though the classification group with the highest number of documented species alive today is arthropods, from a YEC perspective it's unclear how many ancestors these modern species possess. Thus, within the YEC model, the fold-difference between the number of descendants and ancestors might actually be higher for vertebrates than for invertebrates, plants, fungi, and bacteria. Hence, I used vertebrates as a representative example.

3. The technical measure of this is *linkage disequilibrium*. Extended Data Figure 10 of the following paper depicts the current levels of linkage disequilibrium in the human population: 1000 Genomes Project Consortium, "A Global Reference for Human Genetic Variation," *Nature,* 2015, 526:68–74.
4. Standard genetic textbooks contain a much more in-depth treatment of, and explanation for, the processes described in this paragraph.
5. J. Wang et al., "Genome-wide Single-Cell Analysis of Recombination Activity and De Novo Mutation Rates in Human Sperm," *Cell,* 2012, 150:402–412.
6. P.F. Palamara et al., "Leveraging Distant Relatedness to Quantify Human Mutation and Gene-Conversion Rates," *Am. J. Hum. Genet.,* 2015, 97:775–789; A.L. Williams et al., "Non-crossover Gene Conversions Show Strong GC Bias and Unexpected Clustering in Humans," *Elife,* 2015, 25:4; B.V. Halldorsson et al., "The Rate of Meiotic Gene Conversion Varies by Sex and Age," *Nat Genet.,* 2016, 48(11):1377–1384.
7. These calculations were done using the full statistical range of recombination and gene conversion events reported in the following papers, respectively:

   J. Wang et al., "Genome-wide Single-Cell Analysis of Recombination Activity and De Novo Mutation Rates in Human Sperm," *Cell,* 2012, 150:402–412. [recombination rate]

   B.V. Halldorsson et al., "The Rate of Meiotic Gene Conversion Varies by Sex and Age," *Nat Genet.,* 2016, 48(11):1377–1384. [gene conversion rate]
8. The X-Y chromosome pair would add nuance to this statement; for simplicity, I'm ignoring the nuances here.
9. Again, the X-Y chromosome pair would add nuance to this statement; for simplicity, I'm ignoring the nuances here.
10. The 100 different versions of chromosome 1 could produce 5,050 different combinations (not permutations) of chromosome *pairs*. Across the entire genome (i.e., ignoring the nuances of the X-Y chromosome pair), 5,050 ^ 23 = 1.5x10^{85}.
11. L. Orlando et al., "Recalibrating Equus Evolution Using the Genome Sequence of an Early Middle Pleistocene Horse," *Nature,* 2013, 499(7456):74–8.
12. See supplementary tables S10–S11 of the following paper (I rounded the reported numbers): H. Jónsson et al., "Speciation with Gene Flow in Equids Despite Extensive Chromosomal Plasticity," *Proc Natl Acad Sci USA,* 2014, 111(52):18655–60.
13. See any standard population genetics textbook for further discussion of these concepts.
14. See the Ageing Database (http://genomics.senescence.info/species/).
15. Ibid.
16. Ibid.
17. This dataset also includes parameters for some marsupials. Naturally, the parameters for species utilizing this mode of reproduction will look slightly different from the parameters for placentals. But for purposes of calculating population growth curves, the data are still present.
18. See the Ageing Database (http://genomics.senescence.info/species/).
19. For details on the calculations of population growth curves, see the following paper: N.T. Jeanson and J. Lisle, "On the Origin of Eukaryotic Species' Genotypic and Phenotypic Diversity: Genetic Clocks, Population Growth Curves, and Comparative Nuclear Genome Analyses Suggest Created Heterozygosity in Combination with Natural Processes as a Major Mechanism," *Answers Research Journal,* 2016, 9:81–122. Available online at https://answersingenesis.org/natural-selection/speciation/on-the-origin-of-eukaryotic-species-genotypic-and-phenotypic-diversity/.
20. Again, for details on the calculations of population growth curves, see the following paper: N.T. Jeanson and J. Lisle, "On the Origin of Eukaryotic Species' Genotypic and Phenotypic Diversity: Genetic Clocks, Population Growth Curves, and Comparative Nuclear Genome Analyses Suggest Created Heterozygosity in Combination with Natural Processes as a Major Mechanism," *Answers Research Journal,* 2016, 9:81–122, available online at https://answersingenesis.org/natural-selection/speciation/on-the-origin-of-eukaryotic-species-genotypic-and-phenotypic-diversity/.
21. Much of the content of this chapter was based on standard population genetic theory, which can be found in any standard population genetics textbook. I am indebted to both D.J. Futuyma, *Evolution* (Sunderland, MA: Sinauer Associates, Inc., 2013) and M.B. Hamilton, *Population Genetics* (Chichester, West Sussex, UK: Wiley-Blackwell, 2009).
22. The three-step mechanism outlined in this chapter applies equally well to the origin of populations below the species level — for example, to the origin of the "races" or ethnolinguistic groups. Specifically, since I'm proposing that Adam and Eve were created with millions of heterozygous sites, they also could have produced diverse offspring in a single generation.

More immediately under the YEC model, since the Flood of Noah reduced the global population to eight people, from whom the modern human population arose, the origin of the "races" is a post-Flood event. Nevertheless, the post-Flood peoples (like people today) would have retained millions of heterozygous sites, and could have produced diverse offspring in a single generation. Conflicts, wars, famines, and the like would have forced migrations throughout human history, and this migration would have isolated subpopulations of humans, allowing unique genetic (and visible) features to form in these subgroups.

These splinter groups could have repopulated quickly. For example, in the absence of deliberate population culling (e.g., genocide), a starting population of just two people could produce one million offspring in just 200 years. In just 1,200 years, over one *decillion* humans could arise from this couple (one decillion is 1 with 35 zeros after it). See N.T. Jeanson and J. Lisle, "On the Origin of Eukaryotic Species' Genotypic and Phenotypic Diversity: Genetic Clocks, Population Growth Curves, and Comparative Nuclear Genome Analyses Suggest Created Heterozygosity in Combination with Natural Processes as a Major Mechanism," *Answers Research Journal,* 2016, 9:81–122, available online at https://answersingenesis.org/natural-selection/speciation/on-the-origin-of-eukaryotic-species-genotypic-and-phenotypic-diversity/).

With respect to the first post-Flood population splits, the events recorded in Genesis 10–11 describe an initial impetus. Following the Flood, the descendants of Noah's family stayed in one place. Rather than spread out over the earth and fill it, they clustered and attempted to build a tower (Genesis 11:1–4).

Since this was an act of direct rebellion against God's explicit commands after the Flood (e.g., "So God blessed Noah and his sons, and said to them: 'Be fruitful and multiply, and fill the earth,' " Genesis 9:1), God forced them to spread out by confusing their languages.

Based on the genealogies of Genesis 10, this event appears to have been just a few generations after the Flood. For example, the text of Genesis 10 lists some of the male post-Flood descendants of Shem, Ham, and Japheth. The commentary on each of these genealogies, as well as on the genealogies as a whole, ties them to the events of Genesis 11:1–9. For example, after listing the descendants of Japheth: "From these the coastland peoples of the Gentiles were separated into their lands, everyone *according to his language*, according to their families, into their nations" (Genesis 10:5) (emphasis added). See also Genesis 10:20, 31–32.

In addition, in Genesis 10, the number of generations listed under each of Noah's three sons is small. Less than 10 generations of offspring are listed for each son. For Japheth's line, less than 5 generations are recorded. Thus, if the Tower of Babel incident follows the generations listed in Genesis 10, then few generations seem to have passed between the Flood and the Tower of Babel incident.

Consequently, the global population size would have been small — perhaps a few thousand people, or less. The confusion of languages would have further subdivided this small population into even smaller groups. In fact, Genesis 10 describes the language groupings as according to families. Each linguistic group may have been a population of just a handful of people. Since these groups could not speak with one another, they would have married only within their groups.

Coupled with the statistics of small population sizes, this would have pushed each group's heterozygous state toward a more homozygous one. Furthermore, since the exact combination of sites of genetic recombination and gene conversion are unique to each individual, each group would have acquired a unique, more homozygous state. Again, since these groups couldn't have spoken to one another, these unique genetic states would have been isolated from other genetic states. And, again, repopulation would have happened quickly.

With respect to the health consequences of intra-group marriages, these reductions in heterozygosity likely had few immediate consequences that were deleterious for human health. For example, just 10 generations occur between Adam and the Flood (see Genesis 5–9). With just a few generations between the end of the Flood and the linguistic confusion event, very few generations would have passed between Adam and the formation of the first major ethnolinguistic groups. Consequently, very few harmful mutations could have accumulated. (This might be the explanation for why the lifespans of these patriarchs were so long.) Today, hundreds of generations removed from Adam, our mutational load is much higher. Hence, reductions in heterozygosity have much more harmful consequences today than they would have had back then, and inbreeding is culturally and biologically discouraged. The scenario that I just described applies to the oft-repeated objection to Genesis "Where did Cain get his wife?" Since Adam and Eve were the founders of all humanity, Cain must have married his sister, an act which was not prohibited until the time of Moses, thousands of years (and thousands of mutations) later.

Thus, the linguistic confusion event of Genesis 11 would have quickly precipitated the formation of separate, visibly distinct populations that continues to this day.

Following the events of the Tower of Babel, language diversification would naturally have led to the over 7,000 ethnolinguistic groups present today. For example, each post-Babel population-splitting event (war, famine, etc.) would have allowed for languages to naturally diverge. Each event would also have allowed the genetics of these groups to diverge.

Thus, the formation of the major ethnolinguistic groups harmonizes with the YEC model. Repopulating the earth with different ethnolinguistic groups in the 4,500 years following the Flood is mathematically and genetically uncomplicated.

23. With respect to the evolutionary model, much of what follows can be found in standard evolutionary textbooks.
24. N.T. Jeanson and J. Lisle, "On the Origin of Eukaryotic Species' Genotypic and Phenotypic Diversity: Genetic Clocks, Population Growth Curves, and Comparative Nuclear Genome Analyses Suggest Created Heterozygosity in Combination with Natural Processes as a Major Mechanism," *Answers Research Journal,* 2016, 9:81–122, available online at https://answersingenesis.org/natural-selection/speciation/on-the-origin-of-eukaryotic-species-genotypic-and-phenotypic-diversity/.

Chapter 10

1. I re-performed the alignment from chapter 7 using just equid mtDNA genome sequences, not other Perissodactyl sequences. See chapter 7 endnotes for NCBI accession numbers of equid sequences.
2. The mtDNA differences among equid species (calculated after aligning the equid mtDNA genome sequences, as per the previous endnote) were visualized in tree form using MEGA6 (http://www.megasoftware.net/).
3. http://www.megasoftware.net/web_help_7/rh_mid_point_rooting.htm.
4. L. Orlando et al., "Recalibrating Equus Evolution Using the Genome Sequence of an Early Middle Pleistocene Horse," *Nature,* 2013, 499(7456):74–8; J. Huang et al., "Analysis of Horse Genomes Provides Insight into the Diversification and Adaptive Evolution of Karyotype," 2014, *Sci Rep.* 4:4958, *Sci Rep.* 5:14106; H. Jónsson et al., "Speciation with Gene Flow in Equids Despite Extensive Chromosomal Plasticity," *Proc Natl Acad Sci USA,* 2014, 111(52):18655–60.
5. See Supplementary Tables S10–S18 of H. Jónsson et al. (see previous endnote).
6. See Supplementary Table S10 of H, Jónsson et al. (see previous endnote).
7. See Figure 1A of H. Jónsson et al. (see previous endnote).
8. See Supplementary Table S10 of H, Jónsson et al (see previous endnote).
9. See Figure 1A of H. Jónsson et al. (see previous endnote).
10. Compare especially the relative branching order implied by Table 7.1 in chapter 7 to the relative branching order implied by Figure 1A of H. Jónsson et al. (see previous endnote).
11. Calculated from Wilson & Reeder's *Mammal Species of the World,* 3rd ed., https://www.departments.bucknell.edu/biology/resources/msw3/.
12. Think of lining up the number of species per family in all 153 families in a row, and then think of doing so in order from least to greatest. Finding the one value in the middle represents the process of finding the median.
13. This skewed distribution of species per family was true in other groups of creatures, as later data in this chapter show.
14. For Table 10.1, the number of species per family was downloaded from the IUCN Red List (http://www.iucnredlist.org/) on February 24, 2017. The numbers for the column "theoretical number of species with mitochondrial DNA sequences available" were obtained from the NCBI RefSeq database, as of February 23, 2017. The numbers for the column "actual number of species with mitochondrial DNA sequences available" is the result of my own manual curation of the numbers in the "theoretical number of species with mitochondrial DNA sequences available" column. Since the NCBI RefSeq database includes fossil DNA sequences as well as sequences from domestic animals, breeds, and subspecies, I removed these sequences in order to reduce the dataset to strictly wild, still-living species.
15. These results were obtained through my own personal analysis. Sequences were downloaded from the NCBI database (NCBI accession numbers = NC_007441.1, NC_009510.1, NC_010640.1, NC_012096.1, NC_012098.1, NC_012346.1, NC_012706.1, NC_013069.1, NC_013751.1, NC_014044.1, NC_014875.1, NC_015889.1, NC_016421.1, NC_016422.1, NC_016689.1, NC_018603.1, NC_020614.1, NC_020615.1, NC_020616.1, NC_020617.1, NC_020618.1, NC_020619.1, NC_020620.1, NC_020621.1, NC_020622.1, NC_020623.1, NC_020624.1, NC_020625.1, NC_020626.1, NC_020627.1, NC_020628.1, NC_020629.1, NC_020630.1, NC_020631.1, NC_020632.1, NC_020633.1, NC_020656.1, NC_020674.1, NC_020675.1, NC_020676.1,

NC_020678.1, NC_020683.1, NC_020685.1, NC_020686.1, NC_020687.1, NC_020688.1, NC_020689.1, NC_020690.1, NC_020691.1, NC_020692.1, NC_020693.1, NC_020694.1, NC_020695.1, NC_020698.1, NC_020699.1, NC_020701.1, NC_020702.1, NC_020703.1, NC_020704.1, NC_020705.1, NC_020706.1, NC_020707.1, NC_020708.1, NC_020709.1, NC_020710.1, NC_020712.1, NC_020713.1, NC_020715.1, NC_020716.1, NC_020717.1, NC_020718.1, NC_020722.1, NC_020723.1, NC_020724.1, NC_020725.1, NC_020726.1, NC_020727.1, NC_020728.1, NC_020731.1, NC_020732.1, NC_020733.1, NC_020734.1, NC_020735.1, NC_020736.1, NC_020738.1, NC_020741.1, NC_020742.1, NC_020746.1, NC_020747.1, NC_020748.1, NC_020749.1, NC_020750.1, NC_020751.1, NC_020752.1, NC_020788.1, NC_020789.1, NC_020793.1, NC_020794.1, NC_021381.1, NC_023457.1, NC_023542.1, NC_023543.1, NC_024818.1, NC_025563.1, NC_026063.1, NC_026064.1, NC_028161.1). Where appropriate, sequences were manually adjusted to share the same position #1. Sequences were aligned in-house with CLUSTALX2 (http://www.clustal.org/clustal2/) or CLUSTALW-MTV (http://www4a.biotec.or.th/GI/tools/clustalw-mtv). In BioEdit (http://www.mbio.ncsu.edu/BioEdit/bioedit.html) software, all non-standard nucleotides (e.g., N, M, R, Y, B, W, S, V, H, D) were replaced with gaps. Then all columns containing gaps were stripped, after which a tree was created using MEGA6 (http://www.megasoftware.net/).

16. The methodology for creating this timeline, and the methodology for creating speciation timelines for other families in this chapter, is the same. A preliminary version of this methodology was published previously (N.T. Jeanson, "Mitochondrial DNA Clocks Imply Linear Speciation Rates within 'Kinds,'" *Answers Research Journal,* 2015, 8:273–304, available online at https://answersingenesis.org/natural-selection/speciation/clocks-imply-linear-speciation-rates-within-kinds/). To create a timeline of speciation for species within a particular family, I adhered to the following outline:

1) The mtDNA sequences were downloaded from the NCBI database (see endnotes for each family for NCBI accession numbers). Where appropriate, sequences were manually adjusted to share the same position #1. The number of species with mtDNA sequences was compared to the total number of living species within a family, using publicly available sources of mammal (IUCN), reptile (Reptile Database), amphibian (Amphibian Species of the World), bird (BirdLife Taxonomic Checklist, version 9), and fish (Catalog of Fishes) taxonomy.

2) Sequences were aligned in-house with CLUSTALX2 (http://www.clustal.org/clustal2/) or CLUSTALW-MTV (http://www4a.biotec.or.th/GI/tools/clustalw-mtv).

3) In BioEdit (http://www.mbio.ncsu.edu/BioEdit/bioedit.html) software, all non-standard nucleotides (e.g., N, M, R, Y, B, W, S, V, H, D) were replaced with gaps. Then all columns containing gaps were stripped, after which a midpoint-rooted tree was created using MEGA6 (http://www.megasoftware.net/).

4) From the tree, the branch lengths between speciation events were manually copied into Microsoft Excel. These branch lengths were treated as successive temporal events based on their distance from the midpoint root. For example, the branch length from the midpoint root to the first speciation event was copied unaltered. However, subsequent speciation events required a modification of this protocol to reflect a temporal order of speciation. In other words, for speciation events that followed the first species splits, I added the branch length from the midpoint root to initial split, to the branch length from the initial split to the subsequent split. This addition was performed only for speciation events that were linked horizontally on the mtDNA tree.

As a particular example, consider the timing of speciation events in the family Equidae (Figure 10.1B). The timing of the speciation of the African wild ass is straightforward. It involves only a single branch length that traces directly back to the midpoint root — in this case, 0.004825 is the distance.

The timing of the split of the Kiang and Onager is more complex. Several branch lengths must be added together. For example, the split happened at 0.027434 (0.004825 + 0.000658 + 0.021951 = 0.027434).

5) Branch lengths were converted to nucleotides by multiplying the branch lengths (from step 4) by the total length of the mtDNA alignment that was stripped of gaps (i.e., the result of step 3). These nucleotides were plotted against cumulative speciation numbers to produce the graphs visible in the Figures showing the timing of speciation for each family.

6) Using Microsoft Excel, linear regression was performed on these plotted data.

7) Because each species varies in its distance from the midpoint root, statistical uncertainty was added to the time element of each graph as follows: Within a particular family, the species with the longest and shortest branch lengths (both measured as total distance from the midpoint root to tip) were identified. Then the average total distance was calculated with these two data points. The percent difference between the average and either of these data points was calculated. Then each speciation time point (i.e., the nucleotide distance from the midpoint root to each

speciation event) was multiplied by this percentage to derive a statistical uncertainty, which was plotted as horizontal gray bars on either side of each data point.

As a practical example, consider again the timing of speciation events in the family Equidae (Figure 10.1B). The shortest total branch length (from midpoint to tip) is the Kiang; the longest total branch lengths belong to the mountain zebra and the Przewalski's horse. The difference between longest (or shortest) branch and the average (of the longest and shortest branches) is 13% of the average (of the longest and shortest branches). Thus, each speciation data point (i.e., nucleotide distance) in the timeline of Equid speciation (Figure 10.24) was multiplied by 13%, and this 13% difference was applied on both sides of each data point (differences represented by horizontal lines) (Figure 10.24). Because later time points represent longer distances, the horizontal lines are longer than at early time points.

17. These results were obtained through my own personal analysis. Sequences were downloaded from the NCBI database (NCBI accession numbers = NC_023965.1, NC_028592.1, NC_021943.1, NC_023964.1, NC_021944.1, NC_023963.1, NC_023962.1, NC_023961.1, NC_007009.1, NC_024933.1, NC_009747.1, NC_008066.1, NC_009748.1, NC_006901.1, NC_021947.1, NC_021954.1, NC_025201.1, NC_023795.1, NC_027449.1, NC_012670.1, NC_025513.1, NC_027604.1, NC_031156.1, NC_005943.1, NC_026976.1, NC_026120.1, NC_025221.1, NC_002764.1, NC_011519.1, NC_025222.1, NC_028442.1, NC_021956.1, NC_008216.1, NC_020006.2, NC_020007.2, NC_001992.1, NC_020008.2, NC_020009.2, NC_020010.2, NC_008219.1, NC_008217.1, NC_020666.1, NC_018062.1, NC_008220.1, NC_018061.1, NC_015485.1, NC_015486.1, NC_018057.1, NC_008218.1, NC_018059.1, NC_008215.1, NC_020667.1, NC_019802.1, NC_023971.1, NC_023970.1, NC_019580.1, NC_019579.1, NC_019583.1, NC_006900.1, NC_024529.1, NC_019581.1, NC_019582.1). Where appropriate, sequences were manually adjusted to share the same position #1. Sequences were aligned in-house with CLUSTALX2 (http://www.clustal.org/clustal2/) or CLUSTALW-MTV (http://www4a.biotec.or.th/GI/tools/clustalw-mtv). In BioEdit (http://www.mbio.ncsu.edu/BioEdit/bioedit.html) software, all non-standard nucleotides (e.g., N, M, R, Y, B, W, S, V, H, D) were replaced with gaps. Then all columns containing gaps were stripped, after which a tree was created using MEGA6 (http://www.megasoftware.net/).
18. See general outline of methodology in previous endnotes.
19. See general outline of methodology in previous endnotes. NCBI accession numbers = NC_009677.1, NC_009678.1, NC_009685.1, NC_009692.1, NC_011125.1, NC_011358.1, NC_011579.1, NC_012141.1, NC_020637.1, NC_020638.1, NC_020639.1, NC_020640.1, NC_020641.1, NC_020642.1, NC_020643.1, NC_020644.1, NC_020645.1, NC_020646.1, NC_020664.1, NC_021749.1, NC_021751.1, NC_023210.1, NC_024942.1, NC_025516.1, NC_028013.1
20. See general outline of methodology in previous endnotes. NCBI accession numbers = NC_020677.1, NC_020680.1, NC_020681.1, NC_020682.1, NC_020684.1, NC_025271.1, NC_007704.2, NC_018595.1, NC_020700.1, NC_024819.1, NC_008749.1, NC_018358.1, NC_020711.1, NC_011821.1, NC_020719.1, NC_020720.1, NC_024812.1, NC_020721.1, NC_004577.1, NC_004563.1, NC_004069.1, NC_016920.1, NC_020729.1, NC_015247.1, NC_020766.1, NC_016707.1, NC_020739.1, NC_020740.1, NC_007703.1, NC_020743.1, NC_014701.1, NC_020744.1, NC_020745.1, NC_031835.1
21. See general outline of methodology in previous endnotes. NCBI accession numbers = NC_005212.1, NC_028306.1, NC_028300.1, NC_027115.1, NC_028307.1, NC_028308.1, NC_028309.1, NC_028310.1, NC_028314.1, NC_028320.1, NC_028321.1, NC_028322.1, NC_028315.1, NC_028317.1, NC_028318.1, NC_028316.1, NC_028313.1, NC_027083.1, NC_028319.1, NC_014456.1, NC_008450.1, NC_028323.1, NC_028302.1, NC_022842.1, NC_010641.1, NC_010642.1, NC_028303.1, NC_028301.1, NC_028312.1, NC_028304.1, NC_028305.1, NC_028299.1, NC_016470.1, NC_028311.1, NC_010638.1
22. See general outline of methodology in previous endnotes. NCBI accession numbers = NC_005278.1, NC_012051.1, NC_012053.1, NC_012057.1, NC_012058.1, NC_012059.1, NC_012061.1, NC_012062.1, NC_019441.1, NC_019577.1, NC_019578.2, NC_019588.1, NC_019589.1, NC_019590.1, NC_019591.1, NC_020696.1, NC_022805.1, NC_023889.1, NC_032301.1
23. See general outline of methodology in previous endnotes. NCBI accession numbers = NC_027956.1, NC_008093.1, NC_008092.1, NC_024172.1, NC_013445.1, NC_028427.1, NC_013700.1, NC_026723.1, NC_023958.1, NC_027935.1, NC_026529.1, NC_008434.1, NC_023459.1
24. See general outline of methodology in previous endnotes. NCBI accession numbers as follows: Muridae = NC_020758.1, NC_016428.1, NC_016662.1, NC_017599.1, NC_019584.1, NC_019585.1, NC_016060.1, NC_028335.1, NC_014696.1, NC_025670.1, NC_027683.1, NC_027684.1, NC_023263.1,

NC_027932.1, NC_025268.1, NC_025269.1, NC_025270.1, NC_030342.1, NC_025287.1, NC_012387.1, NC_025952.1, NC_010650.1, NC_023960.1, NC_019617.1, NC_028715.1, NC_014698.1, NC_012389.1, NC_014867.1, NC_014855.1, NC_014858.1, NC_023347.1, NC_032286.1, NC_001665.2, NC_012461.1, NC_012374.1, NC_014871.1, NC_011638.1, NC_029888.1, NC_014861.1, NC_014864.1
Cricetidae = NC_025746.1, NC_007936.1, NC_024592.1, NC_025330.1, NC_031802.1, NC_013571.1, NC_027418.1, NC_030330.1, NC_025283.1, NC_013276.1, NC_015241.1, NC_003041.1, NC_008064.1, NC_027945.1, NC_024538.1, NC_016427.1, NC_029477.1, NC_016055.1, NC_033356.1, NC_029760.1, NC_031809.1, NC_013563.1, NC_013068.1, NC_025747.1
Vespertilionidae = NC_002626.1, NC_005436.1, NC_015484.1, NC_015828.1, NC_016872.1, NC_016873.1, NC_021119.1, NC_022474.1, NC_022694.1, NC_022698.1, NC_024558.1, NC_025308.1, NC_025568.1, NC_025949.1, NC_027237.1, NC_027973.1, NC_027977.1, NC_029191.1, NC_029342.1, NC_029346.1, NC_029422.1, NC_029849.1, NC_029939.1, NC_033347.1
Soricidae = NC_024563.1, NC_023950.1, NC_026204.2, NC_027249.1, NC_027247.1, NC_029329.1, NC_027248.1, NC_027245.1, NC_027244.1, NC_027242.1, NC_027243.1, NC_027246.1, NC_006893.1, NC_021398.1, NC_026131.1, NC_003040.1, NC_029840.1, NC_023351.1, NC_025559.1, NC_027963.1, NC_025278.1, NC_025327.1, NC_005435.1, NC_024604.1
Sciuridae = NC_030071.1, NC_025550.1, NC_031210.1, NC_026705.1, NC_026706.1, NC_026442.1, NC_030072.1, NC_031847.1, NC_026443.1, NC_027278.1, NC_030070.1, NC_018367.1, NC_023922.1, NC_023089.1, NC_019612.1, NC_023780.1, NC_002369.1, NC_027283.1, NC_032372.1, NC_032374.1, NC_032373.1, NC_032370.1, NC_032371.1, NC_025277.1, NC_032375.1, NC_032376.1, NC_029325.1, NC_026875.1, NC_031209.1

25. See general outline of methodology in previous endnotes. NCBI accession numbers as follows:
Phasianidae = NC_003408.1, NC_004575.1, NC_007238.1, NC_007239.1, NC_007240.1, NC_010767.1, NC_010770.1, NC_010771.1, NC_010774.1, NC_010778.1, NC_010781.1, NC_011816.1, NC_011817.1, NC_012453.1, NC_012895.1, NC_012897.1, NC_012900.1, NC_013619.1, NC_013979.1, NC_014576.1, NC_015526.1, NC_015897.1, NC_016679.1, NC_018033.1, NC_018034.1, NC_020583.1, NC_020584.1, NC_020585.1, NC_020586.1, NC_020587.1, NC_020588.1, NC_020589.1, NC_020590.1, NC_020591.1, NC_020613.1, NC_022683.1, NC_022684.1, NC_023264.1, NC_023779.1, NC_023939.1, NC_023940.1, NC_024533.1, NC_024554.1, NC_024615.1, NC_024616.1, NC_024619.1, NC_025318.1, NC_026547.1, NC_026548.1, NC_027279.1
Gruidae = NC_020569.1, NC_020570.1, NC_020571.1, NC_020572.1, NC_020573.1, NC_020574.1, NC_020575.1, NC_020576.1, NC_020577.1, NC_020578.1, NC_020579.1, NC_020580.1, NC_020581.1, NC_020582.1, NC_021368.1
26. See general outline of methodology in previous endnotes. NCBI accession numbers as follows:
Geoemydidae = NC_006082.1, NC_009330.1, NC_009509.1, NC_010970.1, NC_010973.1, NC_011819.1, NC_012054.1, NC_014102.1, NC_014401.1, NC_014769.1, NC_015101.1, NC_016685.1, NC_016691.1, NC_016951.1, NC_017875.1, NC_017878.1, NC_017885.1, NC_018793.1, NC_020665.1, NC_020668.1, NC_022857.1, NC_023220.1, NC_023221.1, NC_026024.1, NC_026027.1, NC_027826.1, NC_029183.1, NC_029369.1, NC_031432.1, NC_032297.1, NC_032300.1
Viperidae = NC_007397.1, NC_009768.1, NC_010223.1, NC_011390.1, NC_011391.1, NC_012146.1, NC_013479.1, NC_014400.1, NC_021402.1, NC_021412.1, NC_022473.1, NC_022695.1, NC_022820.1, NC_025560.1, NC_025666.1, NC_026051.1, NC_026052.1, NC_026553.1, NC_029165.1, NC_029166.1, NC_029424.1, NC_029494.1, NC_030181.1, NC_030182.1, NC_030760.1, NC_030781.1
27. See general outline of methodology in previous endnotes. NCBI accession numbers as follows:
Hynobiidae = NC_004021.1, NC_008076.1, NC_008077.1, NC_008078.1, NC_008079.1, NC_008080.1, NC_008081.1, NC_008082.1, NC_008083.1, NC_008084.1, NC_008085.1, NC_008088.1, NC_008089.1, NC_008090.1, NC_008091.1, NC_009335.1, NC_010224.1, NC_012430.1, NC_013762.1, NC_013825.1, NC_020634.1, NC_020635.1, NC_020649.1, NC_020650.1, NC_021001.1, NC_021106.1, NC_023789.1, NC_026032.1, NC_026033.1, NC_026698.1, NC_026853.1, NC_026854.1, NC_032306.1, NC_032307.1
Salamandridae = NC_002756.1, NC_006407.1, NC_015788.1, NC_015790.1, NC_015791.1, NC_015792.1, NC_015794.1, NC_015795.1, NC_015796.1, NC_017870.1, NC_017871.1, NC_027421.1, NC_027505.1, NC_027507.1, NC_028278.1, NC_029231.1, NC_029345.1, NC_032068.1, NC_032308.1, NC_032309.1, NC_032310.1, NC_032311.1, NC_032312.1, NC_032313.1, NC_032314.1, NC_032315.1

28. See general outline of methodology in previous endnotes. NCBI accession numbers = NC_006531.1, NC_006532.1, NC_006543.1, NC_006534.1, NC_006537.1, NC_006538.1, NC_006539.1, NC_002707.2, NC_011575.1, NC_006536.1, NC_006540.1, NC_006541.1, NC_006542.1, NC_006544.1, NC_006545.1, NC_006546.1, NC_006547.2
29. See general outline of methodology in previous endnotes. NCBI accession numbers = NC_024055.1, NC_023948.1, NC_026696.1, NC_027081.1, NC_023522.1, NC_025520.1, NC_024862.1, NC_024284.1, NC_020611.1, NC_024596.1, NC_023521.1, NC_026871.1, NC_022193.1, NC_028342.1, NC_028338.1, NC_023361.1, NC_021768.2, NC_028344.1, NC_028341.1, NC_028340.1, NC_029843.1, NC_022819.1, NC_018052.1, NC_022679.1, NC_028508.1, NC_025778.1, NC_026287.1
30. See general outline of methodology in previous endnotes; see previous endnotes for NCBI accession numbers as well.
31. See Supplementary Table S10 of the following paper (I rounded the reported numbers): H. Jónsson et al., "Speciation with Gene Flow in Equids Despite Extensive Chromosomal Plasticity," *Proc Natl Acad Sci USA,* 2014, 111(52):18655–60.
32. See Supplementary Table S13 of the following paper: H. Jónsson et al., "Speciation with Gene Flow in Equids Despite Extensive Chromosomal Plasticity," *Proc Natl Acad Sci USA*, 2014, 111(52):18655–60. In Supplemental Tables S12–S18, the columns labeled "Homozygous derived" and "Heterozygous" have been inadvertently switched. This was confirmed to me by direct email correspondence with corresponding author Ludovic Orlando.
33. Ibid. See supplementary Table S10.
34. Ibid. See Supplemental Tables 12, 14. Again, in Supplemental Tables S12–S18, the columns labeled "Homozygous derived" and "Heterozygous" have been inadvertently switched. This was confirmed to me by direct email correspondence with corresponding author Ludovic Orlando.
35. L. Orlando et al., "Recalibrating Equus Evolution Using the Genome Sequence of an Early Middle Pleistocene Horse," *Nature,* 2013, 499(7456):74–8; J. Huang et al., "Analysis of Horse Genomes Provides Insight into the Diversification and Adaptive Evolution of Karyotype," *Sci Rep.,* 2014, 4:4958.
36. C. Mei, H. Wang, W. Zhu, H. Wang, G. Cheng, K. Qu, X. Guang et al., "Whole-Genome Sequencing of the Endangered Bovine Species Gayal (*Bos frontalis*) Provides New Insights into its Genetic Features," *Science Reports,* 2016, 6: 19787. For more detailed discussion of this and other numbers derived in this paragraph, see the following paper: N.T. Jeanson and J. Lisle, "On the Origin of Eukaryotic Species' Genotypic and Phenotypic Diversity: Genetic Clocks, Population Growth Curves, and Comparative Nuclear Genome Analyses Suggest Created Heterozygosity in Combination with Natural Processes as a Major Mechanism," *Answers Research Journal,* 2016, 9:81–122, available online at https://answersingenesis.org/natural-selection/speciation/on-the-origin-of-eukaryotic-species-genotypic-and-phenotypic-diversity/.
37. Y.S. Cho, L. Hu, H. Hou, H. Lee, J. Xu, S. Kwon, S. Oh, et al., "The Tiger Genome and Comparative Analysis with Lion and Snow Leopard Genomes," *Nature Communications,* 2013, 4:2433.
38. M.J. Montague et al. "Comparative Analysis of the Domestic Cat Genome Reveals Genetic Signatures Underlying Feline Biology and Domestication," *Proc Natl Acad Sci USA,* 2014, 111(48):17230–17235.
39. H-S. Yim, Y.S. Cho, X. Guang, S.G. Kang, J-Y. Jeong, S-S. Cha, H-M. Oh, et al., "Minke Whale Genome and Aquatic Adaptation in Cetaceans," *Nature Genetics,* 2014, 46 (1):88–92.
40. C. Li, Y. Zhang, J. Li, L. Kong, H. Hu, H. Pan, L. Xu, et al., "Two Antarctic Penguin Genomes Reveal Insights into their Evolutionary History and Molecular Changes Related to the Antarctic Environment," *Gigascience,* 2014, 3 (1):27.
41. Fossil numbers are from my own searches on the Paleobiology Database (http://fossilworks.org/). For more detail and documentation, see the following paper: N.T. Jeanson and J. Lisle, "On the Origin of Eukaryotic Species' Genotypic and Phenotypic Diversity: Genetic Clocks, Population Growth Curves, and Comparative Nuclear Genome Analyses Suggest Created Heterozygosity in Combination with Natural Processes as a Major Mechanism," *Answers Research Journal,* 2016, 9:81–122, available online at https://answersingenesis.org/natural-selection/speciation/on-the-origin-of-eukaryotic-species-genotypic-and-phenotypic-diversity/.
42. For documentation, see the Committee on Recently Extinction Organisms (CREO) (http://creo.amnh.org/) and the IUCN Red List Database (http://www.iucnredlist.org/).
43. The number of mammal species per family was downloaded from the IUCN Red List (http://www.iucnredlist.org/) on February 24, 2017.
44. The number of reptile species per family was downloaded from Reptile Database (http://www.reptile-database.org/data/) on February 23, 2017. At that time, the website listed the most updated version as "24 Dec 2016."

45. The number of amphibian species per family was downloaded from the Amphibian Species of the World database (http://research.amnh.org/vz/herpetology/amphibia/) on February 23, 2017.
46. The number of avian species per family was downloaded from the BirdLife Taxonomic Checklist (version 9; http://datazone.birdlife.org/species/taxonomy) on February 23, 2017.
47. The number of fish species per family was downloaded from the Catalog of Fishes (http://researcharchive.calacademy.org/research/ichthyology/catalog/SpeciesByFamily.asp) on February 23, 2017.
48. See the following paper for a more in-depth treatment of these data, as well as primary references: N.T. Jeanson, "Mitochondrial DNA Clocks Imply Linear Speciation Rates within 'Kinds,'" *Answers Research Journal,* 2015, 8:273–304, available online at https://answersingenesis.org/natural-selection/speciation/clocks-imply-linear-speciation-rates-within-kinds/.
49. Fossil numbers are from my own searches on the Paleobiology Database (http://fossilworks.org/).
50. Some Przewalski's horses possess leg bars. I have observed them in the Przewalski's horses at my local zoo (the Cincinnati Zoo), and one of the zookeepers confirmed my observation.
51. http://www.metmuseum.org/toah/hd/play/hd_play.htm.
52. Again, for an intriguing discussion of stripe formation, see J.B.L. Bard, "A Unity Underlying the Different Zebra Striping Patterns," *J. Zool., Lond.,* 1977, 183:527–539.
53. From the Introduction (p.1) to *On the Origin of Species*: "When on board H.M.S. 'Beagle,' as naturalist, I was much struck with certain facts in the distribution of the inhabitants of South America, and in the geological relations of the present to the past inhabitants of that continent. These facts seemed to me to throw some light on the origin of species — that *mystery of mysteries,* as it has been called by one of our greatest philosophers" (emphasis mine).
54. Wilson & Reeder's *Mammal Species of the World,* 3rd Ed., was accessed electronically at https://www.departments.bucknell.edu/biology/resources/msw3/ on October 7, 2016. From the .csv file of the complete taxonomy, the data were filtered for "SPECIES" or for "SUBSPECIES" via the "TaxonLevel" column, and then filtered for "FALSE" via the "Extinct?" column. See the following reference, especially the endnotes, for a similar analysis: https://answersingenesis.org/reviews/articles/does-biologos-strive-for-dialogue/.

Afterword

1. M.J. Behe, *Darwin's Black Box* (New York: Touchstone, 1996). Not surprisingly, his conclusions have been vigorously debated. For his responses, see later editions to his original work, which contain responses, as well as the following: M.J. Behe, "Self-Organization and Irreducibly Complex Systems: A Reply to Shanks and Joplin," *Philosophy of Science,* 2000, 67(1):155–162; M.J. Behe, "Reply to My Critics: A Response to Reviews of Darwin's Black Box: The Biochemical Challenge to Evolution," *Biology and Philosophy,* 2001, 16:685–709; https://www.trueorigin.org/behe01.php; https://www.trueorigin.org/behe08.php. See also the following for a more complete bibliography: http://www.discovery.org/p/31, http://behe.uncommondescent.com/.
2. I am indebted to A.E. Wilder-Smith and his lecture "Is Man a Machine?" for some of the background to and the concepts in this paragraph and the several following. See also N. Jeanson and B. Thomas, "Unmistakable Evidence for God's Design: Cells Lead the Way," in *Creation Basics & Beyond* (Dallas, TX: Institute for Creation Research, 2013).
3. See the following works for a much more in-depth and scholarly treatment of the biblical issues surrounding theistic evolution: T. Mortenson and T.H. Ury, eds., *Coming to Grips with Genesis: Biblical Authority and the Age of the Earth* (Green Forest, AR: Master Books, 2008); T. Mortenson, ed., *Searching for Adam: Genesis & the Truth About Man's Origin* (Green Forest, AR: Master Books, 2016).
4. Again, see the two works in the previous note for a much more in-depth and scholarly treatment of religious and philosophical coherence of the scriptural view espoused in this book.
5. "Then God saw everything that He had made, and indeed it was very good" (Genesis 1:31).
6. "So God created man in His own image; in the image of God He created him; male and female He created them. Then God *blessed* them, and God said to them, 'Be fruitful and multiply; fill the earth and subdue it; have dominion over the fish of the sea, over the birds of the air, and over every living thing that moves on the earth' " (Genesis 1:27–28) (emphasis added).
7. "But God, who is *rich in mercy,* because of His great love with which He loved us, even when we were dead in trespasses, made us alive together with Christ" (Ephesians 2:4–5) (emphasis added).

8. "And the Lord passed before [Moses] and proclaimed, 'The LORD, the LORD God, *merciful and gracious*, longsuffering, and abounding in goodness and truth, keeping mercy for thousands, forgiving iniquity and transgression and sin, by no means clearing the guilty, visiting the iniquity of the fathers upon the children and the children's children to the third and the fourth generation' " (Exodus 34:6–7) (emphasis added).
9. I don't mean that God would have lacked the attribute of mercy. Rather, God would still have possessed the attribute of mercy, but this facet would not have been directly experienced by humans in their being recipients of it.
10. ". . . the LORD, the habitation of justice . . ." (Jeremiah 50:7).

Acknowledgments

Many thanks to my wife who patiently bore with my many hours of writing and with long conversations about abstract topics, all while maintaining the happiness of our three toddlers. She was an invaluable sounding board, a reader of early drafts of this manuscript, and a great encouragement on this project.

The scientific reviewers of this work provided very helpful critical feedback. Joseph Edward (pseudonym), Jean O'Micks (pseudonym), Rick Roberts, Kevin Anderson, Andrew Fabich, Steve Carter, Dave Menton, Jason Lisle, and Tim Clarey were instrumental in taking this book from its original state to the much-improved state in which it exists today. Any errors which remain are my own and not theirs.

This book would not have been possible without the support of Ken Ham of Answers in Genesis (AiG), and of AiG's research director, Andrew Snelling. Dr. Snelling's guidance in my navigation of the publishing process was a huge help.

Tim Dudley, Laura Welch, and the team at Master Books were incredibly supportive and kind in letting me give so much input into each detail of the publishing process. Randy Pratt caught the vision for this book and has been immeasurably helpful in designing a marketing plan to broaden the reach of this work.

Charis Hsu (pseudonym) took on a monster task. Her patient and skillful creation of the many of the genetics diagrams in this book have imparted a quality and professionalism for which I am extremely grateful.

Intelligent Designs Creative kindly bore with my many ideas on the book cover and then produced the product you now possess. I am deeply indebted.

Laura Strobl of AiG was also very kind in helping me find images and navigate the AiG repository of graphical information.

Steve Austin was gracious in allowing me to use some of his Mount St Helens photos.

The AiG librarian, Walt Stumper, has been an enormous help in promptly tracking down critical papers. His aid to my research cannot be overstated.

As the endnotes demonstrate, much of the scientific basis for this book can be found in a series of technical papers that I published in the *Answers Research Journal*. The initial research for these papers began at the Institute for Creation Research (ICR), with the support of the CEO, Henry Morris III. My collaborative interactions with the scientific staff at ICR were a tremendous boon to the development of my own ideas. Jeff Tomkins in particular was an ever-available colleague, and Jason Lisle was always a helpful sounding board and patient counselor.

I cannot overstate the contribution of Rob Carter at Creation Ministries International. He and I have spent countless hours on the phone and in person crunching numbers, bouncing research ideas off one another, and provoking each other to new hypotheses and pursuits. I have learned much from these interactions.

Several other current and former AiG staff members have provided help for this project:

Tommy Mitchell provided extremely valuable feedback on early drafts, which lead to a massive overhaul of the structure of the book.

Steve Ham and Joe Owen also read early drafts and were instrumental in helping me craft the tone and level of scientific complexity in this book.

Georgia Purdom aided my early discussion of book ideas, and was very helpful as head of the AiG editorial review board. Her willingness to promptly review the final proofs of this manuscript is a kindness for which I cannot sufficiently state my thanks. Again, any errors that remain are my own.

Dan Leitha was always a helpful sounding board for anything artistic, including book cover concepts.

Bodie Hodge was helpful during the early stages of this manuscript, and was another valuable sounding board for ideas.

Dan Zordel and Joel Leinweber provided invaluable advice on how to navigate the logistics of book publishing.

Ron Uebel was extremely supportive and caught the vision for this book. He and Jeremy Ham have been tremendously helpful in developing the marketing side of publishing this book.

Roger Patterson provided critical feedback on early discussions of current rates of speciation. He has also been an ever-present confidant and aid.

The AiG IT department has been extremely helpful in supporting my bioinformatics analyses.

My early discussions with Tim Chaffey were the spark that lit the desire to create a resource on speciation, which eventually lead to this book.

Finally, my friend and colleague, David Sparks, eagerly and enthusiastically provided me with early feedback on this manuscript. I am very grateful for his comments and friendship.

Illustration and Photo Credits

T = top, TL = top left, TR = top right, B = bottom, BL = bottom left, BR = bottom right, CP = Color Plate

Answers in Genesis: p. 68 CP8, p. 188

Charis Hsu: p. 21-22, p. 34 B, p. 37, p. 42, p. 46, p. 47 B, p. 48, p. 49, p. 50, p. 51, p. 54, p. 58, p. 60, p. 67, p. 69, p. 70, p. 71, p. 73, p. 74, p. 94, p. 95, p. 101, p. 103, p. 104, p. 167, p. 206, p. 209

David Menton: p. 75 CP16 (all)

Flickr: p. 172 Aardvark (Heather Paul) CC BY-ND 2.0

Nathaniel T. Jeanson: p. 24, p. 25 (all), p. 47 T, p. 91, p. 97-99, p. 133, p. 146 (all), p. 153, p. 159, p. 170, p. 179-181, p. 184, p. 185 (all), p. 186, p. 191 (all), p. 193, p. 197-201 (all), p. 211, p. 213 (all), p. 214, p. 217-218, p. 220, p. 222-225 (all), p. 247, p. 250-266 (all), p. 270 B, p. 271-272, p. 276 Freestockphotos.biz: p. 172 Tiger (Claudio Gennari) CC BY 2.0

Pixabay.com: p. 116 T, p. 173 Kangaroo, p. 66 CP2

Public Domain: p. 27, p. 29, p. 31, p. 43, p. 68 CP9, p. 72, p. 76, p. 81 CP37, p. 83 CP44, p. 84 CP49, p. 86 CP61, p. 88 CP68, p. 121, p. 134 B, p. 135 (all), p. 138 BL, p.138 BR, p. 145 T, p. 145 B, p. 172 Moose, p. 172 Manatee, p. 173 Pangolin, p. 227, p. 269, p. 270 TL, p. 270 TR

SciencePhoto.com: p. 35, p. 36, p. 53, p. 80 CP31, p. 80 CP32, p. 87 CP64, p. 187, p. 195 TL, p. 195 TR, p. 195 BL, p. 195 BR, p. 216,

Shutterstock.com: p. 109 B, p. 137 (all except bottom left and the right top 3), p. 138 TR, p. 172 Zebra

Steve Austin: p. 92 CP76, p. 93 CP77

Wikimedia Commons - p. 17 (Wellcome Library, London), p. 26, p. 30, p. 32 (Adenosine), p. 33, p. 34 T, p. 45 (Dr Graham Beards), p. 66 CP1 (Alan Wilson), p. 66 CP3 (Steve Sayles), p. 66 CP4 (Alan Vernon), p. 66 CP5 (Minette Layne), p. 66 CP6 (Luxil), p. 77 CP18 (Joachim Huber), p. 77 CP19 (Moongateclimber), p. 77 CP20 (Bardrock), p. 78 CP21 (Stig Nygaard from Copenhagen, Denmark), p. 78 CP22 (Donna Brown), p. 78 CP23 (CHUCAO), p. 78 CP24 (Quartl), p. 79 CP25 (Kabir Bakie), p. 79 CP26 (Eric Kilby from USA), p. 79 CP27 (H. Zell), p. 79 CP28 (Fiver Löcker from Wellington, New Zealand), p. 80 CP29 (Alina Zienowicz), p. 80 CP30 (päts from Mexico), p. 81 CP33 (Harald Zimmer), p 81 CP34 (Jean & Nathalie),

p. 81 CP35 (Darren Swim), p. 81 CP36 (Charles W. Hardin), p. 82 CP38 (Eric Kilby from USA), p. 82 CP39 (Charlesjsharp), p. 82 CP40 (Allie_Caulfield), p. 82 CP41 (David Sifry), p. 83 CP42 (JC i Núria), p. 83 CP43 (Tierpfotografien at de.wikipedia), p. 83 CP45 (Ealdgyth), p. 83 CP46 (Shetta), p. 83 CP47 (Vkarel), p. 84 CP48 (Pete Markham from Loretto, USA), p. 84 CP50 (Randy Stewart), p. 84 CP51 (Fábio Vidigal), p. 84 CP52 (Jean-Pol GRANDMONT), p. 84 CP53 (François Marchal), p. 85 CP54 (Tomasz Sienicki), p. 85 CP55 (Herby), p. 85 CP56 (Leadgold at en.wikipedia), p. 85 CP57 (4028mdk09), p. 86 CP58 (CaptainHaddock), p. 86 CP59 (asibiri), p. 86 CP60 (Justlettersandnumbers), p. 86 CP62 (Chixoy), p. 87 CP63 (Oggmus), p. 87 CP65 (Ikiwaner), p. 87 CP66 (Carlos Delgado), p. 88 CP67 (© BrokenSphere / Wikimedia Commons), p. 88 CP69 (SajjadF), p. 88 CP70 (Miroslav Duchacek (from Czech Republic)), p. 90 CP72 (Yathin S Krishnappa), p. 90 CP73 (Jiří Sedláček - Frettie), p. 90 CP74 (Prabir K Bhattacharyya), p. 92-93 CP78 (Jean-Christophe BENOIST),p. 109 T (kaibara87), p. 112 B, p. 113 Wild Camel (John Hill), p. 116 M (Martin Pot (Martybugs at en.wikipedia)), p. 116 BL (David Iliff), p. 116 BR (John O'Neill), p. 119 (Mfield, Matthew Field, http://www.photography.mattfield.com), p. 131 (H. Zell), p. 132 T (H. Raab), p. 134 T (Eduard Solà), p. 137 BL (Javier Casado Tirado), p. 137 TR#1 (M 93), p. 137 TR#2 (order_242 from Chile), p. 137 TR#3 {Aude), p. 138 TL (Matt), p. 141 T (Blausen.com), p. 141 M (BodyParts3D by DBCLS), p. 141 B (Andrew Z. Colvin), p. 172 Whale (Whit Welles), p. 172 Elephant (Yathin S Krishnappa), p. 172 Lemur (Alex Dunkel (Maky)), p. 172 Mole (Michael David Hill), p. 173 Squirrel (Peter Trimming), p. 173 Bat (Bernard DUPONT from FRANCE), p. 173 Rabbit (DAVID ILIFF), p. 173 Armadillo (www.birdphotos.com), p. 173 Hyrax (Prosthetic Head), p. 173 Elephant Shrew (Joey Makalintal from Pennsylvania, USA), p. 173 Echidna (benjamint444), p. 195 TM (Robbie Rae, Amit Sinha, and Ralf J. Sommer)

Maps in the book use a base image from Shutterstock; source material for each adaptation are referenced in the individual caption information for each map.

Glossary

activation energy — In chemical reactions, energetically favorable processes don't always begin on their own. They sometime require a small "push." This "push" is the energy required to get the process started — the activation energy.

adenine — One of the four nitrogen-containing bases in DNA (or RNA) that is chemically distinct from the other three. The base forms one part of each individual unit of the DNA double helix (or of the RNA strand).

adenosine triphosphate (ATP) — A modified free form of one of the four individual subunits of an RNA molecule. It contains adenine at the nitrogen-containing base position.

allele — One of two or more versions of a particular DNA position.

amino acid — The basic chemical unit of proteins.

amphibians — A group of species typified by frogs, toads, and salamanders.

anaphase — A phase of the cell division cycle in which the dividing cell is beginning to form two distinct cells, and the chromosomes are moving into these newly forming cells. A stage before the final separation into two cells. Anaphase follows metaphase but precedes telophase.

anatomy — The physical structure of a creature — for example, the bones, muscles, intestines, blood vessels, etc. It also includes the three-dimensional relationships of these structures to one another.

arachnids — A group of creatures typified by spiders, scorpions, ticks, and mites.

atom — The smallest unit of an element (elements are things like carbon, oxygen, and nitrogen). Atoms can be combined to make molecules.

bacteriophage — A type of virus that infects bacteria.

base pair — A single unit of the DNA double helix. It consists of two bases, each chemically connected to their respective carbohydrate-phosphate backbone units. The two bases are weakly bound to each other in a form of chemical attraction known as hydrogen bonding.

base pairing — A chemical phenomenon in which the bases of one strand of the DNA double helix find their natural chemical complement in the bases of the other strand of the DNA double helix.

bottleneck — A dramatic reduction in population size from many individuals to few individuals; alternatively, a dramatic reduction in the number of genetic variants, from many variants to few variants.

breed — A domesticated version of a wild species. Breeds can take many forms, and one species might have hundreds of breeds.

Caenorhabditis elegans, C. elegans — The scientific name for a roundworm species that is commonly used in the laboratory.

carbohydrate — A type of chemical, typified by sugars.

catalyze, catalysis — The chemical process of aiding a natural chemical reaction. For example, proteins help catalyze intracellular chemical reactions. Though these reactions are chemically favorable, they go much more quickly with the help of these proteins.

cell — The basic unit of a creature. Sperm and egg are examples of cells. Skin consists of many sheets of linked cells.

cell division cycle — The regular, controlled process by which cells reproduce themselves.

central dogma — The shorthand summary for the typical flow of information in the cell. Typically, information in DNA is transcribed into RNA, which is translated into the chemical language of proteins.

centrifugation — The process of spinning substances (usually liquids) at high speed to separate the contents.

chemical bond — The linkage of two atoms together into a molecule.

chromosome — A condensed form of DNA. In bacteria, a chromosome is more diffuse than in species like human, in which the chromosomes assume a characteristic condensed structure during cell division.

chromosome pair — The condensed form of DNA is a chromosome. In species like humans, DNA comes in two copies. Thus, the two copies of DNA exist in pairs of chromosomes.

class — A classification rank above order but below phylum. For example, bony fish are a class of species, as are reptiles, amphibians, and mammals.

clinical trial — An experimental test of a treatment on human patients.

clone — An offspring that is genetically identical to its parents.

cloning — The process of creating genetically identical offspring.

coccyx — The base of the human spine, typically referred to as the tailbone.

complementary — Genetically, a physically and chemically favorable fit between two bases of DNA or RNA.

conifer — A group of plant species typified by pines and firs.

cross — In genetics, the mating of two individuals.

crustaceans — A group of species typified by lobsters, crabs, barnacles, and shrimp.

cyclin D — A gene encoding a specific protein involved in the regulation of the cell division cycle.

cytosine — One of the four nitrogen-containing bases in DNA (or RNA) that is chemically distinct from the other three. The base forms one part of each individual unit of the DNA double helix (or of the RNA strand).

Daphnia pulex, D. pulex — A crustacean known as the water flea; commonly used in the laboratory.

deductive reasoning — A method of reasoning that starts with a premise and uses rules of logic to draw a conclusion.

descent with modification — Darwin's term for the results of the process of inheritance. Each time reproduction happens, something is preserved ("descent"), while something else changes ("modification").

development — The natural biological process of building a creature from a single cell.

diploid — The state of having a full complement of DNA. For example, the non-reproductive cells of our body have *pairs* of chromosomes; they are diploid. Sperm and egg have half of the DNA content; their chromosomes do not exist as pairs.

DNA — Chemically, an abbreviation for deoxyribonucleic acid. Practically, the physical substance of heredity that typically exists in the form of a double helix.

DNA difference, DNA variant — Typically, a position in the DNA double helix in which individuals possess different base pairs. The difference may or may not be the result of mutation.

DNA sequence — The identity and order of base pairs on a DNA double helix.

DNA sequencing — The process of obtaining the DNA sequence of an individual.

dominance — A genetic phenomenon in creatures that have two different versions of hereditary instructions. When a dominant trait is crossed to a recessive trait, the dominant trait appears in the offspring, while the recessive trait is hidden.

Drosophila melanogaster, D. melanogaster — The scientific name for a fruit fly species that is commonly used in the laboratory.

E. coli — The scientific name for a bacterial species that is commonly used in the laboratory.

ecology — The field of science concerned with environments and their relationships to the creatures that they contain.

ecosystem — The relationships among creatures and their environment.

egg — The female reproductive cell.

embryology — The study of embryos.

ENCODE project — A large study dedicated to experimentally examining as much of the human genome as possible for function.

endemic, endemism — Native to a region — and native to that region alone.

enucleated — Having the nucleus removed

enzyme — A cellular component that catalyzes a chemical reaction. Typically, enzymes are proteins, but RNA can act as an enzyme as well.

equid — Any member of the horse family, Equidae.

ethnolinguistic group — A group of people sharing an ethnicity and language classification.

Eurasia — The combined regions of Europe and Asia.

extant — Still living.

extinct — Dead.

extract (noun) — A substance that has been chemically extracted from another mixture.

F_1 generation — In a genetic cross, the generation following the parental generation.

F_2 generation — In a genetic cross, the generation following the F_1 generation.

falsifiable — Something that can be formally disproven, at least in theory.

family — A classification unit above the level of genus but below the level of order. For example, all living horse, ass, and zebra species belong to the same family.

fauna — The animal content of an area.

fixity of species' geography — The hypothesis that God created species in their present locations. In other words, the hypothesis that species have been fixed in their present locations for as long as they have been in existence.

Flood (when capitalized) — The biblical Flood of Noah described in the Book of Genesis, chapter 6–9.

flora — The plant content of an area.

fossil — The rock remnants of a once-living individual. Dinosaurs are known almost exclusively from their fossils.

gamete — a reproductive cell; for example, a sperm cell or an egg cell.

gene — A hotly debated term in genetics; typically, a section of DNA that is transcribed and eventually translated into protein. But genes can also encode RNA molecules that do no get translated into protein.

gene conversion — In organisms that possess two versions of their DNA sequence, a hybrid of the two versions is passed on to offspring. The process of forming this hybrid DNA involves copying large chunks from each chromosome as well as copying small parts from each chromosome. Gene conversion is the latter.

gene regulation — The cellular process of controlling when and where a gene sequence functions.

generation time — The time from conception to sexual maturity.

genetic diversity — The amount of DNA differences between two copies of DNA; these two copies can be within an individual, between individuals, or between groups of individuals.

genetics — The study of inheritance at the visible, cellular, and molecular levels.

genome — The DNA sequence of a species, individual, cell, or subcellular compartment. Since I have listed at least four possibilities, the specific entity intended is usually specified — i.e., "the nuclear genome," or "the genome sequence of chimpanzees," or "the mitochondrial genome."

genus — The classification rank above the level of species but below the level of family. For example, chimpanzees, gorillas, and orangutans each belong to separate genera.

geology — The study of the earth, especially the rocks it contains.

germ cell — Reproductive cells, such as sperm and egg.

great apes — Among living species, the chimpanzees, gorillas, and orangutans represent the great apes.

guanine — One of the four nitrogen-containing bases in DNA (or RNA) that is chemically distinct from the other three. The base forms one part of each individual unit of the DNA double helix (or of the RNA strand).

haplogroup — A group of individuals sharing a particular type of DNA sequence.
haploid — The state of having a half of the normal complement of DNA. For example, sperm and egg have half of the DNA content of non-reproductive cells; their chromosomes do not exist as pairs.
heredity — The transmission of features from one generation to the next.
heterozygosity — The state of having DNA differences between two copies of DNA.
heterozygous — In organisms that possess two copies or versions of their DNA sequence, the sections of DNA in which these two versions differ are referred to as heterozygous.
Homo sapiens, H. sapiens — The scientific name for our own species.
homozygosity — The state of having no DNA differences between two copies of DNA.
homozygous — In organisms that possess two copies or versions of their DNA sequence, the sections of DNA in which these two versions are the same are referred to as homozygous.
Hox — A particular gene (several of them in some species) that controls multiple aspects of development.
hybridization — In genetics, the mating of two individuals.
in vitro — In the culture dish or laboratory. For example, testing a chemical on freshly isolated liver cells.
in vivo — In the creature itself. For example, testing a chemical by injecting it into the livers of live mice.
inbreeding — Mating between two closely related individuals.
incubator — A laboratory instrument used to control the humidity, temperature, and other factors in a particular environment. Often looks like a mini oven.
indel — A mutation in which DNA base pairs are either inserted (the "in" part) or deleted (the "del" part) in a particular DNA sequence.
independent assortment — The phenomenon where traits are inherited independently of one another. The opposite of linkage.
indigenous — Native to a particular area (but not necessarily to that area alone).
inductive reasoning — The method of reasoning that begins with observations and then tries to draw these observations into a general principle or rule. Similar to the process of elimination.
inheritance — The transmission of features from one generation to the next.
insects — A group of species typified by flies, beetles, butterflies, bees, and ants.
invertebrates — A group of animal species lacking a backbone.
kingdom — The highest level of classification in the Linnaean classification system. Animals all belong to the same kingdom. Plants belong to a separate kingdom, and fungi to a third.
linkage — At the visible level, the phenomenon of co-inherited traits. At the genetic level, the phenomenon of genes or sections of DNA being on the same chromosome. The opposite of independent assortment.
Linnaeus — The father of modern classification.

lyse, lysis — When a cell bursts open, often as a result of another organism (like a virus) growing inside of it.

mammals — A group of species that suckle their young. Humans, elephants, howler monkeys, cattle, camels, cats, and many more species are all classified as mammals.

marsupials — A subgrouping of mammals typified by kangaroos, koalas, and Tasmanian devils.

meiosis — The process of cell division that gives rise to sperm and egg cells. The products of meiosis have half of the DNA content of the parental cells.

messenger RNA — A subcategory of RNA molecules that specifically encode protein sequences.

metabolism, metabolic — Having to do with the chemical transformations that occur in the cell. For example, the extraction of energy from the chemical bonds in sugars, fats, and proteins. Also, the synthesis of molecules like DNA, RNA, protein, and other molecules.

metaphase — A phase of the cell division cycle in which the dividing cell lines up the chromosomes in the middle of the cell. Metaphase comes after prometaphase but precedes anaphase.

mitochondria — A subcellular compartment that functions primarily in energy metabolism and also contains its own DNA.

mitosis — The process of cell division that gives rise to cells other than sperm and egg. The products of mitosis have the same amount of DNA content as the parental cells; the DNA content is not halved.

molecular — Having to do with the molecules of the cell.

molecular biology — The study of life at the molecular level — at the levels of molecules such as DNA, RNA, protein, etc.

molecular clock — The concept of using DNA changes that occur each generation to mark the passage of time.

molecule — At least two atoms chemically bound together. They can be tiny — like an oxygen molecule that consists of two oxygen atoms. Or they can be large — like the DNA molecule that consists of millions of oxygen, carbon, nitrogen, phosphorus, and hydrogen atoms.

monotremes — A subgroup of mammals typified by the platypus.

morphology — The form of a creature — such as the color of a butterfly's wings and the length of an elephant's trunk.

Mus musculus, M. musculus — The scientific name for the house mouse, which is commonly used in the laboratory.

mutagenic — Something that induces mutations.

mutant — A DNA base pair, a section of DNA, or a creature that is different from normal.

mutation — A DNA base pair or section of DNA that is different from normal.

natural selection — The survival of the fittest to reproduce.

Neanderthal — An extinct group of humans.

nested hierarchy — A groups-within-groups pattern, similar to Russian nesting dolls.

New World — Shorthand for the Americas — North America, South America, Central America, and the Caribbean.

non-mutant — A DNA base pair, a section of DNA, or a creature that is normal.

nucleic acid — A type of chemical molecule; DNA and RNA are nucleic acids.

nucleotide — The fundamental subunit of DNA, consisting of one of the four bases (adenine, guanine, cytosine, thymine) attached to a carbohydrate molecule and a phosphate molecule.

nucleus, nuclei — The subcompartment of the cell where nearly all of the DNA resides in plant, fungal, and animal cells.

Old World — Shorthand for the continents of Europe, Africa, and Asia.

oogenesis — The process of producing egg cells.

order — The level of classification above family but below class. Among mammals, carnivores constitute an order, as do elephants.

paleontology — The study of fossils.

Perissodactyla — An order of mammals that includes horses, asses, zebras, rhinos, and tapirs.

petri dish — A plastic dish used in the laboratory for growing cells.

pharyngula stage — A stage of development in which embryos from very diverse creatures have an overall appearance of similarity.

phenotype — The visible characteristics of an organism; in contrast to the genetic characteristics of an organism.

phosphate — A molecule consisting of phosphorus and oxygen atoms.

phylum — A level of classification below kingdom but above class that is defined by body plans. For example, chordates (i.e., vertebrates plus some lesser known species) constitute the phylum Chordata.

physiology — The field of science dedicated to the study of the operation of the organs, systems, and tissues of an individual.

pipet — A laboratory instrument for transferring liquids of a defined volume.

pipet tip — The end of a pipet that actually holds the liquid. It's typically disposable so as to prevent cross-contamination of samples.

primates — A group of creatures typified by the great apes and monkeys.

prometaphase — a stage of the cell division cycle following prophase and preceding metaphase.

pronucleus — In egg cells, the location of egg DNA before the egg DNA compartment fuses with the compartment containing the sperm DNA; once sperm fuses with egg, the sperm DNA is contained in a pronucleus before the two pronuclei join.

prophase — A very early phase of the cell division cycle before chromosomes line up in the center of the cell. Prophase precedes prometaphase.

protein — A common molecule in the cell consisting of amino acids.

protein moonlighting — A cellular phenomenon in which proteins that are thought to do only one task actually perform additional, seemingly unrelated cellular tasks.

protein-coding — A DNA sequence that codes for the amino acid sequence of a protein.

pure-breeding — A breed that consistently produces the same outcome each time.

recessiveness — A genetic phenomenon in creatures that have two different versions of hereditary instructions. When a dominant trait is crossed to a recessive trait, the dominant trait appears in the offspring, while the recessive trait is hidden.

recombination — In organisms that possess two versions of their DNA sequence, a hybrid of the two versions is passed on to offspring. The process of forming this hybrid DNA involves copying large chunks from each chromosome as well as copying small parts from each chromosome. Recombination is the former.

regulatory sequence — A sequence whose function is the regulation of other sequences. For example, DNA sequences controlling the timing and location of the transcription of RNA.

reptiles — A group of species typified by crocodiles, turtles, snakes, and lizards.

retrodiction, retrodictive — The opposite of a prediction. Predictions use current data to project into the future and set expectations. Retrodictions use current data to project into the past and examine whether one model is a better fit to the data than another.

ribosome — The structure in the cell which participates in the translation of a messenger RNA into protein.

RNA — A chemically analogous molecule to DNA. Unlike DNA, it's often single-stranded, rather than a double helix. It performs a tremendous diversity of functions in the cell.

rodents — A group of species typified by mice, rats, beavers, squirrels, chipmunks, porcupines, hamsters, and gophers.

Saccharomyces cerevisiae, S. cerevisiae — The scientific name for baker's yeast, which is commonly used in the laboratory.

Second Law of Thermodynamics — The law which states that, in a closed system, things tend toward disorder rather than order.

segregation — Genetically, it is the independent behavior of maternal and paternal genetic instructions.

SNP — Acronym for single nucleotide polymorphism. SNPs are single base pair differences between two DNA sequences.

solvent — The liquid in which a chemical is dissolved.

somatic cell — A cell of the body, but not a reproductive cell. Skin cells, stomach cells, pancreatic cells, muscle cells, etc. are all somatic cells. Sperm and egg are not.

species — A unit of classification above subspecies but below genus. Difficult to define. However, we intuitively recognize many species by their visible features. For example, we readily spot the difference between the African elephant species and the Asian elephant species.

species fixity — The hypothesis that separate species do not have common ancestors.

sperm — The male reproductive cell.

spermatogenesis — The biological process by which sperm cells are produced.

subspecies — A unit of classification below species. Even less well-defined than species.

telophase — A very late phase of the cell division cycle. The stage in which the two daughter cells begin to separate from one another into distinct entities. Telophase follows anaphase.

terrestrial — On the land. In contrast to aerial or aquatic.

tetranucleotide hypothesis — The hypothesis that DNA was not the physical substance of heredity but, rather, the scaffold for the real genetic material.

thymine — One of the four nitrogen-containing bases in DNA that is chemically distinct from the other three. The base forms one part of each individual unit of the DNA double helix. In RNA, uracil takes the place of thymine.

trait — A feature of an individual, often something visible like eye color, hair color, or number of limbs.

transcription — In the cell, it is the process by which the cell makes an RNA copy of DNA sequence. It is usually just a small section of the entire DNA sequence.

translation — In the cell, it is the process by which the sequence of a messenger RNA is read and decoded into an amino acid sequence.

translocation — The movement of a chunk of one chromosome to another chromosome.

tRNA — A type of RNA that is used in the translation of the messenger RNA into an amino acid sequence.

unit factors — Gregor Mendel's term for the particulate units of inheritance that he documented.

uracil — One of the four nitrogen-containing bases in RNA that is chemically distinct from the other three. The base forms one part of each individual unit of the RNA strand. In DNA, thymine takes the place of uracil.

vertebrates — A group of species possessing a backbone; typified by mammals, reptiles, amphibians, fish, and birds.

vestigial — In evolutionary thought, vestigial structures are leftovers of past evolutionary processes. While they may have performed a function in the ancestors of the species in which they exist, they often no longer do so.

Young-earth creation, young-earth creationist (YEC) — A branch of creation that espouses an age for the earth and universe of just a few thousand years; it also espouses a global Flood on the earth around 4,500 years ago.

zygote — The first cell of development and the product of conception. It results from the union of sperm and egg.